PAT MᶜDADE

Circuits, Interconnections, and Packaging for VLSI

H.B. Bakoglu

ADDISON-WESLEY PUBLISHING COMPANY

Reading, Massachusetts • Menlo Park, California • New York
Don Mills, Ontario • Wokingham, England • Amsterdam • Bonn
Sydney • Singapore • Tokyo • Madrid • San Juan

This book is in the **Addison-Wesley VLSI Systems Series**
Lynn Conway and Charles Seitz, **Consulting Editors**

Library of Congress Cataloging-in-Publication Data

Bakoglu, H. B.
    Circuits, interconnections, and packaging for VLSI.
    Bibliography: p.
    Includes index.
    1. Integrated circuits—Very large scale integration—Design and construction. 2. Electronic packaging.
    I. Title.   TK7874.B345 1990   621.381'73   87–22964
    ISBN 0–201–06008–6

BCDEFGHIJ–MA–943210

FOREWORD

The subject of VLSI systems spans a broad range of disciplines including semi-conductor devices and processing; integrated electronic circuits; digital logic; design disciplines and tools for creating complex systems; and the architecture, algorithms, and applications of complete VLSI systems. The Addison-Wesley VLSI Systems Series comprises a set of textbooks and research references that present the best current work across this exciting and diverse field, with each book providing a perspective that ties its subject to related disciplines.

Microelectronic interconnection and packaging technology is entering a period of essential change. The switching performance of the transistors in $1\mu m$ MOS technologies has overtaken the capability of the traditional in-terconnection technology—dual-in-line packages, stitch-bond wires, and even the long metal wires on chips—to convey such high-speed signals within and between chips. It is becoming widely recognized that the performance, cost, size, power consumption, and reliability of computing systems will depend increasingly on the system designer's understanding of the physical effects of interconnections and their associated circuitry.

Circuits, Interconnections, and Packaging for VLSI by H. B. Bakoglu provides students and practicing system designers with immediate access to timely information about advanced packaging technologies and interconnec-tion issues. Descriptions, characteristics, and illustrations of microelectronic process and packaging technologies are presented before a subject is treated analytically. Thus, the reader is led systematically from direct answers to such factual questions as "What is tab packaging?" to the practical answers to such engineering questions as "How does lead inductance affect signal delay and power distribution?" and, finally, to the analytically correct answers to such fundamental questions as "What is the mechanism by which noise is coupled between adjacent wires?" and "How do I calculate its effects?"

In addition to connecting artifact to analysis, Bakoglu provides a perspective on interconnection and related issues that ranges from current practice to futuristic technologies, and from the scale of the VLSI chip to that of large systems. This book is, to the best of our knowledge, the first comprehensive treatment of this important subject, and we hope and expect that it will make a significant contribution to the VLSI system designer's understanding of the problems of interconnection and awareness of the opportunities of advanced packaging.

Lynn Conway
Ann Arbor, Michigan

Charles Seitz
Pasadena, California

PREFACE

This book focuses on the technology and design of high-speed VLSI circuits and systems. It covers topics such as packaging, interconnections, clock skew, and noise that are not usually addressed in introductory VLSI design books. The book describes the basic principles of these concepts, accompanied by many illustrations and examples. Practical aspects of the covered topics are emphasized, and design guidelines and rules of thumb are presented. This book can be used in an advanced course in VLSI design or by the practicing engineer who wants to enhance his or her knowledge of VLSI device and circuit technologies, wants a better understanding of the design trade-offs in nMOS, CMOS, bipolar, and GaAs ICs, and wants to have better understanding of interconnection and packaging requirements and limitations.

Since the advent of the integrated circuit, the packing density of transistors and their intrinsic gate delay have improved continuously. With submicron device dimensions and nearly a million transistors integrated on a single microprocessor, on-chip and chip-to-chip interconnections are playing a major role in determining the speed, power consumption, and size of digital systems. In this book, interconnections and packaging are analyzed in detail. Coverage also includes scaling of MOS devices, short-channel effects, electromigration, reliability of integrated circuits, sense amplifiers for SRAM and DRAM chips, merged bipolar/CMOS (BiCMOS) circuits, emitter-coupled logic (ECL), cryogenic CMOS, superconductors, and optical interconnections. The emphasis is on issues that are gaining importance in current designs and that are likely to become even more important in the near future.

First, device, interconnection, and packaging technologies are described, and essentials of advanced VLSI device and packaging concepts are reviewed. Then wiring capacitance and resistance, transmission-line concepts, simultaneous switching (dI/dt) noise, clock skew, and high-speed clock distribution are treated in detail. Various specific performance-enhancement techniques are also presented. In addition, a new system-level circuit model that encompasses material, device, circuit, logic, architecture, and packaging parameters is derived. This model is used to compute the clock frequency, power dissipation, and chip and module sizes of central processing units (CPUs). It is also used to compare competing IC (CMOS, bipolar, GaAs) and packaging (wafer-scale integration, thin-film hybrid, multilayer ceramic) technologies.

The chapters in this book form a coherent unit, but they are structured so that any one chapter is sufficiently self-contained to be read independently from the rest.

- **Chapter 1** is a brief introduction to the topics covered in the book.

- **Chapter 2** discusses VLSI device and interconnection technologies. A thorough review of circuit performance and reliability issues resulting from device scaling are presented, such as velocity saturation, mobility degradation, source/drain resistance, subthreshold current, DIBL, hot-carriers, electrostatic discharge, and latch-up. In addition, advanced CMOS, BiCMOS, cryogenic CMOS, MESFET, bipolar, and gallium arsenide (GaAs) MESFET, HEMT, and HBT technologies are described and compared. Interconnection reliability issues, such as electromigration and contact failures, are also discussed.

- **Chapter 3** covers packaging. Material, thermal, and electrical properties of various packaging technologies are comparatively presented. First, low-end electronic packages such as DIP, pin-grid array, and surface-mounted chip carriers are described, and wire bonding, TAB, and flip-chip mounting are compared. Then more sophisticated ceramic and polyimide/ceramic multichip carriers are presented. Recent developments and research topics such as silicon-on-silicon packaging, wafer-scale integration, microchannel cooling, superconductors, and optical interconnections are also discussed in detail.

- **Chapter 4** presents driver and receiver circuits for large-capacitance on-chip and chip-to-chip wires in high-speed VLSI systems. Static and dynamic RAM sense amplifiers are reviewed that minimize the delay associated with large-capacitance bit lines. Later, single- and double-ended CMOS logic circuits with reduced potential swings are presented. Finally, BiCMOS circuits are described that combine the advantages of bipolar and CMOS technologies.

- **Chapter 5** analyzes the effects of parasitic interconnection resistance and presents alternative scaling schemes and repeater circuits that minimize distributed RC delays of VLSI wires.

- **Chapter 6** discusses the transmission line properties of on-chip and chip-to-chip interconnections. Various transmission line driving methods are described and compared.

- **Chapter 7** is devoted to noise. Capacitive and inductive couplings between wires are modeled, and simultaneous switching (dI/dt) noise is analyzed at the chip and board levels.

- **Chapter 8** reviews various clocking disciplines and presents methods for distributing high-speed clock signals with minimal skew at the chip and board levels.

- **Chapter 9** presents a system-level model that combines material, device, circuit, logic, architecture, and packaging parameters to predict clock frequency, power dissipation, and chip and module sizes. This model can be used to optimize the number of transistors integrated on a die and to determine the impact of new device and packaging technologies on system performance.

The contents of this book are based on my doctoral work at Stanford University. The book is significantly expanded to include useful background material in addition to the original thesis work. I extend my appreciation to Professor James D. Meindl for his guidance throughout my years at Stanford. I am also fortunate to have participated in many stimulating conversations with the members of the Stanford Integrated Circuits Laboratory and Center for Integrated Systems. Careful review and suggestions by Robert Sechler of IBM Advanced Workstation Division were valuable in improving the presentation and content of Chapters 6 and 7. I thank the IBM Corporation for supporting the final three years of my doctoral study through an IBM Graduate Fellowship and the Defense Advanced Research Projects Agency, which partly funded my work at Stanford. Finally, I thank my employer, IBM Corporation, for providing an environment in which completing this book was possible, and special thanks go to my wife Marie-Christine for her love, support, and patience.

None of the institutions, companies, or persons named above can be held responsible for the contents of this book. The information presented is believed to be accurate, and great care was paid to assure its accuracy. However, no responsibility is assumed for its use or for any infringement of patents or other rights of third parties that may result from its use. No license is granted by implication or otherwise under any patent, patent rights, or other rights.

The figures in this book were hand drawn by the author, and the book was typeset by the author on an IBM RT using a software package based on LaTeX and TeX.

H.B.B.
Austin, Texas, 1989

CONTENTS

4   INTERCONNECTION CAPACITANCE   134

5   INTERCONNECTION RESISTANCE   194

1

INTRODUCTION

What will limit the performance of future integrated circuits that may contain millions of transistors? Since the beginning of the integrated circuit era, the number of devices per chip has been increased by reducing the minimum feature size, enlarging the chip area, and improving the packing efficiency of the devices [1.1]–[1.3]. Between 1959 and 1983 minimum feature size shrunk 11 percent per year, chip area increased 19 percent per year, and packing efficiency improved by a factor of 2 per decade. These three factors combined produced a one-hundredfold increase in devices per chip every decade [1.2]. How long can this growth continue, and how can the performance potentials of the future ultra-large-scale-integrated (ULSI) circuits be best utilized?

Currently, interconnection- and packaging-related issues are among the main factors that determine the number of circuits that can be integrated on a chip as well as the chip performance. This is true for CMOS, bipolar, and gallium arsenide (GaAs) systems. Interconnections and packaging will gain even more importance as we continue reducing feature sizes of transistors and enlarging chip dimensions, and they may eventually become the dominant performance limiting factors. Consequently, understanding the on-chip and chip-to-chip wiring issues and subtleties of the electronic packaging will be key to harnessing the full potential of future ULSI systems.

The main topics addressed in this book are the interconnection- and packaging-related chip- and system-level electrical performance issues, such as wiring capacitance and resistance, transmission line phenomenon, cross talk between wires, simultaneous switching noise, and clock skew. First the advantages of MOS device scaling are described, and second-order short-channel

1

and reliability issues related to scaling are presented. Then advanced CMOS, BiCMOS, and cryogenic CMOS technologies are described, and CMOS devices are compared with silicon MESFET and BJT, and GaAs MESFET, HEMT, and HBT devices. Later various single chip and multichip packaging technologies are reviewed. Next analytical models for capacitive and resistive delays of interconnections, cross talk, and simultaneous switching noise are developed, and the effects of reduced minimum feature size and increased chip dimensions are calculated. Specific material, process, circuit, packaging, and architecture solutions that reduce the adverse effects of scaling are then proposed and quantitatively evaluated. A system-level circuit model encompassing material, device, circuit, logic, packaging, and architecture parameters is also described. This model is used to calculate the clock frequency, power dissipation, and chip and module sizes of present and future computers. The data obtained from the model are analyzed to investigate the effects of the model parameters on the overall system throughput, to compare various IC and packaging technologies, and to determine the impact of new devices and packages on system performance.

1.1   VLSI Device Technologies

Because of their high packing density, low power dissipation, and high yield, metal-oxide-semiconductor field-effect transistor (MOSFET), and especially complementary MOSFET (CMOS), is the technology of choice for VLSI applications. Basic MOSFET device structure is illustrated in Fig. 1.1.

Miniaturization of MOS transistor dimensions continues to improve circuit speed and packing density. Ideal scaling theory was one of the first formal approaches to shrinking MOS transistors [1.4]. When the MOS devices are scaled in five dimensions—the three physical dimensions, voltage supply level, and doping concentration—the field patterns in the smaller device will be identical to those in the larger one. This leads to greater speed, greater density, and lower power consumption. Scaling improves circuit performance by reducing capacitances and potential swings and at the same time ensures the physical integrity of devices by keeping the electric fields constant. When simple scaling (Table 1.1) is applied, all horizontal and vertical dimensions of transistors (W, L, t_{gox}, X_j) are scaled down by $1/S$ ($S > 1$). Substrate doping (N_{SUB}) is increased by S, and all voltages (V_{DD}, V_{TN}, V_{TP}) are reduced by $1/S$. As a result, internal electric field strengths remain approximately unchanged, and velocity saturation, oxide breakdown, and carrier multiplication problems are avoided. With simple scaling, devices get faster, power dissipation and power-delay products of the devices are reduced, packing density improves, and power dissipation density remains constant. Fast chips that integrate a large number of transistors become possible.

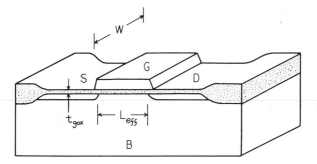

FIGURE 1.1   Metal-Oxide-Semiconductor Field-Effect Transistor (MOSFET). The main dimensions that determine device properties are gate oxide thickness t_{gox}; gate length L, gate width W; and junction depth X_j.

These observations are strong evidence for the advantages of scaling MOS transistors; however, there are a number of second-order effects that require modifications in the simple scaling approach. For example, velocity saturation, mobility degradation, and parasitic source and drain resistances degrade the current drive capability (transconductance) of devices. The silicon bandgap and built-in junction potentials are material parameters and cannot be scaled. This gives rise to short-channel effects. This makes the threshold potential

TABLE 1.1   Scaling of MOS Transistors

Parameter	Scaling Factor
Dimensions $(W,\ L,\ t_{gox},\ X_j)$	$1/S$
Substrate doping (N_{SUB})	S
Voltages $(V_{DD},\ V_{TN},\ V_{TP})$	$1/S$
Current per device $(I_{DS} \propto \frac{W}{L}\frac{\epsilon_{ox}}{t_{gox}}(V_{DD} - V_T)^2)$	$1/S$
Gate capacitance $(C_g = \epsilon_{ox}\frac{WL}{t_{gox}})$	$1/S$
Transistor on-resistance $(R_{tr} \propto \frac{V_{DD}}{I_{DS}})$	1
Intrinsic gate delay $(\tau = \frac{C_g \Delta V}{I_{av}} = R_{tr}C_g)$	$1/S$
Power-dissipation per gate $(P = IV)$	$1/S^2$
Power-delay product per gate $(P \times \tau)$	$1/S^3$
Area per device $(A = WL)$	$1/S^2$
Power-dissipation density (P/A)	1

dependent on transistor width and length, causes degradation in subthreshold behavior, and increases leakage currents.

In addition, to satisfy TTL compatibility and to keep utilizing existing ICs without having to install multiple power supplies in systems, voltage levels have not been scaled down along with feature sizes. This has an advantage; it reduces the gate delay by an additional factor of S (if the devices are not velocity saturated) because it increases the current drive. The power dissipation density also goes up. In addition, keeping the supply voltage constant while scaling device dimensions causes the electric fields to increase, which raises reliability concerns such as oxide breakdown, drain avalanche, and hot-electron injection into gate oxide.

Another parameter that worsens with scaling is the distributed RC delays of the wires. The relative rise in the RC constants of local interconnections (especially the polysilicon gate material) increases circuit delays. In addition, when CMOS circuits are scaled down, parasitic bipolar transistors formed by the n-p-n-p structure of the source-bulk-well-source layers become more likely to turn on and damage the circuit (CMOS latch-up).

These and other second-order scaling issues will be described in detail in Chapter 2. In addition, advanced CMOS, BiCMOS, cryogenic CMOS, MESFET, bipolar, and GaAs technologies are also presented and compared in Chapter 2.

1.2   VLSI Interconnections

Digital integrated circuits contain two basic components: transistors and interconnections. At low integration levels (SSI and MSI), circuit speed, packing density, and yield are determined by transistors, but as more and more devices are integrated on a single die, wires gain importance. Interconnections play an important role in determining the speed, area, reliability, and yield of VLSI circuits.

1.2.1   Interconnection Capacitance

One of the reasons interconnections limit the performance is the wiring capacitance. With increasing chip dimensions, parasitic interconnection capacitance dominates the gate capacitance, and the speed improvement expected from simple scaling does not apply to circuits that drive global communication lines. Simple scaling assumes a reduction in the capacitive loading due to wires. This is true locally when a circuit is connected only to its neighbors, but for circuits that drive long global wires, the capacitive loading actually increases because chips get bigger with time. Since the critical paths that determine the overall performance of anintegrated circuit are usually dominated by global wires, faster local circuits do not necessarily improve the clock frequency. To minimize the negative effects of the long wires, the floor planning and placement of the functional subblocks must be constructed such that long wires are avoided in the critical paths.

In an SSI or MSI chip designed in 7 μm nMOS technology, the gate input capacitance (20 fF per minimum size transistor) dominates the wiring capacitance (200 fF/mm). Here, a typical transistor with $W/L = 10$ has a capacitance of 200 fF, which is equivalent to 1 mm of wire. With a typical MSI die size of a few millimeters on a side, most of the nets will be well below 1 mm. When the transistors are scaled down, the gate input capacitance is reduced. In a 0.7 μm CMOS technology, a minimum-size transistor has 2 fF input capacitance, which yields a typical transistor ($W/L = 10$) input capacitance of 20 fF. This is equivalent to 0.1 mm of wire, which represents a large number of the nets in a 10 × 10 mm VLSI chip. Consequently as the devices are scaled down below 1 μm, the node capacitances will not go down as much as they did in a 7 μm nMOS MSI chip because of the increased role of the wiring capacitance. To enhance system performance, devices driving long lines can be made stronger; however, these buffers do not come for free because they have long internal delays, and they occupy a large area. As a result, large CMOS buffers do not solve all the problems associated with the capacitance of the long wires but significantly reduce them.

Large capacitive loads also increase power consumption. Currently in high-performance CMOS microprocessors, a good portion of the power may be dissipated by the off-chip drivers. In dynamic random-access memories (DRAMs), practically all the power is consumed in driving and sensing long column and row lines and in off-chip drivers. A thin-film hybrid package with dense interconnections and flip-mounted chips or a wafer-scale integrated (WSI) circuit may partially resolve this problem by reducing module-level wire lengths and capacitive loading associated with them; however, global interconnections will still be responsible for a significant portion of the delay in the critical nets and will consume a major portion of the available power.

The power-delay product of long lines can be improved by driver and receiver circuits that reduce the voltage swing at high-capacitance nodes. This can be achieved by a number of static and dynamic circuits that precharge the interconnection to the high-gain region of the receiver and by driver circuits that combine bipolar and CMOS transistors. Such circuits are described in Chapter 4 [1.8]–[1.9].

1.2.2   Interconnection Resistance

In addition to large capacitance loads resulting from long interconnections, the resistance of the lines also becomes a major concern [1.4],[1.10]. Simple scaling of local and global interconnections is summarized in Table 1.2, and the basic interconnection parameters are illustrated in Fig. 1.2.

TABLE 1.2 Scaling of Local and Global Interconnections

Parameter	Scaling Factor
Cross sectional dimensions (W_{int}, H_{int}, W_{sp}, t_{ox})	$1/S$
Resistance per unit length ($\mathcal{R}_{int} = \rho_{int} \frac{1}{W_{int}H_{int}}$)	S^2
Capacitance per unit length ($\mathcal{C}_{int} = \epsilon_{ox} \frac{W_{int}}{t_{ox}}$)	1
RC constant per unit length ($\mathcal{R}_{int}\mathcal{C}_{int}$)	S^2
Local interconnection length (l_{loc})	$1/S$
Local interconnection RC delay ($\mathcal{R}_{int}\mathcal{C}_{int}l_{loc}^2$)	1
Die size (D_c)	S_C
Global interconnection length (l_{int})	S_C
Global interconnection RC delay ($\mathcal{R}_{int}\mathcal{C}_{int}l_{int}^2$)	$S^2 S_C^2$
Transmission line time of flight (l_{int}/v_c)	S_C

Resistance is inversely proportional to the cross-sectional area of the conductor, and capacitance is proportional to the plate area and is inversely proportional to the distance between the plates. Accordingly interconnection resistance and capacitance (neglecting fringing fields) per unit length are ex-

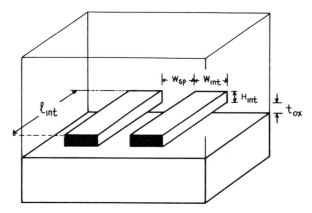

FIGURE 1.2 Basic interconnection parameters

pressed as

$$\mathcal{R}_{int} = \rho_{int} \frac{1}{W_{int} H_{int}}$$

$$\mathcal{C}_{int} = \epsilon_{ox} \frac{W_{int}}{t_{ox}}$$

(1.1)

where ρ_{int} and ϵ_{ox} are the resistivity of the conductor and the dielectric constant of the insulator, W_{int} and H_{int} are the width and thickness of the interconnection, and t_{ox} is the insulator thickness. In order to improve packing density, interconnection dimensions are usually scaled down by the same factor as the transistors, which causes an increase in the distributed RC constant per unit length by S^2

$$\mathcal{R}_{int}\mathcal{C}_{int} = \rho_{int}\epsilon_{ox} \frac{1}{H_{int}t_{ox}} \propto S^2. \tag{1.2}$$

In addition, chip size has increased continously throughout the evolution of ICs. This can be modeled as a chip scaling factor S_C, which represents the increase in the length of one side of a die. If the chip size increases from 5×5 mm^2 to 10×10 mm^2, S_C is 2. As a result, global interconnection length l_{int} increases by S_C, and the distributed RC delay of long lines degrades by $S^2 S_C^2$:

$$R_{int}C_{int} = \mathcal{R}_{int}\mathcal{C}_{int}l_{int}^2 \propto S^2 S_C^2 \tag{1.3}$$

Because the distributed RC delay of an interconnection is independent of driver size, it cannot be reduced by a larger driver. In a given technology (i.e., with given wire and insulator thicknesses), the RC constant of the wire cannot be decreased significantly by making the wires wider. When the width is increased, resistance goes down, but the capacitance increases by the same amount.[1] Consequently the RC constant cannot be reduced by adjusting mask dimensions; it is determined by the technology.

To provide global wires with acceptable RC delays, the top few levels of interconnections must be made wider and thicker than the first few levels. Alternative scaling schemes that supply wider, thicker, and lower resistance lines for long interconnections and repeaters that can minimize the distributed RC delay are described in Chapter 5 [1.11]–[1.12].

---

[1] Actually, the increase in capacitance is less than the reduction in the resistance because the fringing field component of the capacitance remains constant. As a result, distributed RC constant improves somewhat when the wires are made wider. After the wire width is doubled or quadrupled, the parallel-plate component of the capacitance dominates the fringing field component, and further increases in the wire width do not help to reduce the distributed RC delay.

1.2.3 Reliability of VLSI Interconnections

Aluminum is the preferred metal for VLSI interconnections because of its low resistivity, good adherence to silicon (Si) and silicon-dioxide (SiO$_2$) layers, bondability, patternability, and ease of deposition. In addition, aluminum (Al) can be easily purified (it does not contaminate the IC with undesirable impurities), and it is a readily available and low-cost material. In spite of its positive qualities, aluminum interconnections introduce many reliability problems such as electromigration, contact failures, and bad step coverage [1.5]–[1.6].

Electromigration is one of the major interconnection failure mechanisms in VLSI integrated circuits. It is caused by the transport of the metal atoms when they get bombarded with electrons. As they collide with the oncoming electrons, the metal atoms migrate, primarily via grain boundary diffusion, generating electrical opens and shorts that cause the circuit to fail.

Diffusion of silicon into aluminum at the points where Al contacts Si, referred to as contact electromigration, is another major reliability problem related to interconnections. This arises when the silicon atoms migrate into aluminum at the interface area where Al interconnect forms a contact to Si, and the void created by the transported silicon is filled by aluminum. If the resulting "spike" is deep enough, the junction below the contact point can be shorted to the substrate. Interconnection reliability issues are described in Chapter 2.

1.3 Packaging Technologies

Having fast and reliable chips is not sufficient to build superior systems; these ICs must be supported by an equally fast and reliable packaging technology [1.13]. Packaging supplies the chips with signals and power, removes the heat generated by the circuitry, and provides physical support and environmental protection. In general, packaging plays an important role in determining the overall speed, cost, and reliability of the system, and electronic package designers strive for compact packages that supply the chips with a large number of pins and dense wiring capability and a noise-free environment.

As a result of today's densely packed submicron devices, on-chip circuitry has become so fast that a significant portion of the total delay in a processing unit is due to the time required for signals to travel from one chip to another. In high-end systems, 50 percent of the total system delay is usually due to packaging, and by the year 2000 the share of packaging delay may rise to 80 percent [1.14]. In order to minimize this delay, chips must be placed close together. In addition, VLSI chips require many signal and power connections, which can be satisfied only by a dense interconnection network. It is important that the connections between the dice and the board have small parasitic inductances and capacitances. The same applies to board wiring, which also have to exhibit good transmission line behavior, and the wires must be shielded to minimize the line-to-line couplings. The package must distribute a stable power level without excessive switching noise. In addition, the thermal properties of the package and semiconductor should match in order to avoid stress-induced breakage when the parts expand and contract as temperature changes. The package should remove the heat generated by the circuits, and its cost should not be excessive.

A variety of packaging options are reviewed in Chapter 3. Dual-in-line package (DIP) and pin grid array (PGA) are popular in low-cost systems. These chip carriers are mounted on printed circuit (PC) boards by stuffing their pins through the holes drilled on the board and then soldered down with a wave-soldering process. Surface-mounted chip carriers, a more recent package type with a smaller footprint size than DIP and PGA, are also presented. Surface-mounted carriers do not require any through-holes, and they can be mounted on both sides of the board so their board packing density is significantly better than through-hole mounting.

Packing density and system performance can be improved by placing many chips directly on a substrate that provides low-inductance and low-capacitance connections between the chip and signal/power lines and supplies a very dense wiring. Multichip carriers can improve system performance by 20 percent or more [1.14]. High-end computers traditionally employ multilayer ceramic substrates and sophisticated cooling schemes. Silicon substrates with aluminum or copper wiring and ceramic substrates with molybdenum interconnections are described as examples of multichip packages in Chapter 3. The insulator on the silicon substrate can be SiO_2 or polyimide, and the ceramic is usually alumina. It is also possible to use a ceramic substrate with tungsten power distribution network as a base and then fabricate dense and low-resistance copper wires insulated by low dielectric constant polyimide on top of this ceramic base. With proliferation of tape-automated bonding (TAB) and availability of bare dice from IC vendors, direct die mounting on PC boards or on insulated metal substrates may become more common in low-end systems.

Recent developments in superconductivity promise lossless wires at liquid nitrogen temperature or possibly even at room temperature. Issues such as skin effect, limited current density, and brittleness of ceramic superconductors must be resolved before superconductors can be employed in practical applications.

Optical interconnections is another promising alternative, which offers fast, reliable, easy-to-route, and noise-free data transmission [1.7]. Many fiber-optic local area networks are already being used in computer-to-computer communication, and some systems employ optical fibers at the backplane level. The benefits of optical fibers, however, are still undetermined at the lower levels of the packaging hierarchy. In the present, tight alignment requirements, size incompatibility with ICs, and high power consumption limit the applicability of optoelectronic components.

1.4   Transmission Lines

Wires at the board level have always been treated as transmission lines in delay calculations [1.16]–[1.18]. As chip size becomes larger and the wavelengths of the signals become comparable to interconnection lengths, the transmission line properties of on-chip interconnections will also gain importance. Reflections from the discontinuities and the end of the line may become important. Transmission line phenomenon becomes significant at the chip level when the signal rise times go below 150 psec.

Figure 1.3 compares a good transmission line and a typical IC interconnection. Between the strip line and ground plane there is a low loss insulator. Because above the transmission line there is air, which has a lower dielectric constant than the insulator, electric fields are confined between the strip line and ground plane, and the propagation mode is close to ideal transverse electromagnetic (TEM) mode. On the other hand, in IC interconnections almost all the field lines are terminated at the neighboring wires. This causes cross-talk noise and may excite multiple modes of propagation with slightly different transmission speeds and may give rise to dispersion and degradation

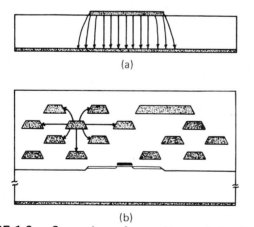

(a)

(b)

FIGURE 1.3   Comparison of a good transmission line and IC interconnections. The strip line in (a) is a good transmission line; its electric fields are confined between the line and ground plane. On the other hand, the electric fields of IC interconnections in (b) terminate at the neighboring lines. This gives rise to cross-talk noise and dispersion of wavefronts.

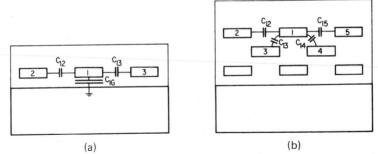

(a)   (b)

FIGURE 1.4   Comparison of coupled noise in single-level and multilevel interconnections. (a) The presence of a dominant capacitance to a stable ground plane ensures less noise coupling between wires. (b) In comparison a multilevel interconnection scheme lacks such a dominant and stable ground plane.

of the sharpness of the waveforms as on-chip rise times go below 150 psec. These second-order transmission line phenomena become significant only if the resistance of the lines is low enough not to dominate the delay and rise time of the signals. To obtain low-loss interconnections, the lines must be sufficiently thick.

Transmission line phenomenon is a combination of capacitive and inductive effects and is determined by the simultaneous solution of electric (capacitive) and magnetic (inductive) fields. Another important issue that becomes significant at even lower frequencies (15–20 MHz) is the noise generated as a result of the capacitive coupling between the lines. This grows as the number of interconnection levels increases. Figure 1.4 illustrates the difference between line-to-line couplings in single-level and multilevel interconnection schemes. As shown in Fig. 1.4(a), a potential swing in line 2 couples to line 1. This coupling is proportional to the capacitance between lines 1 and 2 (C_{12}) and inversely proportional to the total capacitance of line 1 ($C_{12} + C_{1G} + C_{13}$). As a result, a voltage swing ΔV_2 on line 2 will cause a noise pulse of $\Delta V_{1,noise}$ in line 1 as described by the following equation:

$$\Delta V_{1,noise} = \Delta V_2 \frac{C_{12}}{C_{12} + C_{1G} + C_{13}}. \tag{1.4}$$

The same is true for the structure in Fig. 1.4(b), which has three levels of interconnections, but there is a major difference between these two cases. In single-level metal technology, C_{1G} is larger than C_{12} and C_{13}, and the coupled noise between the interconnections is minimal. For the wires in the third and higher levels of the multilevel interconnections, a dominant and stable capacitance to ground plane does not exist, and as a result coupling between the lines increases. This is similar to the noise problems observed in silicon-on-insulator (SOI) and silicon-on-sapphire (SOS) based technologies.

Additionally, the coupling between the wires is made worse by the fact that interconnection thickness is increased as feature size is reduced in order to keep the wire resistance low. This yields interconnection cross-sections that are almost square shaped and increases the wire-to-wire capacitance to a level larger than it would have been if both wire thickness and width were scaled by the same factor.

The silicon substrate also introduces unique problems by behaving as a conductor for capacitive effects and as an insulator for inductive effects because of the peculiar behavior of the magnetic and electric fields in semiconductors [1.19]. These properties are strongly dependent on substrate resistivity and operating frequency and make a metal current return path in close proximity to high-performance interconnections very desirable.

In Chapter 6, transmission line theory as applied to CMOS and bipolar VLSI circuits is presented, and termination methods that prevent reflections while maintaining the low dc power-dissipation property of CMOS circuits are described.

1.5   Noise

As illustrated by the capacitive cross-talk example in the previous section, a significant system-level problem is externally and internally generated noise. When the feature sizes and signal magnitudes are reduced, circuits become more susceptible to outside disturbances. The levels of noise coupled from external sources that used to be acceptable can become intolerable when internal potential swings are reduced. Smaller feature sizes result in reduced node capacitances. This helps to improve circuit delays; however, these nodes also become more vulnerable to external noise, especially if they are dynamically charged.

An example of this problem is the loss of information stored in DRAM cells due to alpha particles. This becomes a problem when the storage capacitance goes below a certain value in a scaled-down DRAM cell [1.27]–[1.28]. It is refered to as *soft errors*. As shown in Fig. 1.5, an alpha particle generates electron hole pairs in the silicon substrate, and these charges can be swept to a nearby dynamically charged node, such as the bottom plate of a DRAM cell or a precharged logic gate, and change its state. The source of these high-energy particles can be any radioactive impurities in the package materials and IC layers (for example, aluminum, ceramic, or plastic) or external sources (for example, radiation from the outer space). To reduce soft-error rates, package and IC materials should be purified from radioactive contaminants as much as possible. Radiation generated by packaging can be screened by protective coatings. Cosmic rays, however, have too much energy to be screened by a protective layer, and radiation from the chip interconnections and substrate

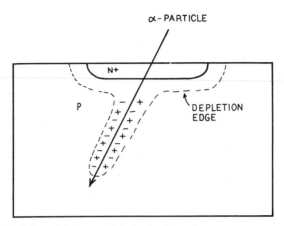

FIGURE 1.5   Alpha-particle-induced soft error

is also impossible to block. Device and circuit innovations, such as creative use of epitaxial layers, trench isolation, and heavily doped buried layers, are continuously developed to minimize the soft-error rates. Soft errors also become significant in static RAM cells as feature size is scaled down. This may cause moving away from the area-efficient four-transistor cell with polysilicon pull-up resistors to a full six-transistor CMOS SRAM cell.

Coupling between neighboring circuits and interconnections and dI/dt noise generated by simultaneous switching of circuits are the most prevalent forms of internal noise. As chip dimensions and clock frequency increase, the wavelengths of the signals become comparable to interconnection lengths, and this makes interconnections better "antennas." In addition, capacitive, inductive, and resistive couplings between neighboring circuits become more significant with higher packing density [1.26].

As the total chip current increases, it also becomes very difficult to control inductive noise in the power lines (dI/dt noise). Board wires, package pins, bond wires, and IC interconnections all have parasitic inductances. When a current that flows through an inductor changes, a voltage fluctuation is generated across the inductor proportional to the inductance $\mathcal{L}$ and the rate of change of the current dI/dt. As a result when the circuits switch on and off, the voltage levels at the power distribution lines fluctuate. This is also refered to as *simultaneous switching noise* because it is most pronounced when many off-chip drivers switch simultaneously. Voltage-level fluctuations are especially important in precharged CMOS and nMOS circuits because they require large current surges at the beginning of the clock periods. In addition, signal levels are decreased with reduced power supply voltage. The combined effect of these could be excessive degradation of the signal-to-noise ratio in high-speed VLSI circuits if they are designed with conventional techniques and structures. Improved power connections with minimal inductance will be essential, and to limit the voltage fluctuations, many such connections may be required. Effective inductance of a power distribution network is inversely proportional to the number of power/ground connections. In addition, decoupling capacitors placed very close to the chip and on-chip decoupling capacitors that occupy minimal chip area may be necessary. On-chip decoupling capacitors may be required especially in DRAM chips where large current surges are generated when bit lines are precharged and sense amplifiers are enabled. Decoupling capacitors momentarily take over the role of the power supply and provide charge to the chip until the potential level on the distribution lines settles down. Chapter 7 addresses these and other noise-related issues.

1.6 High-Speed Clock Distribution

The *clock* is one of the most important signals in synchronous digital circuits because it controls the timing and throughput of the entire system. To achieve high data rates, it is essential to provide a fast storage element (latch, flip-flop, double-latch) and a robust clocking scheme free from race conditions. Various storage element designs and clocking schemes are reviewed and compared in Chapter 8.

Distributing the clock with minimal skew is a major concern in high-speed digital hardware design. Clock skew is due to the variations in the time of arrival of a clock signal at various parts of a digital circuit. This skew is caused by the fact that not all circuits are equidistant from the clock driver. Since the clock must be distributed globally, clock skew is a major concern in digital system design because it may account for 10 percent or more of the system cycle time [1.20]–[1.23].

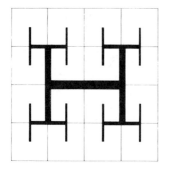

FIGURE 1.6 Symmetric H-clock tree

Based on the concepts developed in the first seven chapters, Chapter 8 presents a symmetric clock distribution tree suitable for high-speed ULSI and WSI circuits, as well as for multichip module and board designs. If all the circuits are equidistant from the clock driver, there will be no skew. Accordingly, in the H-clock tree method, clock signal is delayed by an equal amount before it reaches all the subblocks in the system and, as a result, the skew is reduced. The H-clock tree is illustrated in Fig. 1.6. The clock signal is fed from the center of the tree and reaches to the end points simultaneously. In addition the distributed RC delay of the line is minimized to ensure steep rise times. This technique can be used in large VLSI chips, as well as in module and board layout. Design considerations regarding interconnection RC delay, skin effect, reflections at the discontinuities, cross talk, termination and driving of the clock lines, and electrical design of interconnections are described. In Chapter 8, a design example in which the clock signal is distributed over a 3-inch wafer with a 0.5 nsec rise time and a 0.5 nsec skew is presented [1.24].

1.7   System-Level Performance Optimization

With larger dice and smaller minimum feature sizes, the number of transistors that can be integrated on a die increases by $S_C^2 S^2$, where S_C and S are the scaling factors for the chip and the minimum feature sizes, both greater than 1. Large numbers of transistors require more interconnections and, as a result, chips become limited by interconnections in many ways. First, integration density is constrained by the number of interconnection levels rather than by transistor packing efficiency because of the large number of wires required to interconnect the continuously increasing number of devices [1.29]. The number of circuits is not limited by how many circuit books can be placed on the chip but by how many of them can be wired. Second, power dissipation density is determined by the interconnections because interconnection capacitance becomes a principal component of the total chip capacitance, and the dynamic power required to charge and discharge this capacitance grows with reduced minimum feature size and increased packing density, chip size, and operating frequency. Third, as integrated circuits get faster, the delay due to long wires on the chips, and packaging limits and chip-to-chip propagation delays become the major factors that determine overall performance.

As more and more devices are integrated on a chip, as chip size increases, and as the clock frequency improves, electrical phenomena that were significant only in packaging, such as transmission line properties and inductive couplings, will also become important in chips, and electrical design of chips will start resembling package design. At the same time, with the increased use of multichip packages, denser package wiring, and increased chip input/output (I/O) counts, package design will become more like chip design. This will eliminate most of the distinctions between the two and may yield to merging of the chip and package design under one discipline.

There is a strong relation between the wiring requirements of the design and the machine architecture, organization, and implementation. Techniques that improve system performance, such as pipelining and parallelism, also increase the communication requirements between the sections of the machine and raise the wiring demands. Thus, the main architecture and logic design-dependent performance metric (average number of cycles required per executed instruction), and the main technology- and circuit-dependent performance metric (cycle time) are tied together by wiring supply and demand of the chip and packaging technologies. To minimize the number of cycles required for the execution of an instruction, pipelining and parallelism of the machine are increased, which also stresses the wireability of the system and degrades the cycle time. This yields a trade-off between the cycles per instruction and the cycle time. One can find an optimal design choice in terms of logic implementation and chip partitioning for a given chip and packaging technology that will give a high clock-frequency-to-cycle-per-instruction ratio.

In Chapter 9, these and other system-related issues described in the previous chapters are combined to form a system-level model that encompasses material, device, circuit, logic, packaging, and architecture parameters. This model is used to project the electrical performance of the future high-performance chips and systems and to evaluate the impact of technology and packaging developments on overall system performance [1.30]. Models like this are becoming important as the distinction between chip and package design is reduced and as a wide range of chip and packaging technologies becomes available. Performance evaluation metrics that combine chip, packaging, and architecture parameters will be essential in making choices that will optimize the overall system performance.

2

VLSI DEVICES AND INTERCONNECTIONS

Metal-oxide-semiconductor field-effect transistor[1] (MOSFET) is the technology of choice for VLSI applications because of its high packing density, low power dissipation, and high yield. In this chapter basic principles of MOS transistors are described, and then speed, power, and density improvements brought by scaling are presented. Later, second-order short-channel effects that result from miniaturization are analyzed, including channel-length modulation, velocity saturation, mobility degradation, finite channel thickness, source and drain resistance, dependency of V_T on device geometry, and sub-threshold current. Device reliability issues, such as hot-carrier effects, oxide and junction breakdown, electrostatic overstress, and CMOS latch-up, are also described. In addition, advanced CMOS, BiCMOS, MESFET, bipolar, and GaAs technologies are presented, and interconnection-related material and reliability issues are described, including electromigration, contact failures, and step coverage.

---

[1] The name *metal-oxide-semiconductor* stems from the basic structure of early FETs with a metal gate electrode, silicon dioxide gate insulator, and silicon channel. Since modern devices have polysilicon gate electrodes, referring to them as MOS is a misnomer. As devices are scaled down, however, *metal gate* technology may again become a popular option.

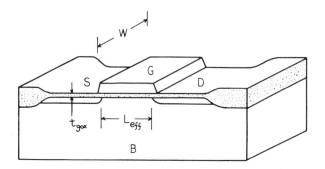

FIGURE 2.1   MOS transistor. The main dimensions that determine device properties are gate oxide thickness t_{gox}, transistor length L and width W, and junction depth X_j.

2.1   Fundamentals of MOS Transistors

An n-channel MOSFET (nMOS) and its basic parameters are illustrated in Fig. 2.1. An nMOS transistor is formed by implanting $n+$ source/drain regions on a p-type substrate. During the normal operation of an enhancement mode nMOS transistor, source/drain junctions are reverse biased, and there is no current flow between the source and drain if the gate potential is low. When the gate potential is raised to a sufficiently high level, it bends the band structure of the semiconductor surface below the gate, referred to as the channel region, and effectively changes the channel from p- to n-type. Then an n-type conducting path is formed between the $n+$ source and drain, and current flows (the switch is on) [2.1]–[2.6]. To a first order, the current through an n-channel MOSFET is given by

$$
I_{DS} = \begin{cases} \mu_n C_{gox} \dfrac{W}{L} \left[(V_{GS} - V_{TN})V_{DS} - \dfrac{V_{DS}^2}{2} \right], & V_{DS} \le (V_{GS} - V_{TN}) \\[2ex] \mu_n C_{gox} \dfrac{W}{L} \left[\dfrac{(V_{GS} - V_{TN})^2}{2} \right], & V_{DS} \ge (V_{GS} - V_{TN}). \end{cases}
$$

$$(2.1)$$

The subscripts D, G, S, and B refer to drain, gate, source, and bulk terminals of the transistor, respectively, and W and L are the width and length of the active transistor area. The effective mobility[2] of the carriers is μ_n, and the

---

[2]The mobility will be symbolized by μ_n for electrons, μ_p for holes, and μ when the equation applies to both electrons and holes.

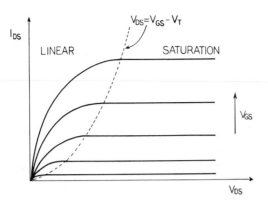

FIGURE 2.2   I-V characteristics of an *n*MOS transistor

gate capacitance per unit area C_{gox} is expressed as

$$C_{gox} = \frac{\epsilon_{ox}}{t_{gox}}, \tag{2.2}$$

where ϵ_{ox} and t_{gox} are the dielectric constant and thickness of the gate oxide. The threshold potential V_{TN} is the value of V_{GS} when the transistor turns on,[3] and for an *n*MOS transistor it is defined as

$$V_{TN} = V_{FB} + 2\phi_F + \frac{\sqrt{2\epsilon_{Si}qN_{SUB}(V_{BS} + 2\phi_F)}}{C_{gox}}, \tag{2.3}$$

where ϵ_{Si} is the dielectric constant of silicon, q is the charge of an electron, N_{SUB} is the substrate doping, and V_{BS} is the substrate (bulk) potential with respect to source. The flatband potential V_{FB} and Fermi potential ϕ_F are

$$\begin{aligned}
V_{FB} &= \phi_g - \phi_{Si} - \frac{Q_{ox}}{C_{gox}} \\[2mm]
\phi_F &= \frac{k_BT}{q}\ln\left(\frac{N_{SUB}}{n_i}\right).
\end{aligned} \tag{2.4}$$

Here $(\phi_g - \phi_{Si})$ is the work function difference between the gate and the semi-conductor, Q_{ox} is the effective fixed charge density at the Si-SiO$_2$ interface, N_{SUB} is the substrate doping concentration, and n_i is the intrinsic carrier concentration of silicon ($n_i = 1.5 \times 10^{10}$ cm^{-3} at room temperature). Thermal voltage k_BT/q is equal to 0.026 V at 300°K, where k_B is Boltzman's constant and T is the temperature in units of Kelvin. The threshold potential

---

[3]The threshold potentials of *n*MOS and *p*MOS transistors will be indicated by V_{TN} and V_{TP}, respectively. For situations that apply to both transistor types, V_T will be used.

can be adjusted by implanting positive or negative ions in the semiconductor region below the gate oxide. This has the the the same effect as Q_{ox} in Eq. 2.4. Ideal current-voltage (I-V) characteristic of an nMOS transistor is plotted in Fig. 2.2. More detailed descriptions of MOSFET device operation can be found in references [2.1]–[2.6].

2.2   MOS Circuits

The basic building block of all MOS digital circuits is the inverter. Figure 2.3 illustrates nMOS inverters. When the input of the inverter is low, the pull-down transistor M2 is turned off, and the output is pulled up high by M1. When the input rises, M2 turns on and pulls V_{OUT} low.

There are two kinds of nMOS devices: enhancement mode and depletion mode. *Enhancement-mode* transistors are normally off and require a positive potential on their gate to be turned on. Enhancement-mode nMOS transistors are good at passing low voltages, but they cannot pass full V_{DD} because both the source and drain ends of the gate are pinched off when the output of the pass transistor reaches $(V_{DD} - V_{TN})$, and the transistor is turned off (Fig. 2.4). As a result the output of the inverter in Fig. 2.3(a), which has an enhancement-mode load, cannot reach V_{DD} unless a separate voltage level $V_{GG} > (V_{DD} + V_{TN})$ is distributed throughout the chip to be connected to the gates of the pull-up transistors. The wiring requirements for this are unacceptably high. On the other hand, if only one power-supply level is distributed, the noise margins and switching speeds of the gates suffer.

Depletion-mode nMOS devices are normally on because they have a negative threshold voltage. The channel region of a depletion-mode transistor is implanted with positive ions that influence V_T the same way as Q_{ox} in Eq. 2.4

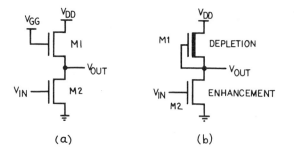

(a)   (b)

FIGURE 2.3 nMOS inverter circuits: (a) enhancement-mode load, (b) depletion-mode load

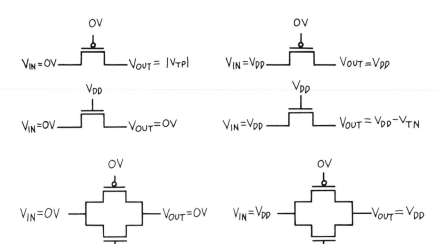

FIGURE 2.4   *p*MOS, *n*MOS, and CMOS pass gates. *n*MOS transistors are good at passing low voltages, but they cannot pass full V_{DD} because both the source and drain ends of the gate are pinched off when the output of the pass transistor reaches $(V_{DD} - V_{TN})$ and the transistor is turned off. *p*MOS transistors, on the other hand, can pass high voltages but not the low ones. A CMOS transfer gate with an *n*MOS/*p*MOS transistor pair passes both low and high voltages well.

does. If a sufficient amount of ion density is established under the gate, the threshold potential of the transistor becomes negative, and it conducts current even with zero gate voltage. As a result, the gate potential of a depletion-mode transistor must be made negative with respect to its source terminal in order to turn it off.

Depletion-mode *n*MOS devices with negative threshold potentials solve the problem of degraded logic high values; they provide inverters with full output swings without requiring two power-supply levels. This improves the noise margins and gate delays. The depletion-mode load device shown in Fig. 2.3(b) has a negative threshold, and its gate is connected to its source. As a result it never turns off; instead it behaves like a mixture between a current source and a resistor. To obtain a sufficiently small logic low value, the on-resistance of the *n*MOS enhancement pull-down M2 must be much smaller than the on-resistance of the depletion pull-up M1. Subsequently, when both M1 and M2 are on, they act similar to a resistive potential divider and provide a low output potential. In an *n*MOS technology where $V_{DD} = +5$ V, $V_{TN} = +1$ V, and $V_{TND} = -3$ V, the following rules are appropriate. The W/L ratio of M2 must be four times the W/L ratio of M1 if the input of the gate is fed from another gate, and $(W/L)_{M2}/(W/L)_{M1}$ must be greater than eight if the gate

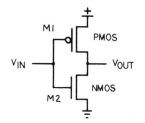

FIGURE 2.5   CMOS inverter circuit

is fed from an nMOS pass transistor, in which case the high value of input is one threshold below V_{DD}. Because of the ratio requirements, the pull-up delay of nMOS gates is four to eight times longer than their pull-down delays. In addition, when their input is low, nMOS gates dissipate dc power because both M1 and M2 are conducting. As a result, the power consumption of highly integrated nMOS VLSI circuits may be as large as 2–15 W.

CMOS circuits with a complementary device pair solve most of the problems associated with E/E and E/D nMOS circuits. In CMOS technology both n- and p-type transistors are available. These two device types are complements of each other. nMOS transistors are formed by diffusing n-type source/drain regions in a p-type substrate, and their gate voltage has to be raised high enough to invert the p-type channel into n-type in order to obtain a conducting layer of electrons between source and drain. pMOS transistors, on the other hand, are formed by implanting p-type source/drain junctions in an n-type semiconductor, and their gate potential must be lowered to induce holes in the channel region, which then carry the current between p+ source and drain. As shown in Fig. 2.40, nMOS and pMOS transistors can be formed on the same wafer by implanting an n-well on a p-substrate. nMOS transistors are fabricated on the p-substrate, and pMOS transistors are placed in the n-well.

CMOS circuits dissipate no dc power because, as illustrated by the CMOS inverter in Fig. 2.5, when the input is high, nMOS transistor turns on and V_{OUT} is pulled low, and when the input goes down, nMOS turns off, pMOS turns on, and V_{OUT} is pulled high. The important fact is that nMOS and pMOS transistors are never turned on simultaneously, and a dc-conducting path is never formed between the power and ground.[4] This eliminates the dc power consumption associated with E/E and E/D nMOS circuits and reduces the heat generation tremendously. This is the most important advantage of

---

[4]Actually both nMOS and pMOS transistors of the inverter are conducting for a brief period during switching when V_{IN} is around the midpoint between V_{DD} and ground. This time duration, however, is usually a small fraction of the system cycle, and the resulting dc power dissipation is small.

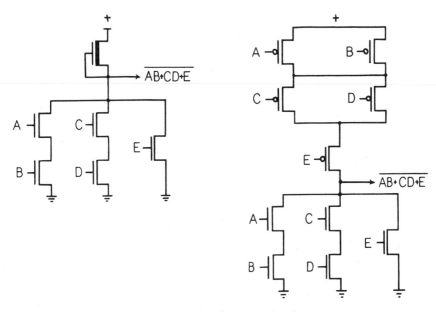

FIGURE 2.6   *n*MOS and CMOS NOR gate circuits

CMOS devices and makes them essential for VLSI circuits where a million transistors may be integrated on a 1×1 cm^2 chip. At such high integration levels, even a very small amount of dc power dissipated by each gate can add up to prohibitively large amounts. In addition, because the pull-up and pull-down transistors are never turned on simultaneously, valid logic "0" values can be achieved without restricting the ratio of the pull-up and pull-down transistors. This gives more balanced rising and falling delays and improves the noise immunity of the circuits. Because *p*MOS passes high voltages well and *n*MOS passes low voltages well, CMOS circuits have excellent noise margins ($V_{OL} = 0$ V, $V_{OH} = V_{DD}$), and the transfer curve of a CMOS inverter is much sharper than that of an *n*MOS inverter. CMOS circuits are also perfect push-pull drivers; they can drive large capacitive loads with equal rising and falling delays and without dissipating any dc power or requiring boot-strapped gates. This gives CMOS a clear advantage in many critical circuits, such as clock buffers and off-chip drivers. CMOS devices also provide excellent transmission gates, which can pass both high and low levels equally well (Fig. 2.4). This is again made possible by the symmetry between the *n*MOS and *p*MOS devices.

The major drawbacks of CMOS circuits are that they require more transistors than simpler *n*MOS gates, and the CMOS process is more complicated and expensive. The CMOS gate shown in Fig. 2.6 requires about twice as many devices as the equivalent *n*MOS gate shown in the same figure. As a result, the total gate capacitance on a given node with the same fan-out will be at least twice as large as in an nMOS circuit. This gets even worse in CMOS transmission gates, which not only require a pair of pass transistors but also the true and complement of the control signal that turn the transmission gate on and off. This implies an additional inverter, which adds two more devices. As a result, a simple single transistor nMOS pass gate transforms into a four-transistor entity in its CMOS version. In many cases, however, a single control signal is shared by multiple pass-transistor pairs, which reduces the overhead (for example, in a 32-bit multiplexor, the same control signal is tied to all 32 transmission gates). In addition the simple single transistor pass gate is still available in CMOS circuits if the threshold drop is acceptable. Similarly, CMOS register, latch, and multiplexor circuits require complementary clock and enable signals to take full advantage of CMOS. This increases the interconnection requirements and transistor counts of the circuits. Finally, the effective mobility of holes is about half the mobility of electrons. To achieve the same current drive, a pMOS transistor must be approximately twice as large as an nMOS device. If the pMOS transistors are made the same size as nMOS transistors, the pull-up delay will be twice as large as the pull-down delay, and the transfer curve of an inverter will not be totally symmetric. When pMOS devices are made twice as large to achieve symmetry, the input capacitance of the inverter becomes three times that of an nMOS inverter. The low power dissipation of CMOS circuits, however, sufficiently dominates the trade-off considerations to establish CMOS as the technology of choice for VLSI. In CMOS circuits the driver sizes can be increased in order to reduce the net delay without having to worry about power dissipation. In nMOS and bipolar circuits, on the other hand, power is a valuable resource that needs to be allocated carefully; as a result one cannot freely power up the nets with large capacitive loadings. This difference makes CMOS superior to nMOS and bipolar in a VLSI design.

Dynamic precharged circuits can be used to reduce the transistor count and delay of CMOS gates. An example of a dynamic logic gate is shown in Fig 2.7. The node V_X is precharged high during ϕ, and when the precharge clock goes away, V_X may or may not be discharged, depending on the value of the inputs. Because of the inverting buffer, the output is zero when the gate is precharged, so the default value of the output is low, and it goes high only if its intended value is high. If such gates feed each other, all the inputs are guaranteed to be zero during and just after precharge, and this prevents unintentional discharging and ensures proper operation. When the precharge clock goes away, the gates discharge one after another similar to falling dominoes; accordingly this circuit family is called *domino logic* [2.7]. Domino CMOS logic is very compact and fast, but it is also sensitive to glitches and charge sharing at precharged nodes. In addition, it can offer only AND/OR gates with no inversion, which is not sufficient to construct any random logic function.

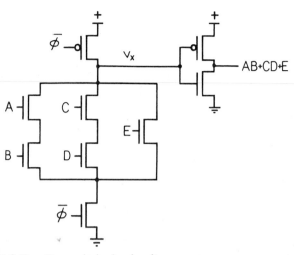

FIGURE 2.7   Dynamic logic circuits

Another important MOS circuit is the single-transistor dynamic random memory (DRAM) cell shown in Fig. 2.8. Here one bit of information (0 or 1) is stored on a capacitor isolated by a pass transistor. Later, depending on the output of the address decoder, one of the many word lines goes high, and the selected memory cell is connected to the bit line. The bit line feeds into a sense amplifier, which reads and amplifies the data stored in the cell capacitor. The data are then passed to off-chip drivers.

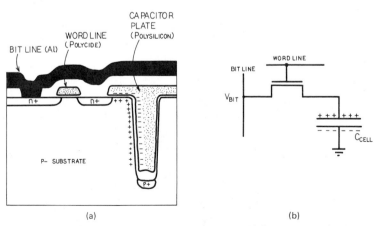

FIGURE 2.8   A DRAM cell with a trench capacitor: (a) cell cross-section and (b) the circuit equivalent of the cell

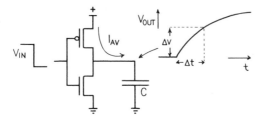

FIGURE 2.9   Delay calculation for a capacitive load driven by an inverter. Delay is proportional to the capacitive load C and the potential swing at the output ΔV and inversely proportional to average current supplied by the driver I_{av}.

2.3   Scaling of MOS Transistors

Miniaturization of MOS transistor dimensions has been and continues to be a popular method to improve circuit speed and packing density [2.8]–[2.16]. As illustrated in Fig. 2.9, circuit delay expressed in its simplest form is proportional to the capacitive load C and potential swing ΔV at the output

TABLE 2.1   Ideal Scaling of MOS Transistors

Parameter	Scaling Factor
Dimensions $(W,\ L,\ t_{gox},\ X_j)$	$1/S$
Substrate doping (N_{SUB})	S
Voltages $(V_{DD},\ V_{TN},\ V_{TP})$	$1/S$
Current per device (I_{DS})	$1/S$
Gate capacitance $(C_g = \epsilon_{ox}\frac{WL}{t_{gox}})$	$1/S$
Transistor on-resistance $(R_{tr} \propto \frac{V_{DD}}{I_{DS}})$	1
Intrinsic gate delay $(\tau = \frac{C_g \Delta V}{I_{av}} = R_{tr}C_g)$	$1/S$
Power-dissipation per gate $(P = IV)$	$1/S^2$
Power-delay product per gate $(P \times \tau)$	$1/S^3$
Area per device $(A = WL)$	$1/S^2$
Power-dissipation density (P/A)	1

node and inversely proportional to the average current supplied by the driver I_{av}:

$$\Delta t = \frac{C \Delta V}{I_{av}}. \tag{2.5}$$

To optimize the circuit performance, the fundamental goal in high-speed circuit design is to reduce the capacitive loading C and potential swing ΔV while keeping the average current drive capability I_{av} high. As described in Chapter 1, ideal scaling improves circuit performance by reducing capacitances and potential swings, and at the same time it ensures the physical integrity of devices by keeping the electric fields constant [2.8]. When ideal scaling (Table 2.1) is applied, all horizontal and vertical dimensions of transistors (W, L, t_{gox}, X_j) are scaled down by $1/S$ ($S > 1$). Substrate doping (N_{SUB}) is increased by S, and all voltages (V_{DD}, V_{TN}, V_{TP}) are reduced by $1/S$. As a result, internal electric field strengths remain approximately unchanged, and velocity saturation, oxide breakdown, and carrier multiplication problems are avoided. With ideal scaling, gate capacitance per unit area ($C_{gox} = \epsilon_{ox}/t_{gox}$) increases by S, current flow in each device (I_{DS}) and parasitic gate capacitance (C_g) decrease by $1/S$, and intrinsic gate delay (τ) improves by $1/S$.

$$
\begin{aligned}
I_{DS} \;&= \mu C_{gox} \frac{W}{L} \frac{(V_{GS} - V_T)^2}{2} \\[6pt]
&\propto 1 \times S \times \frac{1/S}{1/S} \times (1/S)^2 = \frac{1}{S} \\[10pt]
C_g \;&= W L C_{gox} \\[6pt]
&\propto \frac{1}{S} \times \frac{1}{S} \times S = \frac{1}{S} \\[10pt]
\tau \;&= \frac{C_g \Delta V}{I_{av}} \\[6pt]
&\propto \frac{1/S \times 1/S}{1/S} = \frac{1}{S}.
\end{aligned}
\tag{2.6}
$$

Consequently power dissipation per gate (P) is reduced by $1/S^2$, and the power-delay product of devices is scaled down by $1/S^3$:

$$
\begin{aligned}
P \;&= IV \\[4pt]
&\propto \frac{1}{S} \times \frac{1}{S} = \frac{1}{S^2} \\[10pt]
P \times \tau \;&\propto \frac{1}{S^2} \times \frac{1}{S} = \frac{1}{S^3}
\end{aligned}
\tag{2.7}
$$

In addition the packing density of transistors increases by S^2, and power-dissipation density remains constant:

$$\text{packing density} \propto \frac{1}{A} \propto \frac{1}{WL} \propto S^2 \tag{2.8}$$

$$\text{power-dissipation density} = \frac{P}{A} \propto \frac{1/S^2}{1/S^2} = 1 \tag{2.9}$$

As a result, with ideal scaling, devices get faster, power-delay products and power dissipation of the devices are reduced, packing density improves, and power-dissipation density remains constant. Faster chips with larger transistor counts become possible; both the speed and functional capability of ICs improve without increasing the power-dissipation density.

As mentioned in Chapter 1, a number of second-order effects can undo the improvements brought by scaling. For example, velocity saturation, mobility degradation, and increased parasitic source and drain resistances degrade the current drive capability (transconductance) of devices. Nonscalability of the silicon bandgap and built-in junction potentials gives rise to short-channel effects that make the threshold potential vary with transistor width and length, causes degradation in subthreshold behavior, and increases leakage currents. In addition, to satisfy TTL compatibility and to keep utilizing existing ICs without having to install multiple power supplies in systems, voltage levels have not been scaled down along with feature sizes. This increases current drive by S until the carrier velocity is saturated. The result of this is a reduction in the gate delay by an additional factor of S and an increase in power dissipation. It also gives rise to excessive electric fields, which may cause oxide breakdown, drain avalanche, or hot-electron injection into gate oxide. When CMOS circuits are scaled down, parasitic bipolar transistors formed by the *n-p-n-p* structure of the source-bulk-well-source layers become more likely to turn on and damage the circuit (CMOS latch-up). These issues will be described in more detail in the following sections.

As a result of second-order effects, a 2 *μm* device that performs satisfactorily may not be acceptable if it is miniaturized to 1 or 0.5 *μm* following ideal scaling rules. Therefore device structures are improved and ideal scaling rules are modified in order to achieve optimal device performance or because of practical constraints. Despite the differences in their implementation details, however, all MOS miniaturization schemes follow the basic principles of ideal scaling.

2.4   Short-Channel Effects

In this section, short-channel effects are described that negatively affect device performance. The major phenomena that degrade the transistor behavior are channel-length modulation, velocity saturation, mobility degradation, finite channel thickness, source/drain resistance, subthreshold current, dependence of V_T on device geometry, punchthrough, and drain-induced barrier lowering.

2.4.1   Channel-Length Modulation

Simple MOSFET theory assumes that once the drain current is saturated $(V_{DS} = V_{GS} + V_T)$, any increase in V_{DS} will have no effect on I_{DS}. In reality, however, as V_{DS} is increased beyond the saturation value, the depletion edge of the drain moves toward the source, reducing the effective channel length (Fig. 2.10). Accordingly I_{DS} increases as V_{DS} rises, and the transistor acquires a finite output conduction at saturation $(g_m = dI_{DS}/dV_{DS})$. As a result, inverter transfer curve sharpness is degraded, and the noise margins are reduced (Fig. 2.11). The transfer curve loses its sharpness because inverter gain is reduced when saturation output conductance of transistors increases.

2.4.2   Velocity Saturation

In a long-channel nMOS transistor, the electron velocity is given by $v_n = \mu_n \mathcal{E}_x$ where $\mathcal{E}_x$ is the lateral electric field. If the power-supply level is not scaled down with channel length, the electric field along the channel increases, and velocity of electrons eventually saturates at a value determined by carrier scattering (Fig 2.12) [2.17]–[2.19]. For an n-type silicon, the critical field at which saturation occurs is 1.5×10^4 V/cm, and the saturation velocity v_{max} is 10^7 cm/s. The holes in p-type silicon, on the other hand, have a smaller mobility and saturate at a slightly lower velocity. This is why pMOS transistors are not as fast as nMOSFETs. GaAs transistors are faster than silicon devices also because of the higher electron mobility of GaAs. In GaAs, holes also have a smaller mobility, and the difference between electron and hole mobilities is much more pronounced in GaAs than it is in Si. That is why complementary GaAs devices would not be as attractive as silicon CMOS.

The saturation of carrier velocity causes degradation in device transcon-

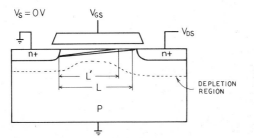

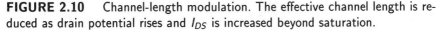

FIGURE 2.10   Channel-length modulation. The effective channel length is reduced as drain potential rises and I_{DS} is increased beyond saturation.

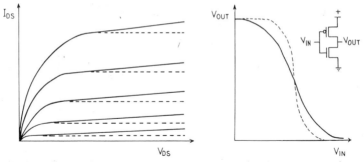

FIGURE 2.11   Effect of finite saturation output conductance on transistor I-V characteristics and inverter transfer curve. Broken lines represent ideal transistor behavior; solid lines are obtained when channel-length modulation effect is taken into account.

ductance. The difference between the drain current at regular long-channel transistors and velocity-saturated short-channel devices can be calculated by the following set of equations. Using the definition of a capacitance $Q = CV$, the amount of charge per unit channel length is expressed as

$$Q_x = W C_{gox}(V_{GS} - V_{TN}). \tag{2.10}$$

Based on the simple drift current equation $(I_x = Q_x v_x)$, the drain current is obtained as

$$I_{DS} = W C_{gox}(V_{GS} - V_{TN})v_n, \tag{2.11}$$

where v_n is the velocity of electrons. In long-channel transistors, $v_n = \mu_n \mathcal{E}_x$, and the pinch-off point occurs at the voltage $V_{DS} = V_{GS} - V_{TN}$. Under these

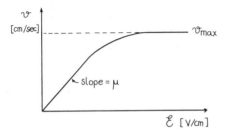

FIGURE 2.12   Velocity saturation in MOS transistors. As the electric field increases, the drift velocity of carriers saturates at a scattering limited value. The slope of the curve at the linear region gives the mobility of the carriers.

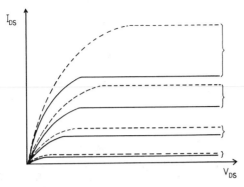

FIGURE 2.13   Effect of carrier velocity saturation on current drive of MOS transistors. Broken lines are the curves predicted for a short-channel device by simple MOS theory, and solid lines are obtained when velocity saturation is taken into account. The transconductance of the transistor is degraded, and as illustrated by the equal spacing of the broken lines, I_{DS} is proportional to $(V_{GS} - V_T)$ rather than $(V_{GS} - V_T)^2$.

conditions, average electric field[5] at pinch-off is $\mathcal{E}_x = (V_{GS} - V_{TN})/2L$, and the familiar saturation drain current of a long-channel device is obtained:

$$I_{DS} = \mu_n C_{gox} \frac{W}{L} \frac{(V_{GS} - V_{TN})^2}{2}. \tag{2.12}$$

In a velocity-saturated transistor $v_n = v_{max}$, and saturation drain current becomes

$$I_{DS} = W C_{gox}(V_{GS} - V_{TN})v_{max}. \tag{2.13}$$

In a long-channel transistor (Eq. 2.12), drain current is proportional to the square of $(V_{GS} - V_{TN})$, and in a velocity-saturated mode (Eq. 2.13), the drain current is linear with $(V_{GS} - V_{TN})$. Consequently, equal gate voltage increments result in equal drain current increments in a velocity saturated device. Also I_{DS} is independent of L in velocity-saturated devices, which implies that once the device is scaled down to such an extent that carrier velocity is saturated, current drive cannot be improved by further reducing the channel length. Velocity saturation can drastically degrade the drive capability of transistors (Fig. 2.13). Velocity saturation can be avoided by reducing the power supply along with channel length.

---

[5]There is a factor of one-half because the electric field is not constant across the channel. The field is higher near the drain, and when it is averaged along the channel, $\mathcal{E}_{xav} = (V_{GS} - V_{TN})/2L$ is obtained.

2.4.3   Mobility Degradation

As described in the previous subsection, when the lateral electric field goes above a certain value, the drift velocity of the carriers saturates. In addition, the electron channel mobility is reduced by the normal electric field because the possibility of carrier scattering from the Si-SiO$_2$ interface increases when the field perpendicular to the channel becomes larger [2.20]–[2.22]. As a result mobility degradation is observed in miniaturized MOSFETs due to an increase in the normal electric field in the inversion layer. The electric field increases because power supply is not reduced with horizontal and vertical transistor dimensions. As the oxide thickness is reduced, a larger electric field is required to establish a given potential difference across the gate oxide. Even when the power-supply level is reduced, the normal field will increase due to the nonscalability of the work function difference between the polysilicon gate and inversion layer.

2.4.4   Finite Channel Thickness

Idealized MOS model assumes the thickness of inversion layer under the gate to be zero. This thickness, however, is on the order of 100 Å. As the gate oxide gets thinner, insulator and inversion layer thicknesses become comparable. Consequently, a significant portion of the gate-to-source potential can drop across the inversion layer. As shown in Fig. 2.14, the gate oxide capacitance C_{gox} is in series with the inversion capacitance C_{inv}. They act as a voltage divider, leading to a reduced effective channel capacitance and charge density. With a nonzero inversion layer thickness, fewer electrons can establish the required voltage difference across the gate and the substrate because there is some potential drop across the inversion layer in addition to the potential drop across the gate oxide. As the oxide thickness is reduced down to the level of the inversion layer thickness, the difference between the actual carrier density and the carrier density with zero inversion layer thickness increases. This causes a degradation in device transconductance because the resulting carrier density under the gate and, subsequently, the drain current are less than predicted by the simple MOS theory [2.23]–[2.24].

2.4.5   Source and Drain Resistance

When transistors are scaled down, their junctions become shallower, and contact openings to these junctions become smaller [2.25]–[2.28]. This results in an increasing parasitic resistance in series with the device, which is expressed as

$$R_S = R_C + R_\square \frac{\Delta x}{W}. \tag{2.14}$$

Here R_S is the series resistance of source/drain, R_C is the contact resistance, $R_\square$ is the sheet resistance per square of source/drain diffusion, and Δx is the

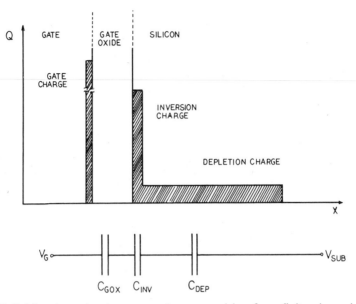

FIGURE 2.14   Inversion layer capacitance resulting from finite channel thickness

distance between the contact and the active transistor area. As illustrated in Fig. 2.15, the effective gate-to-source potential difference V'_{GS} is

$$V'_{GS} = V_{GS} - IR_S, \qquad (2.15)$$

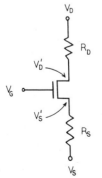

FIGURE 2.15   Modeling of the series resistance of source and drain

and the saturation current is obtained as

$$I_{DS} = \mu C_{ox} \frac{W}{L} \frac{(V_{GS} - V_T - I_{DS}R_S)^2}{2}$$

$$= k(V_{GT} - I_{DS}R_S)^2,$$

(2.16)

where $k = \frac{1}{2}\mu C_{ox}\frac{W}{L}$, and $V_{GT} = V_{GS} - V_T$. For small R_S, the term with R_S^2 can be neglected, and I_{DS} is obtained as

$$I_{DS} = kV_{GT}^2 - 2kV_{GT}I_{DS}R_S,$$

(2.17)

which can be further simplified to obtain

$$I_{DS} = \frac{k}{1 + 2kR_SV_{GT}}V_{GT}^2$$

$$= \frac{k}{1 + 2\frac{R_S}{R_{CH}}}V_{GT}^2$$

(2.18)

$$= \frac{1}{1 + 2\frac{R_S}{R_{CH}}}\mu C_{ox}\frac{W}{L}\frac{(V_{GS} - V_T)^2}{2}$$

where $R_{CH} = (kV_{GT})^{-1}$. As a result the parasitic series resistance causes degradation in device transconductance and reduces its current drive capability by a factor $(1 + 2\frac{R_S}{R_{CH}})$.

The source and drain resistances can be reduced by exposing the diffusion areas to a silicidation process (e.g., titanium or tungsten). Then a layer of low-resistance silicide will form on top of the diffusion area and reduce its overall resistance (Fig. 2.16). This silicidation process can also reduce the resistance of the polysilicon gate in addition to diffusion areas. A sidewall spacer is required to avoid forming undesired shorts between gate and source/drain.

Sidewall spacers can also be employed to build lightly doped source/drain (LDD) structures. Devices with LDD exibit improved short-channel behavior because their junction depths and doping concentrations are smaller at the tip of the source/drain areas. Since the severity of the short-channel effects is determined by the interaction between the gate and source/drain, LDD structure modifies the region in which the interaction takes place in order to minimize the short-channel effects (see Section 2.4.7). LDD transistors also improve hot-electron injection (see Section 2.5.1), breakdown voltages, and gate overlapping capacitances [2.29]–[2.30]. The increased series resistance as a result of lightly doped end points is a concern, but this is minimized by the fact that most of the diffusion area is deep, heavily doped, and covered by a silicide layer.

2.4.6   Subthreshold Current

Simple MOS theory assumes a zero drain current for $V_{GS} < V_T$. In reality, I_{DS} does not drop linearly to zero but decreases exponentially, similar to the

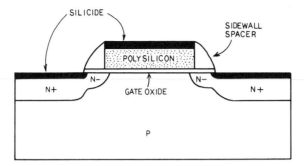

FIGURE 2.16   An LDD MOS transistor with silicided source/drain junctions and silicided polysilicon gate. The sidewall spacer is used to form a lightly doped drain and source structure (the lightly doped area is implanted before the spacer is in place and the latter heavy dose implant is blocked by the spacer). The sidewall spacer also prevents electrical shorts between the polysilicon gate and source/drain diffusion areas during the subsequent silicidation step.

operation of a bipolar transistor. The I-V characteristics of an MOS transistor at two different temperatures are plotted in a logarithmic scale in Fig. 2.17.

The leakage current is governed by a bipolar-like mechanism because when the surface is not inverted, minority carriers and diffusion currents are dominant just like in a bipolar device [2.6], [2.31]–[2.35]. This is unlike MOS transistor operation above the threshold where the surface is inverted and the current flow is dominated by the drift of majority carriers. At the subthreshold region, the inverse rate of decrease of I_{DS} in Volts per decade is given by

$$\left[\frac{d(\log I_{DS})}{dV_{GS}}\right]^{-1} = \left(\frac{k_B T}{q}\ln 10\right)\left(1 + \frac{C_D}{C_{gox}}\right), \qquad (2.19)$$

where C_D is the depletion layer capacitance. The physical basis for Eq. 2.19 is conceptually illustrated in Fig. 2.18. Source $(n+)$, bulk (p), and drain $(n+)$ terminals of the nMOS transistor form an npn bipolar device. The base of this npn transistor, however, is capacitively coupled to the gate terminal, and as a result only a portion of the gate voltage variation is reflected to the base. The capacitive divider is formed by the gate oxide capacitance (C_{gox}) and the depletion layer capacitance (C_D). This determines how much of the gate voltage swing will be seen by the bipolar base and is responsible for the $(1 + \frac{C_D}{C_{gox}})$ term in Eq. 2.19. The rest of the equation is simply a different representation of the familiar bipolar current equation $I_E = I_0 exp\{\frac{V_{BE}}{k_B T/q}\}$. (See Section 2.10 for a detailed description of bipolar device operation.)

Under normal operating conditions, $\frac{k_B T}{q}\ln 10 \approx 60\ mV/decade$, which means that for an ideal pn-junction, every $60\ mV$ increase in the barrier height

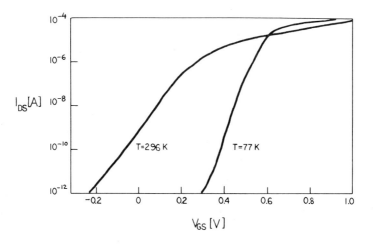

FIGURE 2.17   Subthreshold I-V characteristics of an MOS transistor at room and liquid nitrogen temperatures ($V_{DS} = V_{DD}$). The device turns off more sharply at the lower temperature.

will reduce the current by a factor of ten. This has significant implications for the subthreshold behavior of MOS transistors. If a transistor with $1 + C_D/C_{gox} = 1.4$ conducts 1 μA of current at threshold, it will conduct $1/84 = 0.012$ μA with V_{GS} 60 mV below V_T and 0.00014 μA with V_{GS} 120 mV below V_T. If the threshold voltage is 500 mV, the current will decrease by $500/84 = 6$ decades when V_{GS} is zero, and the subthreshold leakage current will be reduced to 1 $\mu A/10^6 = 1$ pA. The finite rate of change of gate current

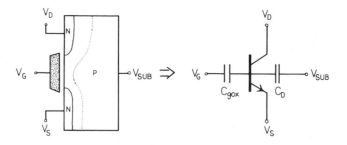

FIGURE 2.18   Equivalent circuit for an MOS transistor at the subthreshold regime. When the device is off, the leakage current between source and drain is determined by a bipolar-like mechanism. The base potential of this bipolar transistor is determined by a capacitive divider formed by the gate and depletion capacitances.

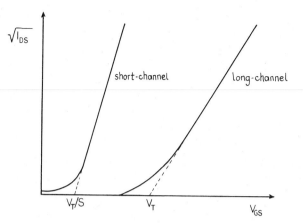

FIGURE 2.19   Subthreshold characteristics of long- and short-channel devices ($V_{DS} = V_{DD}$)

with gate potential poses a fundamental limit on how much the threshold potential can be scaled down. For the circuits to function properly, the device should function close to an ideal switch. When the gate potential is above V_T, it should conduct a significant amount of current, and when the gate potential is 0 V, it should conduct very little current. The ratio of the current with $V_{GS} = V_T$ to the current with $V_{GS} = 0$ V indicates how good a switch the nMOS transistor is. As illustrated in Fig. 2.19, if V_T is reduced too much, the potential differential between 0 V and V_T may not be sufficient to turn off the transistor completely. In this figure, even with a 0 V gate potential, the leakage current of the scaled device is unacceptably high. (Figures 2.17 and 2.19 plot the same quantities but one in a logarithmic scale and the other in a square-root scale to illustrate two separate but related issues.)

The impact of the subthreshold conduction for dynamic circuits and especially for DRAMs is significant because stored charge on a capacitor may be lost due to subthreshold leakage before it is refreshed. Figure 2.8 illustrates a DRAM cell with a trench capacitor. The memory cell consists of a capacitor that stores the information and a pass transistor that selectively connects the cell to the bit line for read and write operations. The gate of the pass transistor is connected to the word line, which is driven by the address decoder. If the cell is not addressed, the pass transistor is off, and the information stored in the capacitor should stay intact. The subthreshold leakage current of the pass transistor, however, will discharge the storage node slowly. In the previous paragraph, one picoampere of leakage was calculated for a 0.5 V threshold potential. This current requires only 50 msec to reduce the potential stored on a 50 fF (typical of 1 Mbit DRAMs) of capacitance by 1 V. As a result all the memory cells in a DRAM must be refreshed continuously by reading the data and writing them back. To ensure reliable DRAM operation, higher dielectric constant materials, insulators with improved dielectric breakdown characteristics that will allow higher electric fields across storage capacitors, and trenches that cut deeply into the substrate to increase effective storage area are investigated as possible solutions. Since subthreshold leakage decreases exponentially with temperature (Fig. 2.17), reducing the operation temperature to liquid nitrogen levels is the most elegant solution but not necessarily the most cost-effective one.

2.4.7   Dependence of V_T on Device Geometry

As the channel length and width are reduced, the threshold potential becomes dependent on L, W, and V_{DS}. The variation of V_T with L in short-channel devices can be qualitatively explained using Fig 2.20. As seen in this figure, there are depletion regions at the reverse-biased source and drain junctions, as well as under the gate area. The amount of charge at the source and drain junctions is determined by the bandgap and built-in junction potentials of silicon (these cannot be scaled down) and also by the reverse-bias potential dictated by the power-supply level (this has not been scaled down as much as the feature sizes). Accordingly when transistor dimensions are reduced, the amount of depletion charge at the source and drain regions becomes comparable to the amount of charge at the intrinsic device area. The threshold potential of the transistor is influenced by the source and drain charges because they "help" the gate to turn on the device. This causes the effective threshold of shorter-channel devices to be smaller than their long-channel counterparts. An alternative way to explain this would be by conceptualizing that some of the depletion charges in the channel region next to the source/drain areas belong to the junction depletion region. As a result the gate does not have to support these depletion charges. One can imagine a trapezoid under the gate that separates the charges that belong to the gate and junctions shown in Fig. 2.20. As feature sizes are scaled down, the depletion charges that belong to source/drain junctions become a larger portion of the total depletion charge, and the effective device threshold decreases (Fig. 2.21). In addition, V_T is reduced when V_{DS} increases (drain-induced barrier lowering, DIBL) because the amount of depletion charge in the reverse-biased drain junction increases, and the gate has to support less charge than it did with a smaller V_{DS}.

Similarly, the threshold potential becomes dependent on transistor width because of the spreading of field lines outside the gate area (Fig. 2.22). Since the gate has to support these "spilled-over" depletion regions in addition to the depletion charges accounted for in the ideal model, the effective threshold potential is larger than the value the ideal model indicates. This difference is larger in narrow-channel devices because the "spilled-over" depletion charges constitute a larger portion of the total charge in narrow devices compared

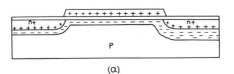

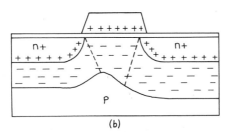

(a)

(b)

FIGURE 2.20   Lateral cross-sections of long- and short-channel *n*MOS transistors. (a) In the long-channel device, the electrical properties of the channel are determined by the gate and bulk silicon, and the junction depths are small compared to channel length. (b) In the short-channel device, on the other hand, source and drain depletion regions encroach under the gate area, junction depths are comparable to channel length, and the threshold potential of the transistor becomes a function of channel length and V_{DS}: Figures (a) and (b) are not drawn at the same scale. If they were, (b) would be much smaller than (a).

to wider transistors (Fig. 2.21). Modern devices with recessed-oxide isolation have complicated structures and exhibit a number of interacting mechanisms that require two-dimensional analysis to determine the net effect of the narrow channel width.

It should be noted that the above explanations are conceptual and simplified. The real nature of short- and narrow-channel effects and drain-induced

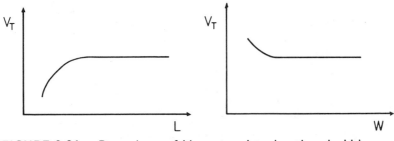

FIGURE 2.21   Dependence of V_T on transistor length and width

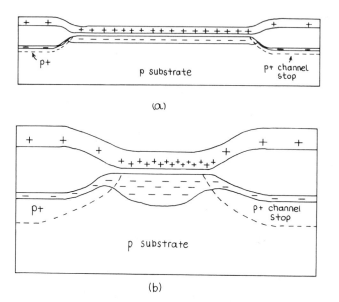

FIGURE 2.22 Cross-sections of long- and short-channel *n*MOS devices along the width of the transistors. (a) In the long-channel device, the electrical properties of the channel are determined by the gate and bulk silicon. (b) In the short-channel device, on the other hand, the depletion charges that spread beyond the gate become a measurable portion of the total charge, and as a result V_T becomes a function of gate width, *W:* Figures (a) and (b) are not drawn at the same scale.

barrier lowering can be best understood with two- and three-dimensional analysis of the band structures, potential fields, and current flow lines of scaled-down MOSFETs.

Among the three effects, dependence of V_T on channel width is the least pronounced one; dependence of V_T on channel length is made more tolerable by the fact that almost all the transistors on a chip have minimum channel lengths, and their current drive capability is adjusted by changing their width. Variation of V_T with V_{DS} is the most troublesome because when the circuit switches, its drain potential will change as part of its function. Increased leakage currents at turned-off devices when drain voltage rises is a major problem in dynamic logic and memory circuits, which depend on reliable storage of charges in isolated nodes. Figure 2.8 illustrates a dynamic node in a DRAM cell. When the gate is turned off ($V_{GS} = 0$ V), the capacitance C is expected to be isolated from the bit line independent of the value of V_{BIT}. As a result of DIBL, however, V_T may be reduced with rising V_{BIT}, and the leakage current may increase and destroy the information stored on the capacitor. This phenomenon is discussed in more detail in the following section.

2.4.8 Punchthrough and DIBL

If the sum of the source and drain depletion layer widths is greater than the channel length of a scaled-down device, the basic MOSFET structure is no longer in effect, and source and drain are shorted together. This is referred to as punchthrough; however, the actual physical phenomenon is more complex than this [2.6], [2.35]–[2.36].

To illustrate the concepts involved in punchthrough and DIBL, surface potential across the MOSFET channel at various drain potentials is shown for a short-channel device in Fig. 2.23. The barrier height at the source end is lowered as drain potential is raised. This increases subthreshold leakage currents. As described in Sections 2.4.6 and 2.4.7, leakage currents can cause failures in dynamic circuits, and especially in DRAMs. To reduce the charge loss, critical pass transistors are not made minimum L. This, of course, increases the delay through the pass transistor. In addition to raising subthreshold leakages, DIBL also increases saturation output conductance of the transistors with a result similar to channel-length modulation (Fig. 2.10). In this case, instead of the effective channel length, the potential barrier at the source end is reduced, and the net effect is an increase in the drain current when V_{DS} rises.

2.5 Device Reliability and Scaling

In this section reliability issues that arise from device miniaturization are described, such as hot carriers, oxide and junction breakdown, electrostatic discharge, electrical overstress, and latch-up.

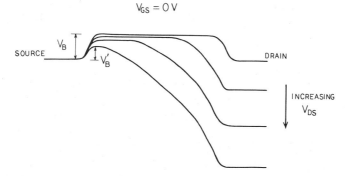

FIGURE 2.23 Drain-induced barrier lowering (DIBL). Surface potential for electrons across the channel is shown for various values of V_{DS} for a turned-off short-channel nMOS device (V_{GS} ¡ V_{TN}). The potential barrier at the source end decreases as V_{DS} rises, which results in an increased leakage current.

2.5.1   Hot-Carrier Effects

As minimum feature size is reduced and power supply remains unchanged, the product of channel length and hot-electron critical field becomes comparable to transistor operating voltages. Then electrons can gain sufficient energy to overcome the interfacial barrier and get injected into the gate oxide [2.37]–[2.40]. The injected electrons can be trapped at the gate and, as they accumulate, the threshold potential shifts upward with time and the transconductance of the devices degrades because the current is proportional to $(V_{GS} - V_{TN})^2$. As the threshold increases, current drive capability of the device is reduced. The injected electrons that are trapped affect the threshold potential the same way Q_{ox} does in Eqs. 2.3 and 2.4. The time dependence of this V_T degradation is especially disturbing because a circuit that passes functional and parametric tests may fail in the field due to threshold shifts.

The hot electrons can originate from the substrate leakage currents, surface channel currents, or impact ionization currents [2.37]. *Substrate hot electrons* are generated as a result of a large potential difference between the gate and the substrate (for example, in bootstrap circuits used in nMOS drivers). The electrons are thermally generated in the bulk near the depletion boundary and accelerate across the depletion region toward the gate Si-SiO$_2$ interface. The electrons that arrive at the surface with sufficient energy may overcome the barrier and enter the SiO$_2$ layer (Fig. 2.24(a)). *Channel hot electrons*

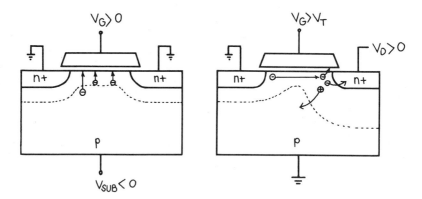

FIGURE 2.24   Hot-electron injection into the gate oxide via substrate leakage and surface channel current. (a) Thermally generated electrons near the depletion region accelerate toward the gate insulator. If they gain sufficient energy, they may enter the oxide region. (b) Regular channel conduction current may also contain electrons with high enough energy to inject into the gate insulator, or they may cause other electrons to inject upon impact ionization.

originate from channel conduction current and impact ionization current near the drain junction. Electrons drifting from source to drain may gain sufficient energy to enter into the gate, or they may collide with the silicon atoms and generate electron hole pairs; the hole becomes substrate current, and the secondary electron may inject into the gate (Fig. 2.24(b)). In addition to threshold shifts, hot carriers can also trigger latch-up because minority carriers generated by ion impaction are injected into the substrate.

Radiation damage to gate oxide (for example, during processing steps such as electron beam or X-ray lithography and e-beam metal evaporation) increases the defect density in the oxide. These defects act as electron traps, and the probability of hot electrons being trapped increases. Process-induced radiation damage up to the metal deposition is annealed during normal high-temperature processing steps, such as diffusion. Damage introduced by the subsequent steps cannot be easily eliminated because annealing at high temperatures ceases to be feasible once the metal is deposited due to the low melting point of aluminum.

To avoid hot electrons, the power-supply level can be scaled down with feature sizes. Alternatively lightly doped drain/source (LDD) structures that tailor the electric fields near the drain can be employed, and the defect densities in gate oxide can be minimized by annealing them at high temperature and pressure. Holes are less likely to inject into the gate oxide than electrons because they are more localized; however, once they enter, they are also more likely to stay there. Overall pMOS transistors are less susceptible to hot-carrier effects than nMOS devices.

2.5.2   Oxide Breakdown

Thin insulators are essential in MOS circuits. When gate oxide thickness is reduced, C_{gox} increases. This improves the coupling between the gate terminal and the silicon surface, and as a result the inversion charge density and drain current increase, as expressed by Eq. 2.1. In effect, transconductance of the devices improves with reduced oxide thickness. Thin dielectrics are also essential in DRAM cells because the cell capacitance is proportional to how thin the insulators can be grown. Cell capacitance is an important parameter because its ratio to total bit line capacitance determines the access time. In addition, circuit noise margins and alpha particle immunity requirements introduce a lower limit to the value of the DRAM cell capacitance and the amount of charge stored on it in order to ensure reliable operation. As a result, the thinner the insulator is, the larger the cell capacitance and the smaller the required cell size. Small cell size minimizes chip area and cost and improves DRAM access time by resulting in short bit lines with smaller capacitance.

If the supply voltage is not scaled down with insulator thickness, the electric field across the gate oxide increases. This may give rise to local breakdown of the gate dielectric, and the rapid discharge (runaway current) of the gate charge through the oxide can generate a large amount of local heating, melting the gate material and insulator. This breakdown mechanism is not very well understood, but it is believed to be due to three failure modes: intrinsic, defect-induced, and wear-out.

The amount of energy dissipated by the runaway current is crucial. The transistor is permanently damaged only if this current has sufficient energy to flow the materials and form a short between the gate and the channel. The time-dependent nature of this failure is also important. When a sufficiently high potential difference is applied across the gate insulator, a tunneling current flows. This starts when $\mathcal{E}_{ox}$ reaches approximately 6×10^6 V/cm. A portion of these charges are trapped by the defects. This alters the electric fields in the oxide, and, with time, a local field in the oxide may reach a critical value ($\mathcal{E}_{crit} \approx 10^7$ V/cm). The resulting avalanche runaway will then destroy the device. Like V_T shifts due to hot electrons, the time-dependent nature of this oxide breakdown mechanism is a concern because devices that are functional after production may later fail in the field, and it is not easy to determine the devices that are prone to early failure.

Burn-in tests that stress devices under high field and temperature are used to eliminate the weak devices that are more likely to experience an early failure. The well-known "bathtub" curve describes time-dependent failure mechanisms (Fig. 2.25). A number of weak devices fail during the initial time period due to defects. Later there is a long time of low failure rates. Finally failure rate increases at the end of the lifetime of the product due to wear-out. The purpose of burn-in is to eliminate the devices with defects by stressing them under high fields and temperatures and pushing them to the flat portion of the failure curve before placing them in the products.

Despite the problems associated with thin oxide layers, 200 Å thick gate oxides are routinely produced today, and it is projected that insulator thicknesses can be reduced to as little as 50–100 Å. In addition, new insulators such as nitrides, which have larger dielectric constants, can support more gate charge at greater insulator thicknesses, and they may also be more resistant to breakdown than SiO_2.

2.5.3 Junction Breakdown

When a sufficiently large electric field is applied across a *pn* junction, the junction breaks down and conducts a very large amount of current [2.6]. The junction breakdown can result from thermal instability, tunneling effect, or avalanche breakdown. *Thermal instability* is a result of reverse-bias current, which heats up the junction. The rising temperature, in turn, increases the reverse current, and a positive feedback mechanism can be initiated, which goes on until the junction is destroyed. Thermal instability is not a primary factor in breakdowns of modern device junctions.

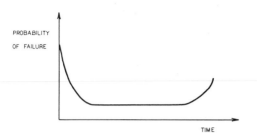

FIGURE 2.25   Failure rates for a time-dependent breakdown mechanism (the "bathtub" curve). A number of weak devices fail during the initial time period due to defects. Later there is a long period of low failure rates. Finally failure rate increases at the end of the lifetime of the product due to wear-out.

Quantum mechanical tunneling is responsible for currents that go straight through the energy barrier across the reverse-biased junction rather than above the barrier (Fig. 2.26). This tunneling current increases with the electric field across the junction and can be destructive at sufficiently large reverse biases. Tunneling current is important in present devices, but *avalanche multiplication* (or impact ionization) is the major mechanism responsible for the junction breakdowns at the collector junction of bipolar transistors and the drain junctions of MOSFETs. If the electric field across the junction is large enough, an electron injected into the depletion region can gain sufficient energy to knock electron-hole pairs from the lattice atoms. These carriers, in turn, can gain energy and generate more electron-hole pairs (Fig. 2.26). The current, as a result, increases rapidly and the junction breaks down. If large amounts of heat are generated, the breakdown can be destructive. Similar to the tunneling effect, avalanche breakdown also worsens with increased electric fields across the junction. (Graded junction profiles result in smaller electric fields than abrupt junctions and can be used to improve the breakdown properties of the junction.) As a consequence of tunneling and avalanche breakdown, depletion widths cannot be smaller than a certain lower limit because the electric field is inversely proportional to the depletion width. Since heavily doped junctions have narrower depletion widths, the maximum doping levels may be limited by junction breakdown. This is significant for scaled-down devices because scaling requires increased doping concentrations and reduced junction depths. Corners of the miniaturized junctions are not as smooth as deep junctions, and this increases the fields at the corners. In addition, if the voltage supply level is not scaled down as aggressively as feature sizes, junction depths, and doping levels, this will worsen the severity of junction breakdown. Despite these difficulties, however, the scaling of modern CMOS devices is not seriously limited by junction breakdowns.

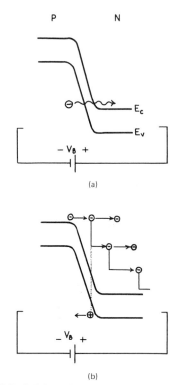

(a)

(b)

FIGURE 2.26   Quantum mechanical tunneling and avalanche breakdown in reverse-biased *pn* junctions: (a) electron tunneling, (b) avalanche breakdown. In these band diagrams, the y-axis is the potential of electrons, and the x-axis is the distance across a reverse-biased junction.

2.5.4   Electrostatic Discharge and Electrical Overstress

Electrical overstress (EOS) and electrostatic discharge (ESD) are major reliability problems closely related to oxide breakdown (Section 2.5.2) and junction failures (Sections 2.5.3 and 2.6.2). Ungrounded conductive objects and people can accumulate static electric charges by induction or by contacting charged insulators. If these objects or people then come into contact with an IC and are discharged through an ESD-sensitive path in the circuit, large momentary currents are generated, and the circuit can be damaged [2.41]–[2.44].

Plastics and other synthetic materials are very good insulators; they accumulate static charges and can retain them for long periods of time. If a charged insulator is in the near vicinity of an insulated conductor, such as a human wearing rubber shoes, large amounts of static charge can be generated on him or her by induction. If that person is holding a pointed metallic object, such as a tweezer, the electric fields and charge densities at that point can reach very large levels, and if the object contacts an IC, the resulting charge

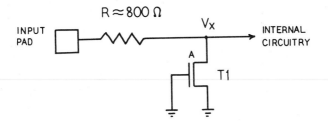

FIGURE 2.27   *n*MOS input protection circuitry

surge can damage the delicate thin oxide layers and shallow junctions. The electrostatic charge can also flow from the IC to an outside ground. If the pins of a packaged IC are insulated, the chip can attain large amounts of charge via induction. If one of the pins is later grounded through a conducting path, the large current surge that discharges the IC can cause damage.

ESD damage can occur during wafer processing, dicing, packaging, testing, board assembly, transportation, and in the field. Wafers are prone to ESD damage during processing. The source of ESD can be the manufacturing equipment or technicians who handle the wafers. Under regular conditions some amount of water condenses on objects and forms a conducting layer on their surfaces. Accumulated charges can be disposed through this thin condensation. The low humidity requirements of wafer-processing environments, however, cause the surrounding objects to collect more charge than they would otherwise and make them more hazardous. Once the dice are packaged, great care is taken to store them in antistatic foam and to use wrist bands that ground the work bench and operator in order to avoid the ICs' coming into contact with a charged object while they are being handled.

Scaling down the devices makes them more susceptible to ESD. When the gate oxide thickness and source/drain junction depth are reduced, their breakdown voltages are also decreased. In addition, miniaturized feature sizes result in smaller device capacitances, and consequently a smaller amount of charge coupled from outside is sufficient to generate large potentials and electric fields in the device.

Robust input devices that protect the internal circuitry from ESD are essential for reliability assurance. A conventional *n*MOS input protection circuit, the so-called lightning arrester, is shown in Fig. 2.27. If V_X is much larger than V_{DD}, T1 turns on due to punchthrough and V_X is clamped. If V_X is one threshold below ground, T1 turns on naturally, and the input potential is again clamped.[6] As a result,

$$-V_{TN} < V_X < V_{Punchthrough}. \tag{2.20}$$

---

[6]If there is no substrate bias, the diode formed by the drain diffusion may be forward biased when input potential goes below 0 V, and the input may be clamped before T1 turns on.

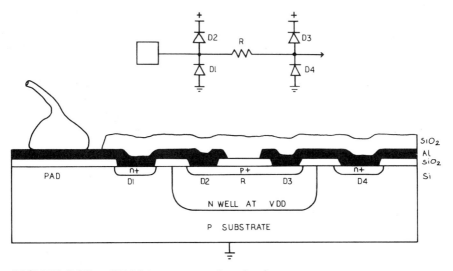

FIGURE 2.28   CMOS input protection circuitry

The resistor R protects T1 by limiting the current and power; it prevents T1 from conducting too large a current and burn.

In CMOS circuits $n+$ and $p+$ source/drain diffusions of nMOS and pMOS transistors provide diodes that can be used to clamp the input node and protect the internal devices. The diode D2 in Fig. 2.28 is formed by the $p+$ diffusion and turns on when the input potential goes 0.7 V above V_{DD}. Similarly, the diode D1 is formed by the $n+$ diffusion, and it turns on when the input potential goes 0.7 V below ground. As a result,

$$-0.7 \ V < V_X < (V_{DD} + 0.7 \ V). \qquad (2.21)$$

It is important that the minority carriers injected into the substrate by the forward-biased diodes do not trigger latch-up (see Section 2.5.5). To prevent latch-up, input circuitry should be surrounded by $p+$ and $n+$ guard rings to collect the injected carriers.

As the technology is scaled down, the input protection devices themselves also need to be protected from ESD. The thin gate oxide of the protection device with the grounded gate in Fig. 2.27 is vulnerable to breakdown at point A. Tying the gate of T1 to ground through a diode or another transistor can reduce the electrical overstress on the input device. During normal operation $(0 < V_X < V_{DD})$, the gates of the protection devices T1 shown in Fig. 2.29 are floating, and the transistor is off. When the input potential goes

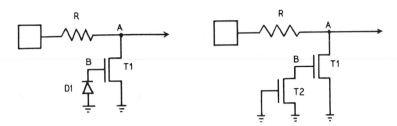

FIGURE 2.29   Improved *n*MOS input protection circuitry. The diode D1 and transistor T2 prevent overstress at node A.

beyond the allowed range given by Eq. 2.20, T1 turns on. At the same time, some of the voltage swing is capacitively coupled from node A to node B by the gate drain overlap capacitance; as a result the thin gate oxide of T1 is not stressed as severely. Later the diode D1 or the transistor T2 is turned on, the charge at the gate capacitance of T1 is discharged, and the input potential is clamped as before.

A hierarchically structured CMOS input protection circuit is described in reference [2.44]. It is shown in Fig. 2.30. Here sturdy but less sensitive protection devices are placed at the input, followed by more fragile devices with smaller clamping potentials. This structure has many positive attributes:

- A floating *n*-well is placed under the pad. This prevents the input pad metal from spiking into the substrate (Fig. 2.31). During the wafer probing, the input pad and the insulator below the pad can be damaged. Later electrical overstress can generate spiking beween the wire bond and the substrate and effectively ground the input pin. The floating well acts as a

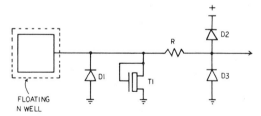

FIGURE 2.30   Improved CMOS input protection circuitry. Sturdy but less sensitive protection devices are placed at the front to filter out gross voltage and current spikes. Later more sensitive devices clamp the smaller spikes that pass through the first stage.

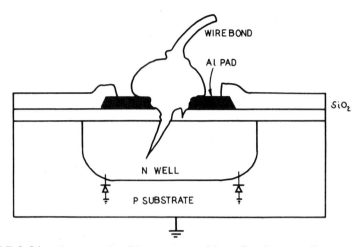

FIGURE 2.31   Input pad spiking prevented by a floating *n*-well

reverse-biased diode and isolates the metal spike from the grounded substrate. In addition, it reduces the capacitance of the input pad because the well capacitance is in series with the pad capacitance.

- A thick oxide transistor T1 and *n*-well enhanced diode D1 form the first step of the protection hierarchy. The thick oxide transistor uses the field oxide as its gate dielectric and *n*-well diffusion as its source and drain (Fig. 2.32). It has a large threshold potential (15–25 V) but is also immune to ESD/EOS. The *n*-well enhanced diode is simply the source/drain diffusion of the thick oxide transistor. It has a very deep and smooth junction with good breakdown properties. These robust devices clamp the gross

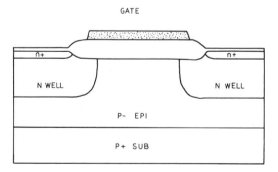

FIGURE 2.32   Thick oxide transistor

voltage surges at the input. If the input is negative by more than V_T of the thick oxide transistor, T1 and D1 turn on. If the input potential goes below 0 V only by V_{BEon}, the diode D1 turns on. If the input is greater than the punchthrough potential of T1, it turns on.

- Diodes D2 and D3 formed by shallow $n+$ and $p+$ source/drain diffusions are used to filter out the smaller energy surges in the input that are beyond the sensitivity of D1 and T1. The diodes D2 and D3 are not as robust as D1 and T1, but the voltage fluctuations that pass by D1 and T1 are not very strong either.

- No additional mask steps are required to build the thick oxide transistor or n-well enhanced diode. This protection circuit can be built using any standard CMOS process.

2.5.5   CMOS Latch-up

All MOS processes produce potentially problematic parasitic bipolar transistors. These bipolar transistors are especially troublesome in CMOS technology because an n-p-n-p structure is formed by the $n+$ source of the nMOS transistor, p-substrate, n-well, and $p+$ source of the pMOS device (Fig. 2.33). This structure has a built-in positive feedback. When it turns on, power and ground are effectively shorted together, large currents are conducted, and the circuit is destroyed. This is refered to as CMOS latch-up [2.45]–[2.49]. Figure 2.34 shows the equivalent circuit for the latch-up structure. The p-n-p transistor is formed by the p-source of the pMOS transistor (emitter), n-well (base), and p-substrate (collector). The n-p-n transistor is formed by the n-well (collector), p-substrate (base), and n-source of the nMOS transistor (emitter). R_{NWELL} and R_{PSUB} are the resistances of the n-well and p-substrate.

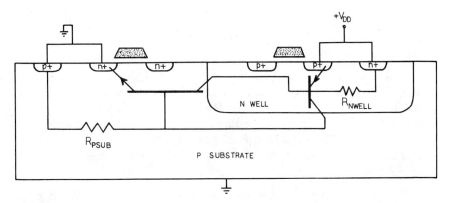

FIGURE 2.33   CMOS latch-up

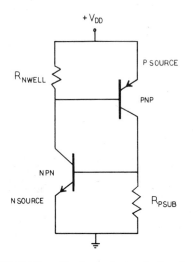

FIGURE 2.34   Equivalent circuit for the latch-up structure

When any one of these two transistors is forward biased, it feeds the base of the other transistor, which in turn feeds the base of the first transistor, and this positive feedback increases the current until the circuit burns out. There are many possible solutions to avoid CMOS latch-up, and in all of them the main goal is to reduce the gains of the bipolar transistors and to minimize the resistances R_{NWELL} and R_{PSUB}. Reduced transistor gain weakens the positive feedback, and smaller resistances rob the undesirable base currents and prevent the bipolar transistors from turning on. Following is a set of guidelines to avoid CMOS latch-up:

- The $n+$ and $p+$ source/drain diffusions of nMOS and pMOS transistors must be kept apart. This increases the base width of the lateral npn device and reduces its gain.

- Guard rings can be placed around transistors ($p+$ guard rings connected to ground around nMOS, and $n+$ guard rings connected to V_{DD} around pMOS transistors). These minimize R_{NWELL} and R_{PSUB} and capture the injected minority carriers before they reach the base of the parasitic bipolar transistors. They also increase the base charge of the parasitic bipolar transistors, crippling their gain. These are usually used only around I/O circuits because of their large area penalty.

- The well and substrate must have periodic contacts (also refered to as "plugs") that tie the p-substrate to ground and n-well to V_{DD}. These contacts must have low-resistance paths to V_{DD} and ground, and they should be placed close to the source/drain of nMOS and pMOS transistors. This minimizes R_{NWELL} and R_{PSUB}.

- Process-based technology improvements can also be employed. A thin epitaxial layer on low-resistivity substrate reduces R_{PSUB}, and the injected minority carriers are collected by the low-resistance substrate. Heavily doped buried layers placed under the p- and n-well in a twin-tub/epitaxial technology reduce R_{PWELL} and R_{NWELL}. Deep trench isolation between n- and p-well increases the effective base width and total base charge, reducing transistor gains. These technology improvements can drastically reduce the occurrence of latch-up, and unlike the previous remedies, they do not degrade circuit packing density; instead they increase process complexity and cost.

2.6 Materials and Reliability Issues in VLSI Interconnections

Aluminum is the preferred metal for VLSI interconnections because of its low resistivity, good adherence to Si and SiO_2 layers, bondability, patternability, and ease of deposition (as a result of its low melting point). In addition, aluminum can be easily purified so that it does not contaminate the IC with undesirable impurities, and it is a readily available, low-cost material. In spite of its positive qualities, aluminum interconnections introduce many reliability problems, such as electromigration, contact failures, and step coverage, which are described in the following subsections [2.50]–[2.54]. Electrical properties of interconnections such as wiring capacitance, resistance, and transmission line properties are covered in Chapters 4, 5, and 6, respectively.

2.6.1 Electromigration

Electromigration is one of the major interconnection failure mechanisms in VLSI integrated circuits [2.55]–[2.57]. It is caused by the transport of the metal atoms when an electric current flows through the wire. This migration is a result of the interaction between the aluminum atoms and the electron current. As the atoms collide with the drifting electrons, the metal atoms are transported primarily by grain boundary diffusion. Because of its low melting point, aluminum has a large grain boundary self-diffusion constant, which increases its electromigration liability. When the metal atoms are displaced, the line may eventually break, and undesirable opens may be formed (Fig. 2.35(a)), or two neighboring lines can get shorted if material accumulated as a result of electromigration forms a bridge between the lines (Fig. 2.35(b)).

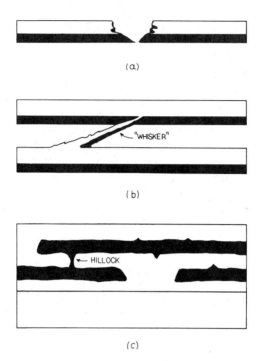

FIGURE 2.35 Electromigration-related failure modes. (a) A broken line caused by removal of metal molecules with electromigration. (b) An undesired short circuit caused by a whiskerlike structure formed by accumulation of metal molecules transported by electromigration. These "whiskers" are observed to grow as long as 100 ⁻m: (c) A "hillock" that goes through the insulator between the two adjoining interconnection levels causing a short circuit.

    Another phenomenon closely related to electromigration is the hillock formation, which can short two lines in two separate wiring levels (Fig 2.35(c)). The hillock formation is caused by the large difference between the thermal expansion coefficients of aluminum and the surrounding silicon and silicon dioxide layers. During the temperature cycling of regular IC processing, the stresses caused by this mismatch give rise to grain boundary movements and induce hillock generation. Electromigration properties are closely related to the stresses applied on the interconnections. For example, both electromigration and hillock generation are strongly influenced by the thickness of the passivation layer placed on top of the chips as the last layer to protect the circuits from outside harm. The thickness of this layer affects the stresses generated on the interconnection layers.

    Electromigration-induced mass transport phenomena increase with cur-

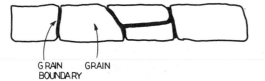

GRAIN GRAIN
BOUNDARY

FIGURE 2.36 Bamboo structure in a metal wire. When the grain size is comparable to wire width and thickness, electromigration is reduced because there are fewer grain boundary paths through which metal molecules can migrate.

rent density and temperature, and the mean time to failure has been observed to fit the following empirical equation [2.58]

$$
MTTF \propto
\begin{cases}
\dfrac{W_{int}H_{int}}{J^n} e^{E_A/k_B T}, & (H_{int} < 5000\ A) \\[2em]
\dfrac{W_{int}}{J^n} e^{E_A/k_B T}, & (H_{int} > 5000\ A)
\end{cases}
\tag{2.22}
$$

where W_{int} and H_{int} are the width and thickness of the interconnection, J is the current density, E_A is the activation energy, k_B is Boltzman's constant, and T is the temperature in $°K$. The constant n varies between 1 and 3 and is observed to be approximately 2. As the minimum feature size is scaled down, MTTF degrades rapidly because W_{int} and H_{int} are reduced and J is increased.

There are a number of ways to minimize the electromigration rate. If the grain size is comparable to interconnection width and thickness, the electromigration rate is reduced because the self-diffusion path through grain boundaries is eliminated [2.59]. An aluminum line with large grains can assume a bamboo-like structure as illustrated in Fig. 2.36. A more common way to reduce electromigration is the addition of alloying elements, which block the grain boundary diffusion path by precipitating at the grain boundary. Addition of 2 to 4 percent of copper to aluminum is shown to provide significant electromigration resistance. When silicon is alloyed with copper, however, it becomes difficult to dry etch, it corrodes more easly, its resistivity increases, and hillocks are not completely eliminated. Layered films, a recently developed technology in which layers of Al/Si are sandwiched between titanium or tungsten layers, promise to reduce electromigration, eliminate hillock formation, and produce low-resistivity lines that can be dry etched [2.60]–[2.62]. These layered structures can have lifetimes in excess of 1000 hours at 250°C while supporting a current density of 1.5×10^6 A/cm^2 [2.61].

2.6.2   Contact Electromigration

Migration of silicon into aluminum at the points where Al contacts Si, referred to as contact electromigration, is another major reliability problem related to interconnections [2.63]–[2.64]. This arises when the silicon atoms diffuse into aluminum at the interface area where Al interconnect forms a contact to Si and the void created by the migrating silicon is filled by aluminum. If the resulting "spike" is deep enough, the junction below the contact point can be shorted to the substrate (Fig. 2.37).

Solubility of silicon in aluminum is an important parameter that affects spiking. At temperatures below 500°C, the solubility of Al in Si is negligible, but the solubility of silicon in aluminum varies from 0.25 to 1 percent between 400°C and 500°C (temperatures common for IC processing steps during and after metal deposition, such as annealing). As a result Si will dissolve in aluminum at the contact areas. Another important factor that influences the severity of contact electromigration is the diffusion coefficient of Si in Al because the volume of aluminum that will be saturated by silicon is determined by how far the silicon atoms can diffuse in aluminum. During a 30-minute annealing at 400°C to 500°C, the diffusion length of silicon in a thin film aluminum ranges from 25 to 55 μm. The thickness of the thin native oxide at the contact area at the time of aluminum deposition is also important. This SiO_2 layer can be 10-20 Å thick and is reduced to Al_2O_3 by aluminum. This reduction process results in low-resistance ohmic contacts and is also responsible for the good adhesion of Al to SiO_2. If the native oxide is thicker, it tends to be consumed faster at certain areas, and as a result silicon is also consumed more under these pinholes, and deep "spikes" are more likely to be formed. In addition, due to preferential formation of spikes at certain crystal directions, (100) substrates are more vulnerable to junction spiking than (111) substrates.

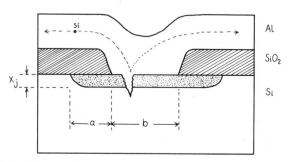

FIGURE 2.37   Junction spiking as a result of silicon diffusion into aluminum

The MTTF for this failure mechanism is described by the following empirical equation [2.58]:

$$MTTF \propto \frac{ab^2 X_j^2}{I^2} e^{E_A/k_B T},\qquad (2.23)$$

where a is the distance from the contact cut edge to the diffusion edge, b is the contact width, X_j is the junction depth, I is the total current through the junction, E_A is the activation energy, k_B is Boltzman's constant, and T is the temperature in °K. As the device technology is scaled down, a, b, and X_j are reduced, and contact electromigration becomes more severe (Fig. 2.37).

Contact failures can be reduced by adding 1 to 2 percent of silicon in aluminum before it is deposited. In this way, Al is saturated with Si before any silicon atoms from the junction can diffuse into aluminum at the contact area. This, however, generates new problems due to precipitation of the Si that was added into Al. Silicon tends to precipitate at the boundaries and surfaces, which increases contact resistances and makes bonding to aluminum more difficult. At highly scaled technologies, a diffusion barrier and contact material between Al and Si may be required. A thin layer of polysilicon above the junction forms a good contact and serves as a supply of Si to saturate aluminum, and as a result no silicon from the junction is consumed. A metal such as titanium (Ti) or tungsten (W) may also serve as a barrier and contact material. It is also possible first to put down a good contact material such as a thin layer of TiSi₂ or PtSi, then to place a good diffusion barrier such as TiW (titanium and tungsten alloy) or separate Ti and W layers, and finally to top this with a low-resistance and electromigration-safe structure. This triple good-contact/good-diffusion-barrier/good-conductor structure solves all the major problems (Fig 2.38). In a different but similar solution, all the junctions

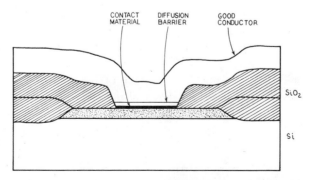

FIGURE 2.38   Barrier material to prevent contact electromigration. A thin layer of polysilicon, titanium, or tungsten can act as an effective barrier against contact electromigration.

are exposed to a silicidation process similar to the one shown in Fig. 2.16. After the contact openings are etched, a nitride layer is formed at the contact area to serve as a diffusion barrier. Finally aluminum interconnection is deposited.

2.6.3   Step Coverage

Less fundamental but still a real problem is step coverage, which is caused by the shadowing effect of the surface topography during metal deposition (Fig. 2.39). This thinning of the line at the steps can cause opens and cracks and generates electromigration-prone and high-resistance points due to reduced cross-sectional area. There are a number of possible remedies:

- Aluminum deposition systems that rotate the wafers above the aluminum source rather than keeping them stationary minimize the shadowing effect by balancing the arrival direction of aluminum atoms.

- Planarization techniques that decrease the roughness of the surface topography can reduce the steps. In addition, if etching methods that avoid steep contact openings are employed, step coverage problems will be minimized.

- Conductive "studs" that fill the via openings before aluminum is deposited eliminate the need for aluminum to fill the vias and minimize the step coverage problem when accompanied by a planarization technique.

- Experimental CVD aluminum deposition technology can be the perfect solution for step coverage [2.65].

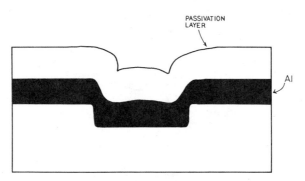

FIGURE 2.39   Step coverage in VLSI interconnections. Thinning of the lines at the steps can cause opens and cracks, generates electromigration-prone spots, and increases the effective resistance of the line.

2.7   Advanced CMOS and BiCMOS Technologies

Figure 2.40 illustrates a hypothetical advanced twin-tub CMOS technology. The process starts with a heavily doped $p+$ substrate on which a thin and lightly doped p-type epitaxial layer of silicon is grown. Low-resistivity $p+$ substrate improves latch-up immunity by reducing R_{PSUB} because minority carriers injected into the epilayer are collected by the low-resistivity substrate and never get a chance to turn on the parasitic bipolar transistors. Separate p- and n-well structures implanted into the lightly doped epitaxial layer enable independent optimization of nMOS and pMOS transistors for best performance. Devices employ sidewall spacers and lightly doped drain and source implants to minimize short-channel effects. Both the polysilicon gate and source/drain implants are covered by silicides (for example, $TiSi_2$ or WSi_2) to reduce parasitic resistances.

Separate $n+$ polysilicon gates for nMOS and $p+$ polysilicon gates for pMOS transistors can be used to optimize the work function difference between the gate conductor and silicon substrate. This reduces the required threshold implant dose for pMOS devices and avoids buried channel transistors; however, the gate electrodes of nMOS and pMOS transistors cannot be connected directly because a diode would form at the interface of the $n+$ and $p+$ polysilicon. If separate $n+$ and $p+$ polysilicon are used, metal jumpers and the associated contacts that degrade packing density are required. A silicidation process that will short out the $p+$ and $n+$ poly will conserve area; however, keeping the boron dopants in the $p+$ poly and the phosphorus dopants in the $n+$ poly may be challenging. A good compromise could be to replace the polysilicon gate with a self-aligned metal gate technology that employs a

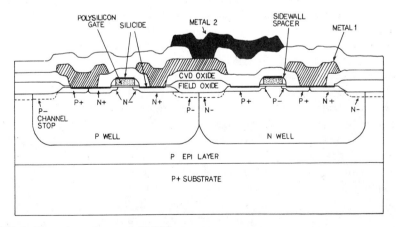

FIGURE 2.40   An advanced CMOS structure

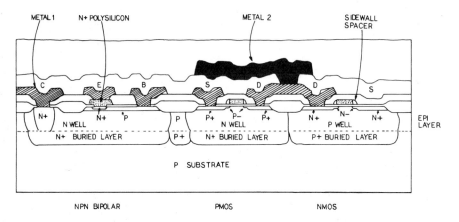

FIGURE 2.41   A BiCMOS technology. Bipolar and CMOS transistors are merged in a single technology, which brings new possibilities to digital, memory, analog, and high-power circuits.

metal with a work function value between the work functions of $n+$ and $p+$ polysilicon.

The process in Fig. 2.40 also uses a barrier metal (for example, TiW) to avoid contact electromigration and spiking (see Sec. 2.6.2). The figure shows only two of the three or more levels of aluminum wires. The interconnections may have a layered structure in which Al layers are sandwiched between thin layers of Ti or W to minimize electromigration and hillock formation (see Sec. 2.6.1).

As illustrated in Fig. 2.41, bipolar and CMOS devices can be fabricated on the same substrate to form a technology that offers the advantages of both types of devices. (See Section 2.10 on bipolar transistors.) Low on-resistance, high current drive capability, and superior matching of bipolar transistors, combined with low power dissipation, high packing density, and dynamic storage capability of CMOS transistors, open up new possibilities for digital, memory, analog, and high-power circuits.

A hypothetical BiCMOS process illustrated in Fig. 2.41 starts with a p-type substrate on which $p+$ and $n+$ buried layers are implanted and a thin layer of epitaxial silicon is grown. Buried layers reduce the collector resistance of bipolar transistors and minimize the possibility of latch-up by increasing the base charge of vertical bipolar transistors (thereby reducing their gain) and by also reducing the resistance of the paths to n-well and p-well plugs (thereby making the collection of injected minority carriers easier). CMOS transistors are built in the n- and p-wells similar to the advanced CMOS technology described above. An npn bipolar transistor is built in the n-well, which forms the collector of the transistor. P-type base and $n+$ emitter are implanted in the well. The emitter contact is done via $n+$ polysilicon, which improves the gain of the bipolar device. This is a complicated BiCMOS technology that has buried and epitaxial layers and independently optimized nMOS, pMOS, and npn bipolar devices for best performance.

Actually it is possible to obtain a BiCMOS technology from an n-well CMOS process without adding any new masking steps. N-well is used as the collector, threshold adjustment implant becomes the p-base, and $n+$ source/drain of the nMOS transistor serves as the emitter. Of course, the performance and latch-up immunity of this simpler technology cannot match the first one; however, some of these ideas can be used to save processing steps in the more advanced BiCMOS structure. Attaining a low collector resistance is difficult in a simple BiCMOS structure with no buried layer. In Chapter 4, circuit aspects of BiCMOS will be analyzed.

2.8   Low-Temperature CMOS Technology

The performance and reliability of CMOS devices are enhanced at low temperatures as a result of improvements in subthreshold slope, carrier mobility, junction capacitance, leakage current, interconnection resistance, electromigration, and latch-up [2.72]–[2.79]. Lowering the temperature makes up for some of the adverse effects of scaling and improves some of the nonscalable device parameters. Liquid nitrogen is used to cool the CMOS circuits because it is inert (it does not react with, harm, or contaminate the devices), and inexpensive, and its boiling temperature (77°K) is close to an optimum for CMOS circuits. When operated at liquid nitrogen temperature, CMOS circuit speed improves by a factor of two to three, and as a result CMOS devices can achieve switching speeds comparable to bipolar transistors. It is predicted that the speed of liquid-nitrogen-cooled CMOS devices will surpass bipolar circuits at 0.5 μm design rules [2.76]. When this is combined with the high integration density, superior yield, and low power dissipation of CMOS, a promising picture emerges. The following are the major performance and reliability advantages of cryogenic CMOS circuits.

a. Improved intrinsic gate delay. Devices speed up at lower temperatures because of the improvements in mobility and parasitic capacitances. Both the saturation velocity and carrier mobility increase at low temperatures. Mobility improves because at lower temperatures carriers have less thermal energy and do not scatter as much. Because of their higher mobility, low-temperature MOS devices enter velocity saturation at a smaller potential difference than room temperature devices, and as a result liquid-nitrogen-cooled transistors can achieve peak currents at lower voltage supply levels. The junction capacitance improves because of the freeze-out effect. At low temperatures, dopant atoms hold on to their extra electrons or holes, and depletion widths are increased. This reduces parasitic source and drain junction capacitances. Improved carrier velocity combined with reduced junction capacitances results in faster devices.

b. Steeper subthreshold slope. In classical scaling theory, temperature is held constant and the device turn-off characteristics do not scale favorably. As a result it is not possible to scale down the threshold potential of the transistor and still maintain low subthreshold currents. This limits how much one can reduce the threshold potential and voltage supply level. At lower temperatures the transistors turn off much more sharply, and as a result V_T and V_{DD} can be reduced without increasing subthreshold leakage currents (see Section 2.4.6). Improved subthreshold slope combined with enhanced carrier mobility provides high-speed CMOS devices at low voltage-supply levels. Lower power-supply levels, in turn, improve the short-channel effects, reliability, and power dissipation.

c. Improved interconnection resistance. Resistance of both metal and silicide lines is decreased at low temperatures because the carriers have less thermal energy and thus a lower scattering rate. (The resistance of diffusion regions increases because of freeze-out.) At liquid nitrogen temperature, the resistance of aluminum wires improves by a factor of five to six, and the resistance of $TiSi_2$ is reduced by a factor of four [2.78]. This is significant for submicron technologies where the resistance of global interconnections may limit the chip performance at room temperature and above (see Chapter 5).

d. Reduced leakage currents. Junction leakage currents at reverse-biased source/drain junctions as well as subthreshold currents are reduced at low temperatures. This is because the reverse-bias current is proportional to $exp\{-\frac{V_{bias}}{k_B T/q}\}$, which decreases exponentially as temperature is lowered.

e. Elimination of CMOS latch-up. The current gain of bipolar devices degrades at low temperatures because of the bandgap narrowing and Schockley-Read-Hall recombination effect. As a result the gains of the parasitic bipolar devices in CMOS circuits are reduced and latch-up is virtually eliminated.

f. Improved reliability. Electromigration and most of the wear-out and failure mechanisms follow an Arrhenius-type relationship where failure rates improve exponentially with reduced temperature (failure rate $\propto e^{-E_A/k_B T}$). As a result device and chip reliability improves at lower temperatures.

g. Reduced thermal noise. Thermal noise improves with reduced temperature, but 1/f noise remains unchanged.

There are also some adverse effects of reducing the temperature:

a. Hot-electron device degradation. Carriers that are injected into the gate are more likely to be trapped at low temperatures. As a result threshold shifts and transconductance degradation of cryogenic CMOS devices are worse than room-temperature devices. This may be prevented by scaling down the power supply level or using LDD structures.

b. Carrier freeze-out. The same carrier freeze-out effect that improves parasitic junction capacitances also increases diffusion resistances. This is especially significant for the depletion-mode devices in nMOS E/D circuits. Depletion devices are formed by implanting n-type impurities at the channel region that release electrons and produce a normally on device. At low temperatures, the donors at the channel are not ionized, and no current flows. The acceptor boron also has severe freeze-out problems because it has a small atomic number and requires more energy to ionize. This makes it harder to control the threshold of pMOS devices and increases the resistance of the p-well. As a result of carrier freeze-out, parasitic source and drain resistances may increase, and the transconductance of transistors may degrade. This is especially significant in LDD devices because lightly doped source and drain extensions are affected more strongly by freeze-out.

c. Temperature cycling. As the packaged chips are repeatedly cooled to liquid nitrogen temperature ($77°$K) and brought back to room temperature ($300°$K), all material interfaces are repeatedly stressed. This may worsen the reliability of some of the most failure-prone components: wire bondings and solder balls that provide electrical connections to the dice.

d. Testing of the chips. If the circuits are designed such that they cannot perform at room temperature due to critical timing requirements or some other reason, they have to be tested at low temperatures. This requires complicated test fixtures, especially for wafer-level tests. Wafers may have to be mounted on liquid-nitrogen-cooled metal plates during testing.

e. Containment of the low-temperature chamber. Keeping the chamber that houses the chips at liquid nitrogen temperature and at the same time providing electrical communication with the outside is challenging. The holes that the wires are fed through must be well isolated. In addition, the heat transferred by the wires will be significant because, besides being good electrical conductors, all metals are also good thermal conductors. In metals both heat and electricity are conducted via electrons, and there is a close correlation between electrical and thermal conductivity of a metal.

f. Cooling and packaging. Gifford-McMahon-type refrigerators can be used to provide 1–400 W of heat removal at $77°$K [2.80]. The basic operating principles are the same as a typical refrigerator. Helium gas is compressed at room temperature, and the heat generated by compression is removed by air or water. Later the compressed gas is expanded while pushing a piston or forcing a turbine blade. The work the helium does against the piston or the blade lowers its temperature. This expanded and cold gas is then piped through room-temperature nitrogen gas and cools it. Helium is then returned to the compressor ready to go through the same cycle again. The liquid nitrogen generated by the refrigerator can be used as a $77°$K heat bath in which the packaged electronics is immersed, or it can be circulated through a metal cold plate on which the chips are mounted, or the liquid nitrogen can be allowed to boil and evaporate from the surface of the chips and cool them directly.

g. High cost. The cost and reliability of cryogenic refrigerators are being improved; however, they are still expensive.

2.9   MESFET and JFET Technologies

Metal-semiconductor field-effect transistors (MESFETs) and junction field-effect transistors (JFETs) are based on the same basic principles as MOS-FETs; they modify the conductivity of a channel between two heavily doped source and drain [2.6]. They are voltage-controlled resistors. In MOSFETs, the channel region is capacitively coupled to a gate, and the potential variation in the gate modifies the band structure and carrier concentration of the channel. In JFETs, the gate capacitor is replaced by a junction diode, and the depth of the depletion region under this diode determines if the thin n-channel between $n+$ source and drain is conducting or is pinched off. In MESFETs, a Schottky-barrier diode takes the place of the junction diode (Fig. 2.42). MESFETs offer lower channel resistance and lower IR drop along the channel. They are also easier to fabricate than JFETs because formation of the metal-semiconductor barrier does not require as high a temperature as does JFET gate diode diffusion. MESFETs are considered an alternative for MOSFETs because they do not require thin gate oxides. They also achieve better carrier mobility because the current flows in the bulk rather than at the surface, and they exhibit better short-channel properties. MESFETs also exhibit excellent radiation hardness and lack problems associated with hot electrons and oxide breakdown because they do not have a gate oxide. On the negative side, MESFETs cannot support dynamic circuits because they lack the capacitively isolated gate structure of MOS transistors, and the gate currents of MESFETs can add up to large values at high-integration densities. Also because of the limited range of Schottky-barrier heights available, MESFETs do not have a readily available complementary transistor pair like

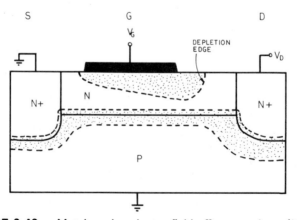

FIGURE 2.42   Metal-semiconductor field-effect transistor (MESFET)

CMOS. At the present, yield problems and difficulties in controlling the device thresholds make MESFET technology difficult.

2.10   Bipolar Transistors

Bipolar junction transistors (BJTs) form a class of devices different from MOSFETs, MESFETs, and JFETs. Bipolar devices are faster, but they require a more complicated processing sequence, take up more area, and consume more power. In the following subsections, basic principles of bipolar transistors are presented, their advantages and disadvantages with respect to CMOS circuits are discussed, and advanced bipolar circuits and processes are described.

2.10.1   Fundamentals of Bipolar Devices

Field-effect transistors (MOSFET, JFET, MESFET) are known as majority carrier devices because, for example, in an nMOS transistor, the current is conducted by electrons via a drift current from the $n+$ source to $n+$ drain through the channel region (which is inverted to n-type by applying a positive potential on the gate). As a result the current always travels in n-type regions, the carriers are always electrons (majority carriers), and the drift current established by the potential difference between the source and drain determines the device operation. In contrast, both majority and minority carriers and drift and diffusion currents play important roles in the operation of a bipolar junction transistor shown in Fig. 2.43. The intrinsic portion of the transistor, which determines the basic device properties, is shown in Fig. 2.44. The

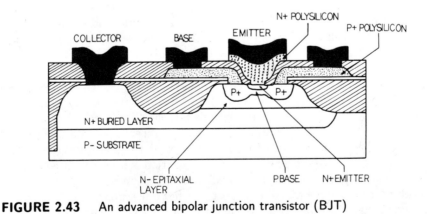

FIGURE 2.43   An advanced bipolar junction transistor (BJT)

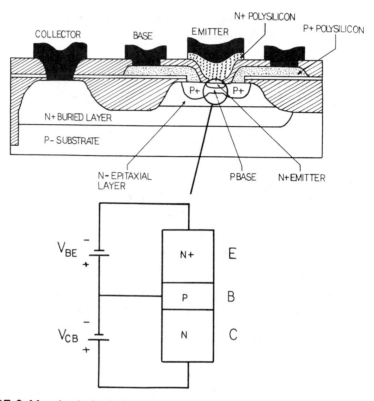

FIGURE 2.44 Intrinsic device region of a BJT

emitter and collector of a BJT correspond to source and drain of a MOSFET, and the BJT base corresponds to the MOSFET channel region. The operation conditions of these regions in bipolar and MOS transistors, however, are fundamentally different. In an n-type MOSFET, source and drain are always reverse biased, and conduction is established by inverting the p-type channel to n-type. In an npn BJT, the base always remains p-type, and the transistor is turned on by forward biasing the emitter-base junction.

In a typical application, the emitter-base junction is forward biased, and the electrons (majority carriers) are injected from emitter to base (where they are minority carriers because the base is p-type). The injection of electrons from emitter to base is primarily controlled by diffusion; electrons diffuse from a region where there are plenty of them (emitter) to a region where the electron density is sparse (base). The emitter-base junction is forward biased, but the base-collector junction is reverse biased. This creates an electron concentration gradient across the base where there are more electrons near the emitter junction than there are near the collector junction (Fig. 2.45). As a result a

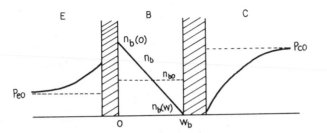

FIGURE 2.45   Majority and minority carrier concentrations in an *npn* bipolar junction transistor in the forward active mode. The emitter-base junction is forward biased, and the collector-base junction is reverse biased. The shaded areas are the depletion regions. The equilibrium concentrations of the minority carriers in the emitter, base, and collector regions are p_{e0}; n_{b0}; and p_{c0}: The current is mainly determined by the slope of the electron concentration across the base $n_b(x)$:

diffusion current of electrons is formed across the *p*-type base region toward the collector. The reverse biasing of the collector also generates an electric field across the collector-base depletion region. Accordingly once the electrons reach the reverse-biased base-collector junction, they are swept toward the positive voltage at the collector by a drift current established by the electric field across the depletion region of the junction. (For a detailed description of BJT device operation see [2.2] and [2.6].)

When the transistor is in the active region (emitter-base junction is forward biased and base-collector junction is reverse biased), collector current is primarily controlled by the electron concentration gradient across the base (Fig. 2.45). Neglecting second-order effects such as recombination of minority carriers at the neutral base region and at the emitter-base and base-collector depletion regions, the collector current is obtained as

$$
\begin{aligned}
I_C &= qAD_b \nabla n_b' \, |_{x=0} \\
&\approx qAD_b \frac{n_b'(0) - n_b'(W_b)}{W_b},
\end{aligned}
\tag{2.24}
$$

where q is the charge of an electron, A is the effective area of the emitter-base junction, D_b is the diffusion coefficient of electrons in the base, $n_b'(x)$ is the concentration of *excess* electrons along the base, and W_b is the base width. Using the diode equation and taking into account the fact that emitter-base is forward biased and base-collector is reverse biased, electron concentrations at the edge of emitter and collector depletion regions are obtained as

$$
\begin{aligned}
n_b(0) &= n_{b0}e^{V_{BE}/V_{TH}} \\
n_b(W_b) &= n_{b0}e^{V_{BC}/V_{TH}} \ll n_b(0).
\end{aligned}
\tag{2.25}
$$

Excess minority carrier concentration is given by

$$
\begin{aligned}
n_b'(0) &= n_b(0) - n_{b0} = n_{b0}(e^{V_{BE}/V_{TH}} - 1) \\
n_b'(W_b) &= n_b(W_b) - n_{b0} = n_{b0}(e^{V_{BC}/V_{TH}} - 1) \ll n_b'(0).
\end{aligned} \tag{2.26}
$$

Here n_{b0} is the equlibrium concentration of electrons in the base determined by $n_{b0} = n_i^2/N_A$ where n_i is the intrinsic carrier concentration of silicon $(1.5 \times 10^{10}$ cm^{-3} at $300°$K) and N_A is the base dopant concentration. V_{TH} is the thermal voltage (not to be confused with the threshold potential of MOS transistors) $V_{TH} = k_B T/q = 0.026$ V at $300°$K. Using equations 2.24 and 2.26, the collector current is obtained as

$$
I_C = \frac{qAD_b n_{b0}}{W_b}(e^{V_{BE}/V_{TH}} - 1). \tag{2.27}
$$

Up to this point, it is assumed that $I_E = I_C$ and $I_B = 0$. In reality, however, there is a nonzero base current due to recombination of electrons injected from emitter with the holes in the base, diffusion of holes from base to emitter (reverse injection current), and diffusion of holes from base to collector. The base and collector current are related by

$$
I_C = \beta_f I_B, \tag{2.28}
$$

where β_f is the forward current gain of the BJT.

2.10.2   Advantages and Disadvantages of BJTs

A number of important observations can be obtained from Eq. 2.27. Collector current is inversely proportional to base width W_b just as the drain current of an MOS transistor is inversely proportional to channel length L (see Eq. 2.1). Channel length in MOS transistors is a horizontal dimension defined by photolithography. The base width, on the other hand, is a vertical dimension and is defined by diffusion of dopants. As a result W_b of BJTs can be made much smaller and controlled much more accurately than L of MOSFETs. This gives bipolar devices superior current drive and fan-out capability. In addition BJT is an exponential device (the collector current in Eq. 2.27 increases exponentially with V_{BE}), and MOSFET is a square-law device (the drain current in Eq. 2.1 is proportional to V_{GS}^2). Consequently, a change in V_{BE} has a greater effect on the output of a bipolar device than a change in V_{GS} has on the output of a MOSFET, which means that BJTs have higher gain than MOSFETs. In addition, the base terminal of a BJT is directly coupled to the base region. MOSFET gate, on the other hand, is capacitively coupled to the channel region. This increases the efficiency of coupling in bipolar devices and enhances the effect of V_{BE} on I_C compared to the effect of V_{GS} on I_{DS}. In MOS devices, the gate oxide thickness is continuously reduced to increase C_{gox} and to improve the coupling between the gate terminal and the channel region. In bipolar devices, the base terminal is coupled to the base region as tightly as possible; it is directly tied to it. (See section 2.4.6.) As a result of the combined effect of these advantages, bipolar transistors can switch much more quickly than their MOS counterparts.

Bipolar devices also exhibit excellent matching property because their turn-on potential ($V_{BEon} = 0.7$ V) is determined by the bandgap of the silicon. This is unlike the threshold potential of an MOS transistor (see Eq. 2.3), which is determined mainly by gate oxide thickness and threshold adjustment ion implantation and therefore cannot be controlled as tightly as V_{BEon} of a BJT. As a result, bipolar transistors provide not only higher gain but also better matched differential amplifiers, which can sense very small potential swings. It is important to have well-matched devices in an amplifier because a differential pair amplifies the mismatches between the two gain devices, as well as the potential difference between their inputs. This has consequences in both logic and memory circuits. Bipolar logic gates with a differential current switch such as emitter-coupled logic (ECL) and current-mode logic (CML) can be designed with very small potential swings, which results in small gate delays. In memory circuits, MOS sense amplifiers have an intrinsic noise due to threshold mismatches among its transistors and as a result cannot sense as small potential swings as bipolar amplifiers can.

Despite the superiority of their raw speed, bipolar devices are not as suitable for VLSI as MOSFETs for many reasons. High-speed circuits based on BJTs dissipate large amounts of power. To take full advantage of their speed, bipolar ECL and CML gates are always kept in a state where the transistors are active and conduct current. This is necessary because when the devices go into deep saturation or completely turn off, their switching speed is degraded. If thousands of these gates are integrated on a single chip, the power dissipation can easily exceed 10–20 W. This requires special cooling techniques and increases the cost of the system tremendously. In ECL circuits, power dissipation, gate switching speed, and chip gate count are traded off against each other.

The trade off goes on as follows. First, system packaging sets an upper limit to chip power consumption. Once the chip power dissipation is fixed, the number of gates placed on the chip (integration density) determines the allowable average gate power dissipation. The average power dissipation per gate is simply the total allowable chip power dissipation divided by the number of gates. Power dissipation of a gate defines its current drive, which in turn determines its switching speed. As a result, chip power dissipation, clock frequency, and chip gate count are traded against each others. A final factor that should be considered is chip-to-chip delays. A chip-crossing delay is usually a factor of four to ten larger than an internal gate delay. If the integration level is very small and system requires many chips, overall system performance will be limited by chip crossings. As a result the number of gates per chip cannot be reduced to improve gate delays without degrading the system speed. The limited integration density of bipolar chips may force some critical paths to be partitioned among multiple chips, causing the path delay to increase. Consequently, by confining a critical path to a single chip, a submicron CMOS-based system may obtain a shorter cycle time than bipolar even with the relatively longer gate delays of CMOS. (This will be analyzed in more detail in Chapter 9.)

Bipolar transistors do not have a good complementary pair like CMOS devices. It is hard to build good *pnp* devices, and, even if good *pnp* BJTs could be fabricated, because the BJT base is not capacitively isolated like the MOSFET gate is, the base currents could add up to unacceptably large levels at high integration densities. An additional disadvantage is the lack of dynamic circuits in bipolar technology. The capacitively isolated gate structure of a MOSFET is advantageous in memories (for example, DRAMs), as well as in logic circuits (for example, domino logic). Because nodes precharged to a high or low potential are guaranteed to stay at that level for many clock cycles, dense and very fast MOS circuits can be constructed using precharging. CMOS transmission-gate circuits that provide compact and speedy implementations for certain functions (for example, multiplexors) do not have a counterpart in bipolar because bipolar transistors are *not* bidirectional. In addition, MOS-specific digital devices such as EPROMs, EEPROMs, and CCDs can be built based only on the MOS capacitor.

The packing density of high-speed bipolar chips suffers from the complicated circuits required to achieve fast switching speeds and small potential swings. As shown in Fig. 2.48, a two- or three-input ECL gate requires many transistors, resistors, and reference potentials. Bipolar chips also have lower yields than MOS chips because of the complicated BJT device structure. The most critical parameter of bipolar transistors is the base width, which must be very small to achieve high switching speeds. As a result the base diffusion is the most critical processing step. Small defects in the silicon wafer can cause "pipes" through which the emitter is shorted through the base to the collector. Accordingly bipolar chips exhibit lower yields and cost more than their MOS counterparts. Finally bipolar transistors do not benefit from scaling as much as MOSFETs. Bipolar transistors are vertical devices, and reduction of horizontal mask dimensions improves the parasitic capacitances but does not affect the intrinsic device performance significantly [2.9],[2.81]. In addition, because the turn-on potential $V_{BEon} = 0.7$ V of a BJT is determined by silicon bandgap, the power supply cannot be scaled down as freely as in MOS circuits, where the threshold potential, in theory, can be made much smaller than 0.7 V.

2.10.3   Bipolar Circuits

Bipolar transistors make good voltage and current amplifiers because a small change in the base-emitter potential and base current can result in a much

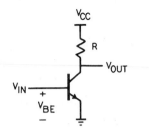

FIGURE 2.46   Simple bipolar transistor inverter

larger change in the collector-emitter potential and collector current. A simple bipolar transistor inverter is shown in Fig. 2.46. When input potential V_{IN} is small, the base-emitter junction is reverse biased, and the transistor is in the cutoff region; no current flows, and the output V_{OUT} is equal to V_{CC}. When input is increased such that V_{BE} is greater than the turn-on potential of the transistor, BJT enters the active region and starts to conduct current. Then small increases in V_{BE} result in large increases in collector current, and the output potential decreases rapidly until both the emitter and collector junctions are forward biased and the transistor is in saturation. When the transistor is deeply saturated, its switching speed degrades because, before it is turned off, all the saturation charge in the base has to be compensated.

In high-speed circuits, a Schottky-barrier diode is formed between the base and collector by extending the metal base contact over on the collector region (Fig. 2.47). When the base-to-collector voltage goes above the turn-on potential of a Schottky diode, the diode turns on, and the base current is channeled to the collector through the diode. As a result the collector voltage gets clamped at a value determined by $V_{CE} = V_{BEon} - V_{Schottky}$, and the bipolar transistor never saturates [2.82]. For this to work properly, the Schottky barrier height of the diode must be less than V_{BEon} of the silicon collector-base diode. (For aluminum/n-silicon Schottky diode $V_{Schottky} \approx 0.4$ V, and V_{CE} gets clamped at $(0.7 - 0.4) = 0.3$ V. The transistor is not saturated because typical saturation potential is 0.1 V.)

The fastest bipolar gates are built using differential current switches. Examples of such circuits are ECL and CML gates shown in Figs. 2.48 and 2.49. In ECL and CML gates the values of the pull-up resistors R_L and I_{DC} of the current source are selected such that the transistors in the current switch never saturate. This keeps the switching delay very small. In addition, taking advantage of the well-matched device parameters of bipolar transistors, ECL can be built with output potential swings as small as 0.5–0.6 V. This reduces the gate-to-gate propagation times. As a result ECL circuits with sub-100 psec gate delays and toggle frequencies in excess of 1 GHz are available [2.83]–[2.86]. ECL is not only the fastest but also the most power-hungry

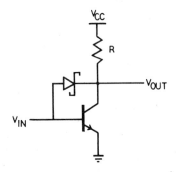

FIGURE 2.47 Inverter with a Schottky clamping diode. The diode clamps the collector-base voltage and prevents the transistor going into saturation. This improves circuit switching speed because less base charge has to be removed to turn the transistor off.

logic family because an ECL gate conducts a current of I_{DC} independent of its state. In addition, to minimize the delay, current I_{DC} has to be large. As a result ECL chips dissipate in excess of 10 W, even at moderate integration densities. As can be seen in Figs. 2.48 and 2.49, the main difference between the ECL and CML circuits is the addition of an emitter follower in the output stage of the ECL gate. This buffers up the capacitive loading at the output

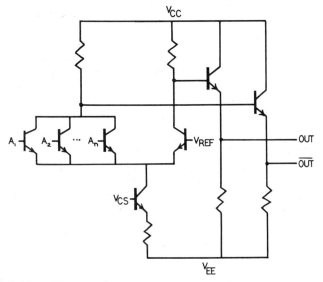

FIGURE 2.48 Bipolar ECL gate

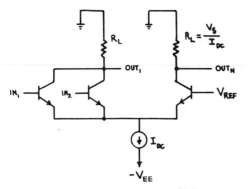

FIGURE 2.49   Simplified model of a CML gate

of the gate from the logic tree. Consequently, total delay can be minimized by separately optimizing the current source value of the logic tree and the current source value of the emitter follower output stage.

2.10.4   Advanced Bipolar Technologies

The most significant recent developments in bipolar technology have been the introduction of self-aligned devices, polysilicon emitter and base contacts, and improved device isolation techniques. These reduce the transistor area and minimize parasitic resistances and capacitances in order to improve circuit speed.

In advanced ECL circuits, base resistance is a major speed-limiting parasitic component. To minimize the base resistance, a self-aligned polysilicon base contact is employed that can get very close to the intrinsic device region (Fig. 2.43). In addition, the $p+$ polysilicon can be used as a diffusion source to dope the extrinsic base region heavily in order to minimize the base resistance. In this way, a lightly doped and narrow intrinsic base with a low hole injection rate to emitter can be obtained simultaneously with a deeper, more heavily doped, and lower resistance extrinsic base area that provides a good electrical contact to intrinsic base. A variant of the BJT in Fig. 2.43 is a bipolar transistor with a sidewall base contact structure [2.89]–[2.90]. As illustrated in Fig. 2.50, sidewall base contact transistors are formed in a mesa-etched substrate, and $p+$ polysilicon deposited next to the device establishes a base contact with minimal resistance and capacitance.

In scaled devices, emitter diffusion and emitter contact also become critical. To achieve high speed, a narrow base is required. And to obtain a well-controlled and narrow base width, the emitter diffusion must be shallow. When the emitter depth is reduced, however, the hole current injected from base to emitter increases, and the forward current gain β_f of the device degrades.

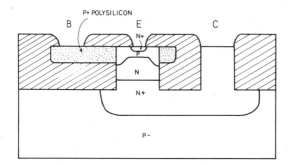

FIGURE 2.50   Bipolar transistor with a sidewall base contact

This effect appears when the hole diffusion length becomes comparable to emitter depth. Then the reverse hole current injected into the emitter is no longer determined by the diffusion constant of holes in the emitter but by the emitter depth and the recombination properties of holes at the silicon/contact interface As shown in Fig. 2.51, with a deep emitter junction, holes injected from base to emitter diffuse inside the emitter and recombine with the electrons; the carrier concentration distribution and hole current are independent of the emitter depth and contact properties. If the emitter depth is less than the diffusion length of the holes, not all the holes have a chance to recombine with the electrons in the emitter, and as a result the emitter depth and contact strongly influence the hole profile and current. The current is proportional to the slope of the hole concentration shown in Fig. 2.51. If a metal contact is used, the recombination rate is very fast, and the undesirable hole current is large. If $n+$ polysilicon is used to contact the emitter, most of the holes continue diffusing into the polysilicon rather than recombining at the interface, and hole current is minimized. As a result polysilicon contact acts as a natural extension of the emitter, minimizes the reverse hole current, and improves the gain of the transistor [2.91]–[2.94].

Fully recessed oxide isolation (Fig. 2.43) or trench isolation is used to increase the packing density and to minimize parasitic emitter, base, and collector capacitances. The reduced capacitances improve the cutoff frequency of the transistors, and smaller gate areas may shorten the global wires. These all contribute to speeding up the circuits.

2.11   Gallium Arsenide Transistors

Gallium arsenide is attractive because of its high electron mobility and high peak electron velocity [2.95]–[2.99]. At high electric fields, saturated electron velocities in GaAs and Si are approximately the same, but GaAs devices re-

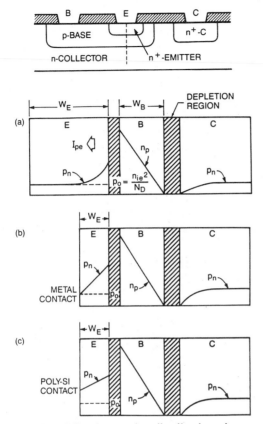

FIGURE 2.51   Minority carrier distributions in *npn* bipolar transistors [2.94] (Reprinted by permission of Gary L. Patton)

quire less potential difference to reach saturation. As a result, with lower power supply and less power dissipation GaAs transistors can achieve higher speeds than silicon devices. In addition, GaAs has a larger bandgap than silicon; therefore it has a low intrinsic carrier concentration and forms a semi-insulating substrate. This creates a structure similar to silicon-on-insulator (SOI) and silicon-on-sapphire (SOS) substrates and reduces the parasitic interconnection capacitances. This gives a significant speed advantage to GaAs, and according to some designers this is even more important than the higher electron mobility. The consequences of this on cross-talk noise should be carefully evaluated (see Chapter 1). GaAs devices are more immune to radiation because of the semi-insulating substrate and lack of gate oxides. They can also operate at a wider temperature range than silicon devices. These make GaAs very attractive for military and aerospace applications. Since GaAs is a direct

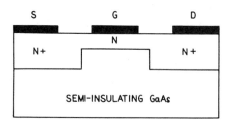

FIGURE 2.52   Gallium arsenide metal-semiconductor field-effect transistor (MESFET)

bandgap material, it can also be used to build light-emitting diodes (LED) and lasers. This makes it possible to integrate digital and optical circuits on the same chip.

On the negative side, hole mobility of GaAs is much smaller than that of silicon, and the thermal conductivity of GaAs is also not as high as silicon. In addition, GaAs chips exhibit lower yields and integration levels because GaAs technology is not as mature as silicon-based technologies. Gallium arsenide wafers are not as defect free as silicon, and they also break easily. As a result, silicon and GaAs currently fulfill different needs and form a complementing rather than a competing technology pair.

There are three major GaAs transistor types: metal-semiconductor field effect transistor (MESFET), high electron mobility transistor (HEMT), and heterojunction bipolar transistor (HBT). The MESFET is the most common and mature GaAs technology. As shown in Fig. 2.52, GaAs field-effect transistors (FET) are formed by implanting a thin layer of n channel and $n+$ source/drain regions on a semi-insulating substrate [2.100]. The gate is a Schottky diode formed by the metal deposited on top of the n-channel. Just like in silicon MESFET, by modifying the potential on the gate, the depth of the depletion layer under the Schottky diode is controlled, and the transistor is turned on and off. If the depletion layer extends across the n-channel all the way to the semi-insulating substrate, no current flows. Otherwise the transistor is on. In depletion-type FETs (DFET), the channel is normally on, and negative potential has to be applied on the gate to increase the depletion layer depth to turn the device off. Enhancement-mode FETs (EFET), on the other hand, are normally off and require a positive potential on the gate to reduce the depletion depth and conduct current. DFET devices supply larger currents and, accordingly, fast circuits can be built using only D-type devices. Slower but lower power circuits can be constructed by using D-type loads and E-type drivers (similar to depletion/enhancement nMOS gates). The major challenge in GaAs MESFET ICs is controlling the variation of device thresholds across the chip.

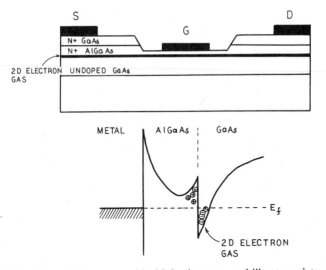

FIGURE 2.53 Gallium arsenide high electron mobility transistor (HEMT)

The high electron mobility transistor (HEMT), also known as modulation-doped FET (MODFET), is shown in Fig. 2.53. HEMT is based on a heterojunction structure formed by a heavily doped AlGaAs and an undoped GaAs layer [2.101]. AlGaAs is heavily doped to provide the electrons, and GaAs is left undoped to avoid scattering from dopant ions and to maximize carrier velocity. Because of the bandgap difference between AlGaAs and GaAs, the electrons are confined in the energy dip created by the heterojunction, when they move from the $n+$ AlGaAs layer to the GaAs layer. This forms a two-dimensional electron gas at the heterojunction, which supplies the device with carriers. Since the GaAs layer is undoped, the carrier scattering is minimized, and electrons can attain very high speeds. The carrier mobility can be further improved by cooling the devices down to, for example, liquid nitrogen temperature ($77°K$).

Gallium arsenide MESFET and HEMT are majority carrier devices, and their operation is similar to silicon nMOS, CMOS, MESFET, and JFET devices. The heterojunction bipolar transistor (HBT), on the other hand, is a minority carrier device and operates like a silicon bipolar transistor [2.102]–[2.103]. Gallium arsenide HBT differs from silicon BJT because its emitter and base are made of different semiconductors; the emitter is AlGaAs, and the base and collector are GaAs. The device structure and band diagram of a GaAs HBT is shown in Fig. 2.54. Since AlGaAs has a wider bandgap than GaAs, electrons can easily diffuse from emitter to base, but holes face a potential barrier. Consequently, the base can be heavily doped to reduce

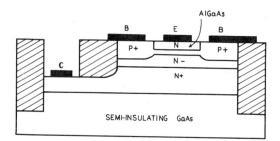

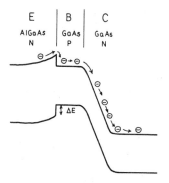

FIGURE 2.54   Gallium arsenide heterojunction bipolar transistor (HBT)

its resistance without impairing electron injection efficiency from emitter to base (reverse hole injection into the emitter is minimized by the barrier in the valence band). GaAs HBT serves very well for applications such as super-computers, where superior circuit speed is the main concern, and high power dissipation and cost are acceptable. (Silicon HBTs can also be constructed by using layers of silicon and germanium. Such transistors with 20 GHz toggle frequencies have been reported.)

Both GaAs HEMT and HBT benefit from the advances in molecular beam epitaxy (MBE) techniques. Molecular beam epitaxy makes it possible to lay down layers of molecules almost one layer at a time, so the thicknesses and dopings of heterojunction structures of HEMT and HBT can be controlled precisely. In an MBE apparatus the substrate is placed in a high-vacuum chamber, and the molecular or atomic beams of Al, Ga, or As, as well as beams of dopants such as Sn or Be, are directed on to the substrate. Col-limated beams of molecules are emitted from heated evaporation cells, and shutters in front of the cells control the deposition time. In this way deposition rates can be controlled very closely. Because the sample is held at a relatively low temperature ($\approx 500°C$), extremely abrupt junctions can be formed. At low deposition rates, layers with thicknesses comparable to lattice constants can be fabricated. These processing techniques are essential for making good heterojunctions for transistors and lasers.

2.12   Summary

CMOS is the most popular VLSI technology because of its low power dissipation, high packing density, wide noise margins, and circuit versatility. The device speed and power consumption of CMOS circuit can be improved greatly by miniaturizing horizontal and vertical device dimensions and reducing the power-supply level.

Second-order phenomena, such as velocity saturation, mobility degradation, finite channel thickness, and source/drain resistances, degrade the device transconductance and reduce the benefits of scaling. In addition, increased subthreshold currents and drain-induced barrier lowering give rise to leakage currents. This becomes problematic in dynamic circuits, especially in DRAMs.

Miniaturized devices also have reliability concerns. When trapped in the gate oxide, hot carriers cause the threshold voltage to drift. This becomes a major problem when the power-supply level is kept constant while the feature sizes are being scaled down. Thin oxides and shallow junctions are susceptible to oxide and junction breakdowns. This is especially important at the chip I/O where high voltages and currents can be momentarily generated by induction or direct contact to charged outside objects. Another reliability problem that is also more frequent in the chip I/O is CMOS latch-up. This is a result of the existence of a positive feedback loop formed by the parasitic bipolar transistors in CMOS technology. Voltage spikes and injection of minority carriers can cause this feedback loop to be activated, which causes the power and ground lines to be shorted together and overheat and eventually damage the circuitry. Care must be taken to keep apart the $n+$ and $p+$ source/drain diffusions and to provide periodic substrate and well contacts to prevent triggering latch-up.

Integrated circuits are built from two basic components: transistors and interconnections. Aluminum is the most popular interconnection metal because it has excellent material and electrical properties. The resistance of aluminum wires becomes significant at submicron design rules. The major interconnection relibility issues are electromigration and contact failures. Electromigration is caused by transportation of aluminum under high current densities. This gives rise to electrical opens and shorts in the wires. Adding impurities in Al or forming layered structures can reduce the occurrence of electromigration. Shallow contacts can fail when silicon diffuses into the interconnection. Various barrier metals can be used to stop Si from diffusing into Al.

Advanced CMOS technologies employ epitaxial layers and separate n-

and p-wells to optimize nMOS and pMOS devices independently and to reduce latch-up susceptibility. Sidewall spacers and lightly doped drain/source structures are used to minimize short-channel effects. Polysilicon gate and source/drain diffusion areas are silicided to reduce parasitic resistances. In the future, to optimize the work function difference between the gate and the substrate, separate $n+$ and $p+$ polysilicon gates for pMOS and nMOS devices may be used, or a self-aligned refractory metal gate technology may be substituted for polysilicon.

The performance and reliability of CMOS devices are enhanced at liquid nitrogen temperature as a result of improvements in subthreshold slope, carrier mobility, junction capacitance, leakage current, interconnection resistance, electromigration, and latch-up. Lowering the temperature makes up for some of the adverse effects of scaling and improves some of the nonscalable device parameters.

Bipolar and CMOS devices can be fabricated on the same silicon substrate to take advantage of the strong points of both devices. Bipolar transistors switch much faster and can drive much larger currents than MOS devices. In addition, BJTs have excellent matching capability and provide very good differential amplifiers. Despite their high speed, bipolar ICs are not as widely used for VLSI applications because of their high power dissipation and lower yield compared to MOS ICs.

Gallium arsenide is attractive because of its high electron mobility, radiation hardness, and wide operation temperature ranges. In addition, because GaAs is a direct bandgap material, both digital and optical devices can be integrated on the same chip. Major GaAs device types are metal-semiconductor FET (MESFET), high electron mobility transistor (HEMT), and heterojunction bipolar transistor (HBT). Currently GaAs technology is not mature enough to compete with silicon for high-volume VLSI applications and instead serves as a niche technology for high-speed supercomputers and digital and microwave circuits in radiation-hardened aerospace systems.

3

PACKAGING
TECHNOLOGIES

Packaging supplies the chips with wires to distribute signals and power, removes the heat generated by the circuits, and provides them with physical support and environmental protection. In high-speed computers, packaging plays an important and sometimes dominant role in determining the performance, cost, and reliability of the system. With today's small feature sizes and high levels of integration, the speed of the on-chip circuitry is so fast that a significant portion of the total delay in a processing unit comes from the time required for a signal to travel from one chip to another. Currently, 50 percent of the total system delay of a high-end computer is due to packaging delays, and the share of packaging may rise to 80 percent by the year 2000 [3.1]. In order to minimize this delay, chips must be placed close together. In addition, VLSI circuits with large numbers of gates require many signal and power connections. These wires should exhibit good transmission line behavior. They also should be shielded to minimize the couplings between the wires and to ensure signal integrity. High-speed and densely packed circuits generate large amounts of heat that have to be removed efficiently.

Let us try to define the "perfect" electronic package every designer strives for. The "perfect" package is compact, and it supplies the chips with a large number of signal and power connections, which have minute capacitances, inductances, and resistances. The wiring on the package is very dense, and the interconnections are good transmission lines with low capacitance and

TABLE 3.1 Electronic Packaging Requirements

Speed
- Short chip-to-chip propagation delays
- High bandwidth (transmission frequency)

Pin count and wireability
- Large I/O count per chip site
- Large I/O count between the first- and second-level package
- Dense wiring

Size
- Compact size

Noise
- High-quality transmission lines (large Z_0, low resistance)
- Low noise coupling among wires
- Power distribution with low inductance to minimize simultaneous switching noise
- Power distribution with low resistance to achieve small IR drops

Thermal and mechanical
- High heat removal rate
- A good match between the thermal expansion coefficients of the dice and the chip carrier

Test, reliability, and cost
- Easy to test
- Easy to fix
- Easy to modify (engineering changes)
- Easy to manufacture (quick turn-around time)
- Highly reliable
- Low cost

high characteristic impedance. This package provides a stable power-supply level even when the circuits are operating at high speeds and many output drivers are switching simultaneously. In addition, the thermal properties of the package and the semiconductor chips match well. This avoids stress-induced cracks and failures when the parts expand and contract as the temperature varies. The package also can remove large amounts of heat generated by the circuits. The reliability record of this package is perfect, and it costs much less than the chips it carries. Table 3.1 summarizes the properties of a high-quality electronic package [3.2].

In the following sections dual-in-line package (DIP), pin grid array (PGA), surface-mounted chip carrier (SMC), direct chip attach (chip on board), ceramic multichip carrier, silicon-on-silicon hybrid, and wafer-scale integration will be described and evaluated on how well they meet the requirements of the "perfect package." First, the DIP and PGA, which require printed circuit (PC) boards with through-holes, are reviewed. Then SMCs are presented. They are similar to DIP and PGA, except SMCs do not require boards with through-holes and therefore can be packed much more densely. Finally substrates that can hold many dice—ceramic multichip carrier, silicon-on-silicon hybrid, and wafer-scale integration—are described. Multichip packages improve the packing density and performance significantly by mounting the dice directly on a substrate via low inductance and capacitance bonds and by supplying a very dense interconnection network. Cooling techniques for removing the heat generated by integrated circuits are described next. Packaging materials with high heat conductivities, heat-sinks that increase the effective cooling area of the package, and forced circulation of air or water are some of the techniques described that can improve the heat removal rate of a package. Superconductors and optical interconnections are also reviewed from a packaging point of view. Superconductors promise zero-resistance wires, and optical interconnections offer a noise-free, easy-to-route, high-speed alternative to electrical wires.

3.1   Packaging Hierarchy

The packaging hierachy of a digital computer is illustrated in Fig. 3.1. After the processing steps and wafer-level testing are completed, semiconductor wafers are diced, and chips are placed on a carrier. The carrier may hold only one, a few, or hundreds of dice. The single-chip carrier can be a plastic package with fewer than 100 pins or a ceramic package with more than 300 pins. The multichip carrier can be a silicon substrate that holds as many as ten dice or a larger ceramic module that supports hundreds of chips and provides thousands of pins. Depending on the cost and performance requirements, therefore, the chip carrier can take many forms. The packaged chips are then placed on PC boards, which are connected to other boards by connectors and electrical or optical cables. PC boards also come in many types with varying number of wiring and ground planes, and with a range of wiring pitches.

In the past, the packaging hierarchy contained more levels. Dice were mounted on individual chip carriers, which were placed on printed circuit cards. Cards then plugged into a larger board, and the boards were cabled into a gate. Finally, the gates were connected to assemble the computer [3.16]. Today higher levels of integration make many levels of packaging unnecessary, and this improves the performance, cost, and reliability of the computers. Ideally all circuitry (including multiple CPUs, caches, I/O processors, and main memory) one day may be placed on a single piece of semiconductor.

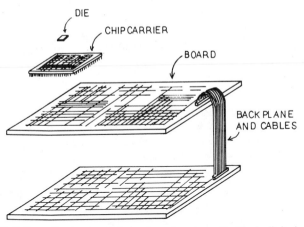

FIGURE 3.1 Packaging hierarchy of a hypothetical digital computer. The physical image depicted in this figure is more representative of a mainframe but the same components and hierarchy exists in almost any computer system.

3.2 Properties of Packaging Materials

Thermal and electrical properties of various metals, semiconductors, and package substrate materials are summarized in Table 3.2. Metals are good conductors of heat and electricity because, in metals, both heat and electricity are transferred via free electrons. This makes metals ideal interconnect and heat sink materials. Silver has very small thermal and electrical resistivities, but problems associated with corrosion, electromigration, incompatibility with high-temperature processing, and high cost make it undesirable as an interconnect or heat sink material. Copper is just as good a heat and electrical conductor as silver and costs much less. Copper, however, has the same corrosion problem as silver and does not adhere well to insulating substrates. Still, copper is more stable than silver. Copper is a harmful contaminant for semiconductor devices. Depositing a thin layer of undercoating metal such as nickel, titanium, or tungsten can ensure adhesion and prevent diffusion into the substrate. Copper can be plated by another metal, such as gold, to avoid corrosion and diffusion. Aluminum has very good metallurgical properties, but its electrical resistivity is higher than silver and copper, making it less desirable for high-speed packaging applications. Molybdenum is not as good an electrical conductor as silver, copper, or aluminum, but it has a high melting point, which makes it suitable for ceramic substrates with multilevels of interconnections. Ceramic substrates must go through processing temperatures in excess of 1500°C, but aluminum and copper melt at 660°C and 1100°C,

TABLE 3.2 Thermal and Electrical Properties of Semiconductors and Packaging Materials

Material	Thermal Conductivity W/cm-K	Coefficient of Thermal Expansion $10^{-6} \times °K^{-1}$	Dielectric Constant	Electrical Resistance $\mu\Omega$-cm
Metals				
Silver	4.3	19.0		1.6
Copper	4.0	17.0		1.7
Aluminum	2.3	23.0		2.8
Tungsten	1.7	4.6		5.3
Molybdenum	1.4	5.0		5.3
Semiconductors				
Silicon	1.5	2.5	11.8	
Germanium	0.7	5.7	16.0	
Gallium arsenide	0.5	5.8	10.9	
Insulating substrates				
Silicon carbide (SiC)	2.2	3.7	42.0	
Beryllia (BeO)	2.0	6.0	6.7	
Alumina (Al_2O_3)	0.3	6.0	9.5	
Silicon dioxide (SiO_2)	0.01	0.5	3.9	
Polyimide	0.004		3.5	
Epoxy glass (PC board)	0.003	15.0	5.0	

respectively. Molybdenum also has a coefficient of thermal expansion close to that of alumina, which reduces thermal stresses between the interconnect and insulating substrate. Similar to molybdenum, tungsten also has a relatively high electrical resistance, its coefficient of thermal expansion matches well with alumina, and it is compatible with temperatures required for ceramic processing (the melting point of tungsten is 3410°C).

Silicon carbide (SiC), aluminum nitride, beryllia (BeO), and alumina (Al_2O_3) are some of the ceramics used in electronic packaging. In comparison with other ceramics, SiC has a thermal expansion coefficient closer to that of silicon, and as a result less stress is generated between the dice and the substrate during temperature cycling. In addition, it has a very high thermal conductivity. These two properties make SiC a good packaging substrate and a good heat sink that can be bonded directly on silicon dice with little stress generation at elevated temperatures. Its high dielectric constant, however, makes it undesirable as a substrate to carry interconnections. Alumina and BeO have properties similar to SiC. Alumina is the most common ceramic substrate. Its thermal conductivity is not as good as SiC or BeO, but is still much

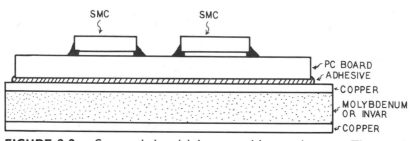

FIGURE 3.2   Copper-clad molybdenum and invar substrates. The metal substrate is used as a stiffener to restrict the thermal expansion of the PC board. In this way, the thermal expansions of the board and chip carriers are matched.

better than plastic packages or PC boards. A disadvantage of alumina and other ceramic substrates is their high dielectric constants, which result in large interconnection capacitances.

As alternatives to ceramics, both SiO_2 and polyimide have dielectric constants about one-third that of alumina. Polyimide, an organic polymer, is preferred over silicon dioxide in packaging because thick layers of polyimide do not exhibit any stress-induced cracking, it is self-planarizing, and it can be deposited more easily than SiO_2. Care should be taken, however, to make sure that polyimide does not contain contaminants that may be harmful to ICs. Epoxy glass is the most common insulator used in PC boards; unfortunately it has a very low thermal conductivity, and its coefficient of thermal expansion is much larger than that of silicon or ceramic packages. This causes stress problems, especially for surface-mounted ceramic parts that lack the support of through-hole pins that ceramic PGAs have. Leadless ceramic chip carriers have been observed literally to pop out of PC boards at elevated temperatures.

Thermal mismatch problems can be solved by incorporating a metal core inside the PC board or by using a metal core board as a base and mounting a regular multilayer PC board on top of the metal core board (Fig. 3.2). The two popular structures that solve the coefficient of thermal expansion mismatch problems are copper-clad invar and copper-clad molybdenum. Molybdenum has a coefficient of thermal expansion close to that of alumina. Molybdenum is also very rigid and prevents the PC board glued on top of it from expanding too much. The coefficient of thermal expansion of copper-clad invar can be adjusted to be exactly the same as the ceramic package mounted on top of it or the same as silicon if chips are directly mounted on the board. Even the temperature difference between the substrate and dice can be compensated by making the coefficient of thermal expansion of the copper-invar-copper substrate slightly greater than that of silicon in order to match their thermal expansions exactly. The advantage of copper-clad molybdenum over copper-clad invar is its higher thermal conductivity and rigidity.

3.3   Through-Hole Mounting

Through-the-board mounting technology using DIPs has been the traditional low-cost packaging choice, and it still remains the most widely used package type. The DIP is a rectangular package with two rows of pins on its two sides. An intermediate step in the fabrication of a DIP is illustrated in Fig. 3.3. Here a die is shown bonded on the lead frame and, in the next step, chip I/O and power/ground pads are wire-bonded to the lead frame, and the package is molded in plastic. Because of the low thermal conductivity of the plastic molding, DIPs have moderate heat-removal capability. (See Section 3.8 for more details on cooling.) The completed DIP is shown in Fig. 3.4. During board assembly, DIPs are mounted in the holes drilled through the PC board and soldered down.

The PC board is manufactured by stacking layers of thin copper sheets and epoxy fiberglass insulating layers on top of each other. The copper sheets are patterned and etched to form the signal and power/ground planes. After all the layers are stacked and pressed together, precision holes are drilled through the board and plated with copper (Fig. 3.5). This copper plating forms the connections between separate layers. There are no dedicated via structures

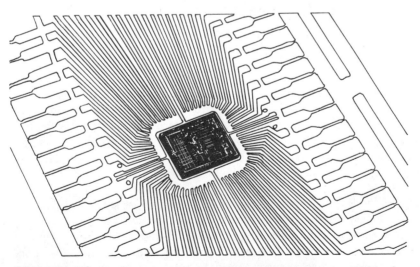

FIGURE 3.3   Intermediate step in DIP fabrication. Here a chip is shown die bonded to a metal skeleton (lead frame). In the next step, chip pads will be wire bonded to the pins, which are the two rows of conductors on both sides of the die. Then the entire structure will be encapsulated in plastic molding.

FIGURE 3.4   Through-hole mounted packages. Dual-in-line (top) has a row of pins on both its sides. Pin grid array (bottom) can support higher lead counts because its pins are not at the periphery but are placed on its entire bottom surface [3.8]. (Photographs courtesy of Motorola, Inc., © 1985 IEEE)

to make connections between wiring levels; through-holes serve that purpose. Finally DIPs are inserted through the holes, and to bond the package on the board, the entire bottom surface of the board is dipped in a special solder solution that adheres only to the metal pins, the copper-plated walls of the holes, and the contact areas surrounding the holes. This is referred to as wave soldering.

Through-holes form a sturdy support for the chip carrier and resist thermal and mechanical stresses caused by the variations in the expansions of

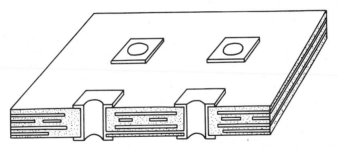

FIGURE 3.5 Printed circuit board with through-holes. Layers of epoxy glass and copper are stacked and pressed together. Precision holes are drilled through the PC board and plated with copper. Copper-plated through-holes form electrical connections between the pads and PC board wires, as well as connections between wires in two separate planes.

components at raised temperatures. The packing density of DIPs, however, is degraded by many factors. For example, DIPs can get excessively large at pin counts greater than 64, and through-holes limit the board packing density by blocking the lines that might have been routed below them and by making it impossible to mount chips on both sides of the board in a way that would increase the board density (Fig. 3.6(a)). This results in long interconnections

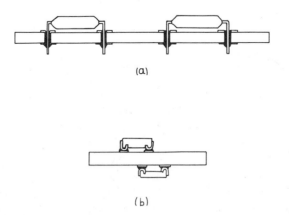

FIGURE 3.6 Through-hole and surface mounted chip carriers on a PC board. (a) DIPs mounted on a through-hole PC board block the passage of wires under them, and packages can be placed on only one side of the board. (b) In surface mounting, wires are not obstructed by pins that go through the board, and chips can be placed on both sides of the board.

at the board level, which, combined with the large capacitance and inductance of DIP pins, degrade system speed [3.4]–[3.9].

A pin grid array (PGA) has leads on its entire bottom surface rather than only at its periphery. This way it can offer a much larger pin-count (300 or more) than a DIP can (maximum of 64). PGAs are available in cavity-up and cavity-down versions (see Fig. 3.27). A cavity-down PGA is shown in Fig. 3.4. Here a die is mounted on the same side as the pins facing toward the PC board, and a heat sink can be mounted on its backside to improve the heat flow. When the cavity and the pins are on the same side, the total number of pins is reduced because the area occupied by the cavity is not available for brazed pins. The mounting and wire bonding of the dice is also more difficult because of the existence of the pins next to the cavity. In cavity-up PGA, pins and chip are on the opposite sides of the square package. Cavity-up PGA is easier to handle and offers a dense pin array on its entire bottom surface, but it cannot dissipate as much power as a cavity-down PGA because the backside of the die faces the board and cannot be directly bonded to a heat sink. High pin count and large power dissipation capability of PGAs make them attractive for minicomputer, workstation, microprocessor, and ASIC packaging.

A PGA usually utilizes a multilayer ceramic substrate (for example, alumina), which has a high dielectric constant ($\epsilon_r = 9.5$) and results in large lead capacitances. In addition, the thermal expansion coefficient of alumina is smaller than that of a typical PC board, and this generates stresses at raised temperatures. The ceramic substrate also increases the cost of PGAs. To support a large number of pins and to provide dense interconnections between the pins and the chip, however, a ceramic substrate is necessary. An additional benefit of ceramic packages is their superior thermal conductivity and hermeticity. Hermetically sealed packages keep the moisture and contaminants away from the chips, and minimize corrosion- and contaminant-related failures. This makes them indispensable for military and aerospace applications, where high reliability and continuous operation are crucial.

As an alternative to ceramic substrate, epoxy fiberglass PC board material is also used in PGAs. These packages are cheaper, match the thermal properties of the PC board better, and have smaller parasitic lead inductances and capacitances, but they cannot dissipate as much power because their thermal conductivity is lower and they cannot be hermetically sealed. The thermal expansion mismatch between the silicon die and epoxy glass substrate must be carefully managed. To improve the heat removal rate, through-holes drilled in the substrate can be filled with metal, and these metal pillars can be attached to similar studs on the PC board to provide an efficient thermal path to a heat sink [3.4].

3.4   Surface Mounting

Surface mounting solves many of the shortcomings of through-the-board mounting. In this more advanced technology, a chip carrier is soldered to the pads on the *surface* of a board without requiring any through-holes (Fig. 3.6(b)). The smaller component sizes, lack of through-holes, and possibility of mounting chips on both sides of the PC board improve the board density. This reduces parasitic capacitances and inductances associated with the package pins and board wiring, and as a result system speed is enhanced. In addition, the manufacturing process is easier to automate because placing chips on the surface of a PC board is simpler and faster than stuffing their pins in through-holes. Of course, there is an initial cost of moving from through-hole-mounting to surface-mounting equipment. Due to the limited observability of the chip leads, testing of the surface-mounted boards for electrical and mechanical defects is more difficult, especially for double-sided boards. Thermal expansion mismatches also become more serious for surface-mounted parts that lack the support of pins that go through the board. (See Table 3.3 for a comparison of various package types.)

The plastic small-outline IC (SOIC) package (Fig. 3.7) has gull-wing-shaped leads with 50 mil spacings, compared to 100 mil lead spacings required for through-hole-mounted DIPs and PGAs. SOIC packages usually have small lead counts (8–28) and are used for discrete, analog, and SSI/MSI logic parts.

A variety of surface-mounted plastic and ceramic chip carriers (CC) is used in applications that require larger lead counts. For example, a flatpack can have densely spaced flat leads on its all four sides. Flatpacks can support

FIGURE 3.7   Small-outline IC package with gull-wing leads [3.8] (Photograph courtesy of Motorola, Inc. © 1985 IEEE)

FIGURE 3.8 Flatpack [3.8] (Photograph courtesy of Motorola, Inc. © 1985 IEEE)

more than 100 terminals, and the lead spacing can be as small as 0.65 mm, compared to 2.54 mm pin spacings of DIPs and PGAs. A low lead count flatpack with leads on only two sides is shown in Fig. 3.8.

Plastic-leaded chip carriers (PLCC), such as gull-wing and J-leaded chip carriers, are also available. Figure 3.9 shows a PLCC with J-shaped leads. PLCCs are offered with lead counts in excess of 124 and lead spacings of 1.27 mm (50 mil). Figure 3.10 compares soldering techniques associated with various through-hole and surface-mounted packages. J-leaded chip carriers pack denser and are more suitable for automation than gull-wing leaded carriers because their leads do not extend beyond the package. In addition, J-leaded PLCCs can be easily inserted into sockets. Sockets are used extensively for PGAs and SMCs, especially for high-cost chips. Boards with gull-wing leaded carriers, on the other hand, are easier to test because their leads are more accessible.

FIGURE 3.9 Plastic leaded chip carrier (PLCC) with J-shaped leads [3.8] (Photograph courtesy of Motorola, Inc. © 1985 IEEE)

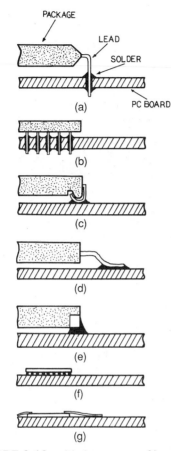

FIGURE 3.10   Various types of integrated circuit packages and mounting techniques: (a) Dual-in-line package (DIP), (b) pin grid array (PGA), (c) J-leaded chip carrier, (d) gull-wing-leaded chip carrier, (e) leadless chip carrier, (f) flip-chip mounting with collapsible solder balls, and (g) tape-automated bonding (TAB) applied directly on the PC board

Leadless ceramic chip carriers (LCCC) take advantage of the same multilayer ceramic technology as the PGAs, but instead of brazing an array of pins on the substrate, the conductors bonded to chip pads are left exposed around the package periphery to provide contacts for surface mounting (Fig 3.11). Lead spacing in LCCCs can be as small as 0.75 mm (30 mil), and terminal counts can reach 124. Dice in leadless chip carriers are mounted in cavity-down position, and the backside of the chip faces away from the board, providing a good heat-removal path. The ceramic substrate also has a high thermal conductivity. These make LCCCs suitable for applications where circuits dis-

FIGURE 3.11 Leadless ceramic chip carrier (LCCC) [3.8] (Photograph courtesy of Motorola, Inc. © 1985 IEEE)

sipate large amounts of power and surface-mounting technology is desired. In addition, LCCCs can be hermetically sealed, which makes them attractive for military and aerospace electronics. Of course, LCCCs have the same thermal expansion mismatch and high-cost problems associated with all other ceramic-substrate-based packages. The mismatch problem is even more serious in LCCCs because they lack the support of through-hole pins. Copper-invar-copper or copper-molybdenum-copper substrates can be used to match the thermal expansions of the chip carriers and PC board. It is also common to place LCCCs in sockets that are mounted on the board.

The properties of through-hole and surface-mounted IC packages are summarized in Table 3.3, and a sampling of packages is shown in Fig. 3.12. All the packages described and the lead counts and spacings quoted in this chapter and in Table 3.3 are typical of those in mid-to-late 1980s. The trend is toward larger lead counts, smaller lead spacings, reduced footprint sizes, improved heat-removal rates, and a greater variety of packages. There are already surface-mount PGAs with 50 mil pin spacings and with pin counts that reach and go beyond 400 [3.10].

3.5 Die Attachment Techniques

The electrical connections between the chip pads and package can be accomplished by one of three ways: wire bonding, tape-automated bonding, or controlled collapse bonding.

FIGURE 3.12   Sampling of IC packages: leadless ceramic chip carrier (top left), dual-in-line package (top right), flatpack (middle row left), plastic-leaded chip carrier (middle row center), small-outline IC package (bottom left), ceramic-leaded chip carrier (bottom center), and pin grid array (far right) [3.8] (Photograph courtesy of Motorola, Inc. © 1985 IEEE)

3.5.1 Wire Bonding

Wire bonding has been used for many years to provide electrical connections between the chip and its package. In this technique, the backside of the die is first glued down on the lead frame with a good thermal conductor (diebond). Later a wire bonding machine, which operates much like a sewing machine, attaches individual wires between the chip bonding pads and the lead frame. Figure 3.13 illustrates a wire bonded chip. The wires are usually aluminum because Al is a good conductor and does not sag like some softer metals do (for example, gold). As the number of I/O and power/ground terminals increase, wire bonding them without causing shorts between the wires becomes challenging. In addition, since the pads are bonded one at a time, wire bonding a high-I/O-count chip takes a long time. This may cause the cost of wire bonding to become prohibitively large. In spite of these difficulties, however, machines exist that can wire bond chips with over 100 terminals. Electrical properties of bonding wires are not perfect either. The irregular structure of bonding wires makes accurate calculation of their electrical parameters difficult. The large parasitic inductance of the bonding wires (approximately 5 nH) is a major problem. The effective inductance can be minimized by bonding with multiple wires and/or by using a thicker bonding wire. Using more than

TABLE 3.3 Various Types of IC Chip Carriers (typical of mid-to-late 1980s)

Package Type	Package Material	Typical Lead Spacing	Typical Maximum Lead Count
Through-hole mounting			
Dual in-line package	Plastic	2.54 mm (100 mil)	64
Pin grid array	Ceramic	2.54 mm (100 mil)	300
Surface mounting			
Small-outline IC package	Plastic	1.27 mm (50 mil)	28
Flatpack	Plastic	1.00 mm	64
		0.80 mm	80
		0.65 mm	100
Leaded chip carrier	Plastic or	1.27 mm (50 mil)	124
(J- or gull-wing-leaded)	ceramic		
Leadless chip carrier	Ceramic	1.27 mm (50 mil)	124
		1.00 mm (40 mil)	
		0.75 mm (30 mil)	
Direct die mounting			
Flip-chip mounting	Ceramic, polyimide, or silicon	0.25 mm (10 mil)	300
Tape-automated bonding	Polymer	0.10 mm (4 mil)	300

two or three wires does not pay off because of the mutual inductances between the wires.

3.5.2 Tape-Automated Bonding (TAB)

In TAB technology, a chip with its attached metal film leads is placed on a multilayer polymer tape similar to a 35 mm photographic film (Fig. 3.14). This film then can be fed to automatic test and assembly machines that place the die on chip carriers or directly on PC boards.

TAB typically places solder bumps on diced chips (Fig. 3.15). (There are also schemes using solder-bumped TAB.) Later these bumped chips are aligned with copper leads fabricated in multilayer polyimide tapes, and the solder is reflown to form the bonding. Because all the terminals are connected simultaneously (also referred to as mass bonding or gang bonding), the process is very fast. In addition, the bonding leads can be placed close together because they are fabricated in a solid supporting structure. TAB reduces the

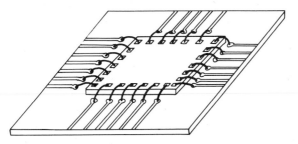

FIGURE 3.13   Wire bonding

length of the connections between the chip and the package and minimizes the lead inductance and capacitance. The minimum pitch in wire bonding varies from 0.16 mm to 0.30 mm. TAB, on the other hand, can offer lead spacings as small as 0.08 mm to 0.12 mm. In addition, the characteristic impedance of TAB leads can be controlled more accurately than the impedance of wire bonding. On the negative side, an individual TAB frame needs to be designed for every chip, and this requires large volumes to justify the cost. Standard

FIGURE 3.14   Tape-automated bonding (TAB). The metal film leads on a polymer tape are shown. Later a chip is placed at the center and attached to the leads. (Photograph courtesy of Motorola, Inc.)

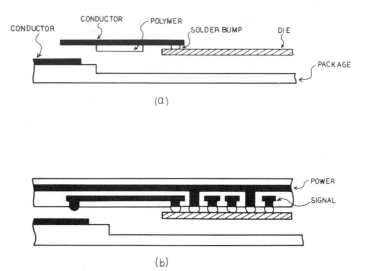

FIGURE 3.15   TAB assembly. (a) A simple TAB structure attaches to the chip via the pads and solder bumps at the periphery. (b) In an area TAB assembly, the pads and solder bumps are distributed over the entire surface of the chip. In area TAB, the polymer tape contains two levels of conductors: one for signals and another primarily for power distribution.

TAB frames, however, can be established for ASICs and other low-volume parts. The reliability of placing an organic material like polyimide in a hermetically sealed cavity also should be evaluated. (This concern may be unwarranted because polyimide has been used as an interlayer dielectric in ICs for may years with great success. It has also been used as a coating to reduce radiation-induced soft errors in DRAMs.)

A new version of TAB, referred to as *area TAB*, borrows a good idea from controlled collapsible bonding (which will be described in the next subsection). In area TAB, solder bumps are distributed over the entire surface of the chip. As a result, I/O and power/ground terminals are not constrained to be placed around the chip periphery. The main advantage of this is the large number of I/O that can be supported by attaching the TAB tape to the entire surface of the chip rather than just the periphery. In addition, on-chip signal and power lines can be routed more efficiently because they are not forced to go to the periphery. This saves die area and results in smaller parasitic resistances, capacitances, and inductances.

Wire bonding will be inadequate for high-pin-count chips, and TAB techniques will be required to connect the chip signal and power pads to package pins. TAB can be used to mount dice on individual chip carriers (to eliminate wire bonding), as well as to attach dice directly on modules or PC boards (referred to as chip on board). Direct TAB mounting on boards can support 500 or more leads per chip [3.10]. But TAB requires complex mounting equipment that is currently too expensive for many users. With time, TAB and direct die mounting may become more common in low- and middle-range digital systems, such as personal computers, workstations, and minicomputers. The thermal mismatch between the dice and PC board remains to be a major concern for chip-on-board technology. This becomes even more troublesome as die size increases.

3.5.3   Flip-Chip Mounting

The length of the electrical connections between the chip and the substrate can be further minimized by placing solder bumps on the dice, flipping the chips over, aligning them with the contact pads on the substrate, and reflowing the solder balls in a furnace to establish the bonding between the chips and the package. This method provides electrical connections with minute parasitic inductances and capacitances (less than 1 nH of inductance and less than 1 pF of capacitance). In addition, the contact pads are distributed over the entire chip surface rather than being confined to the periphery as in wire bonding and most TAB technologies. This saves silicon area, increases the maximum number of I/O and power/ground terminals available with a given die size, and provides more efficiently routed signal and power/ground interconnections on the chips. Thermal resistance of the package, however, may increase because the chip is not attached to the package with a die bond, and the thermal path from the chip to the package is limited to the solder balls unless a thermal contact is established from the backside of the die (examples will be described for ceramic multichip carriers in Section 3.6.2).

Inspection of the solder bonds is also difficult because they are hidden between the chip and the substrate. In addition, thermal expansion mismatch between the semiconductor chips and the substrate can cause strains at the bumps and may lead to eventual failure. The farther the contacts are from the center of the chip, the greater the generated mechanical strain will be. As a result solder bumps must be kept near the center of the die as much as possible, especially in large-die-size VLSI chips [3.5]. When care is taken to avoid the possible pitfalls, however, flip-chip mounting provides a large number of low-capacitance and -inductance electrical connections between the die and the substrate.

3.6   Direct Die Mounting (Hybrids and WSI)

Direct die mounting (chip on board) eliminates the chip carrier and places dice directly on a board or module. Dice can be attached to the substrate using wire bonding, TAB, or flip-chips with solder bump connections. This technique is used widely in mainframe computers (flip-chip mounting on ceramic substrates), and in high-performance special-purpose thick film hybrid parts for military and commercial applications (mostly wire bonding to ceramic substrates). Direct die mounting is also used in consumer products such as electronic calculators, video games, compact disk players, video recorders, and video cameras to achieve small size and light weight. In consumer products, the I/O count is usually low, and wire bonding is used to form electrical connections between the chip and the substrate.

3.6.1   Insulated Metal Substrate

In low-cost consumer products, direct die attachment on aluminum-based metal substrates is very attractive. As shown in Fig. 3.16, the base is formed by an aluminum substrate a few millimeters thick. On top of the aluminum, 10–30 μm of insulating resin, such as epoxy or polyimide, is spun. Then ground vias are etched, and copper 15–75 μm thick is deposited and patterned. Finally, chips are attached, wire bonded, and encapsulated in a protective coating.

Single-level wiring on the substrate keeps the cost down because no precise alignment is required. Wire bonding is also inexpensive. In addition, wire bonds can be used as a second-level wiring by forming jumpers between the copper wires on the substrate.

The aluminum substrate acts as an excellent heat sink and provides a perfect ground plane for power distribution. In addition, by folding the aluminum substrate like a book, an effective shield can be formed against any electromagnetic emission generated by the chips. Since both the substrate and the bonding wire are aluminum, there is no thermal mismatch between them. It is important that the die attach can stand the thermal expansion differences between the silicon chips and the aluminum substrate.

This insulated metal substrate technology is suitable for consumer elec-

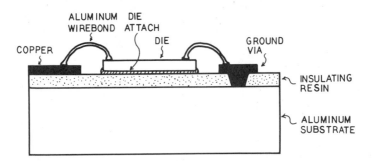

FIGURE 3.16   Insulated metal substrate

tronics and high-power applications. To meet the high interconnection demands of low- and middle-end digital systems, the substrate has to provide multilevels of wires. Increased total insulator layer thickness resulting from multilayer wiring, however, reduces the heat-sinking efficiency of the metal substrate. In addition, larger chip sizes of VLSI circuits introduce more stringent requirements for the thermal matching between the dice and substrate. To reduce the thermal mismatch, aluminum can be replaced by molybdenum, which is more expensive but has a coefficient of thermal expansion closer to that of silicon. Copper-clad invar is another alternative. Then the mismatch between aluminum wire bonding and substrate becomes important. TAB or flip-chip mounting can be used to replace the wire bonding. Flip-chip mounting eliminates the die attach and increases the thermal resistance of the path between the chip and the substrate. Nevertheless a variant of the insulated metal substrate technology is a good candidate for packaging of future low- and middle-end computers.

3.6.2   Ceramic Multichip Carriers

Depending on the application, multilayer ceramic substrates can be used to support one, a few, or more than 100 chips. Here the idea is to eliminate one level of packaging, the single-chip carrier, and place many chips on a substrate that can provide smaller inductance and capacitance electrical connections among the dice than that provided by traditional single-chip carriers and PC boards. Usually the dice are mounted on a multilayer ceramic substrate via solder bumps, and the ceramic substrate offers a dense interconnection network. The advantages of the multichip module over the single-chip carrier are numerous. The multichip module minimizes the chip-to-chip spacing and reduces the inductive and capacitive discontinuities between the chips mounted on the substrate by replacing the die-wirebond-pin-board-pin-wirebond-die path with a much superior die-bump-interconnect-bump-die path. In addition, narrower and shorter wires on the ceramic substrate have much less capacitance and inductance than the PC board interconnections.

The ceramic multichip carrier developed for the IBM 3081 mainframe (1981) is an elegant example of a complex multichip module [3.12]–[3.13]. The so-called thermal conduction module (TCM) consists of a 9×9 cm^2 alumina substrate that may contain up to 33 levels of molybdenum conductors. At the top, there is a bonding level on which chips are flip-mounted via solder bumps. The next five layers are for redistribution and have a 0.25 mm pitch that matches the chip contact pad spacing. The redistribution layers are required to connect the chip bonding pads to the signal wires, because the x- and y-signal wires are not sufficiently dense to connect directly to the bonding pads in the small area in which they are concentrated. This is also referred to as the *escape problem*. Sixteen of the following layers are x- and y-signal planes with a controlled impedance of 55 Ω and a line pitch of 0.5 mm. This yields 320 cm

FIGURE 3.17   TCM ceramic multilayer substrate with flip-mounted chips [3.12] (Copyright 1982 by International Business Machines Corporation; reprinted with permission.)

of signal wiring per cm^2 of TCM, and typically there is 130 m of signal wiring per TCM. Between each pair of signal planes is a voltage reference plane (total of eight). Three power distribution planes (for two supply levels and ground) bring the number of layers up to 33 $(1 + 5 + 16 + 8 + 3 = 33)$.

The chips on the TCM are surrounded by external pads for engineering changes. This is important because designing and fabricating many versions of a complex multichip module during system debug would considerably increase the time to bring up the system. Engineering change pads enable modifications in the connectivity without requiring a new TCM design. The center-to-center spacing between the chips is 300 mils (7.6 mm), and a typical 3081 TCM carries up to 133 chips. The multilayer ceramic substrate with its flip-mounted chips and engineering change pads that surround the dice is shown in Fig. 3.17.

The bottom of the module contains an array of 1800 brazed pins to make connections to a high-performance PC board; 1300 of these pins are for signals, and 500 are for power. Although the number of pins on a TCM is an order of magnitude greater than that of a typical chip carrier, the spacing between the TCM pins is 100 mils, the same as in a typical PGA. The bottom view of two TCMs that shows the dense array of brazed pins is in Fig. 3.18. These pins fit into low-insertion force connectors on the board. Because of the large number of pins on the module, it is important that these connectors require a very low insertion force. Because of the high cost of the chips and package, it is essential that the TCMs are demountable from the board and fixable.

A spring-loaded piston contacts the back of each chip to provide a thermal conduction path to a water-cooled heat exchanger [3.14]–[3.15]. In addition, the entire housing is filled with helium because it has a thermal conductivity

FIGURE 3.18   Bottom view of two TCMs (Photograph courtesy International Business Machines Corporation.)

an order of magnitude greater than that of air. A cross-section of the TCM is shown in Fig. 3.19. Here the cooling housing as well as multilayer ceramic package are cut open to show the internal details. From top to bottom, cold plate with water circulation channels, cooling housing with spring-loaded pistons, MLC substrate with flip-mounted dice, multilayers of wires and ceramic insulator layers, and the base plate are all visible. A blown-up view of the

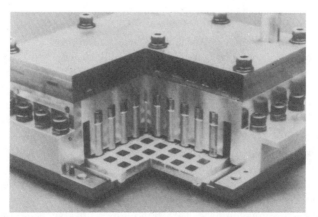

FIGURE 3.19   Cross-section of the IBM TCM [3.12] (Copyright 1982 by International Business Machines Corporation; reprinted with permission.)

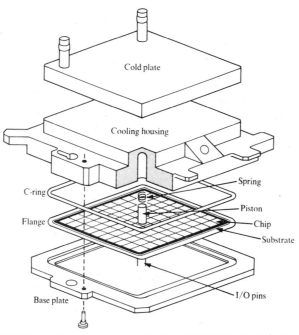

FIGURE 3.20   Blown-up view of the IBM TCM [3.12] (Copyright 1982 by International Business Machines Corporation; reprinted with permission.)

TCM is in Fig. 3.20. The entire package can remove up to 300 W of power, and each chip can dissipate a maximum of 4 W. This corresponds to a power density of 4 W/cm^2 at the module level and 20 W/cm^2 at the chip level. (See Section 3.8 for more details on cooling of IBM TCM.)

Nine TCMs plug into an epoxy glass PC board that measures 60 × 70 cm^2 [3.16]–[3.17]. The PC board has 18 layers of copper, 6 of which are signal planes and 12 are power and reference planes. The signal wires are 80 ± 10 Ω transmission lines. The power bus on this PC board can supply 600 amperes of current with a maximum power-supply level fluctuation of 15 mV. Decoupling capacitors are placed on the backside of the board to ensure noise-free power distribution. The board also supports more than 2000 connections for cables that provide communication between the board and other subassemblies. A board contains one of the dual CPUs of IBM 3081. A full-featured 3081 has 26 modules and requires three PC boards; two are CPUs and one is memory. Figure 3.21 shows two of these boards before the TCMs are mounted on them.

In a more recent development, thin film copper lines with polyimide interlayer dielectric are integrated with ceramic substrates [3.18]. The NEC SX Supercomputer (1985) uses 1000-gate current mode logic (CML) gate arrays

FIGURE 3.21   IBM printed circuit boards. Each board has sockets for nine TCMs. (Photograph courtesy International Business Machines Corporation.)

with 250 ps gate delays as logic elements, 1 Kbit bipolar CML static RAMs with 3.5 nsec access time as cache memory and vector registers, and 64 Kbit CMOS static RAMs with 40 nsec access time as main storage. A 10×10 cm^2 multilayer ceramic substrate is used for distributing the power, and dense thin film signal lines with polyimide insulating layers are fabricated on top of this ceramic substrate. The NEC SX supercomputer achieves a 6 nsec machine cycle and can execute 1300 million floating-point operations in 1 second (1.3 Gigaflops).

The chips are first placed in the so-called flipped TAB carriers (FTC). A logic chip is TAB mounted facing a multilayer ceramic substrate, which has 176 TAB outer lead bonding pads on its upper surface to accommodate the chip and 169 solder bumps arranged in a 13×13 array on its bottom surface to contact the multichip module. After the logic chip is TAB mounted on this ceramic substrate, the entire structure is encapsulated in a low-thermal-resistance copper-tungsten alloy cap, and the backside of the flip-mounted chip is also die bonded to the cap to reduce its thermal resistance. The cap measures 12×12 mm^2 and has a matrix of holes on its bottom surface to allow access to the solder bumps on the ceramic substrate. Four CML RAM chips are mounted in a similar 14×14 mm^2 FTC.

Thirty-six of these FTCs are placed on a high-density multichip package that provides fine pitch and high-speed interconnections and a large number of I/O terminals (Fig. 3.22). The basic substrate is a 2.75 mm thick 10×10 cm^2 alumina. This ceramic substrate includes tungsten metal layers for power distribution and has 2177 I/O pins brazed on its bottom surface on

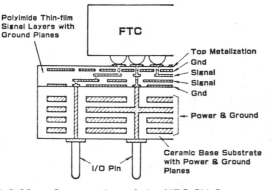

FIGURE 3.22   Cross-section of the NEC SX Supercomputer multichip package [3.18]. (© 1985 IEEE)

a 2.54 mm (100 mil) staggered grid. On top of this alumina substrate, four polyimide insulating layers and five thin film conductor layers are deposited. Two of the metal layers are for signals, two are ground layers to minimize cross talk, and a top layer provides contacts arranged as a 0.8 mm grid for the flipped TAB carriers. Interconnection width is 25 μm, and the pitch is 75 μm. Vias with 50×50 μm holes provide connections between x- and y-signal layers. The 10×10 cm^2 substrate can supply a maximum of 110 m of signal lines. Engineering change pads are also included to simplify the system debug and bring up process as well as testing of the multichip package. This module achieves high speed and large I/O count by combining low-dielectric-constant polyimide insulating layers and densely packed thin film interconnects with high-dielectric-constant alumina substrate that can provide a large decoupling capacitance for power distribution and support a large pin count.

Metal studs are placed on top of the FTCs to remove the large amounts of heat generated by the CML circuits. There is one stud for every FTC, and these studs are placed in holes machined in a heat-transfer block, which, in turn, is attached to a water-cooled cold plate. The studs are bonded to the FTCs by a thermal compound after the entire structure is fitted together. The total thermal resistance between a chip and the cooling water is 4.5°C/W. By this method, a module can dissipate up to 250 W of power. (See Section 3.8 for more details on cooling of NEC SX.)

The multichip packages are mounted on a high-density multilayer board by zero insertion force connectors. Twelve modules fit on a board with a 545 × 469 mm^2 area and 4.9 mm thickness. Six signal planes, eight power/ground planes, and two surface layers (one on each side) are provided by the board. High-speed coaxial cables can be attached to the bottom end of the module pins via zero insertion force connectors to supplement the PC board wiring. These coaxial cables also provide connections between boards; a special board-to-board connection area is not necessary. The propagation delay in this high-speed cable is 3.8 ns/m.

3.6.3 Silicon-on-Silicon Hybrid

A silicon substrate also can be used as an interconnection medium to hold multiple chips as an alternative to ceramic substrates [3.24]–[3.28]. This is referred to as silicon-on-silicon packaging or, sometimes, as hybrid wafer-scale integration. Thin film interconnections are fabricated on a wafer, and separately processed and tested dice are mounted on this silicon substrate via wire bonding, TAB, or solder bumps (Fig. 3.23). Using this technique, chips fabricated in different technologies (silicon CMOS or bipolar, and GaAs MESFET or HBT) can be placed on the same hybrid package. The silicon substrate can also potentially contain active devices that serve as chip-to-chip drivers, bus and I/O multiplexers, and built-in test circuitry [3.30].

The silicon substrate usually has two layers of aluminum or copper interconnections, and the interlayer dielectric is either SiO_2 or polyimide. Copper is preferred over aluminum because of its lower resistance ($\rho_{Cu} = 1.7\ \mu\Omega\text{-}cm$ and $\rho_{Al} = 2.8\ \mu\Omega\text{-}cm$). Because the substrate usually does not contain any transistors, the qualities of aluminum that make it desirable for ICs are not as important here. Polyimide is preferred over SiO_2 because of its lower dielectric constant ($\epsilon_{r\,Polyimide} = 3.4$ and $\epsilon_{r\,SiO_2} = 3.9$) and self-planarization property. Polyimide is applied on the silicon wafer in liquid form by spinning; it provides a very flat surface, ensuring good step coverage for the second-level metal. In addition, polyimide is more elastic than SiO_2 and, as a result, it does not have as serious stress-induced cracking and peeling problems as thick silicon

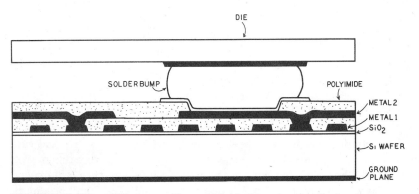

FIGURE 3.23   Silicon-on-silicon hybrid. Two levels of aluminum or copper interconnections are fabricated on a silicon wafer. Polyimide is used as the intermetal insulator. Later separately fabricated chips are flip-mounted on the wafer using solder balls.

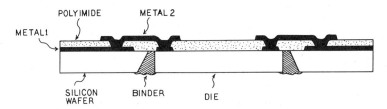

FIGURE 3.24   Silicon-in-silicon hybrid. The dice are stuffed through chemically etched holes in the wafer and glued down by polyimide or epoxy binding. A final level of metal interconnects the chips and the global lines on the wafer.

dioxide layers do. Once the global interconnections are fabricated on the silicon substrate, separately manufactured and tested chips are mounted on it by wire bonding, TAB, or solder bumps.

Figure 3.24 illustrates an alternative way of mounting dice on semiconductor substrates [3.29]. Here tested dice are mounted through the holes etched in a silicon wafer and then interconnected using global aluminum wires fabricated on the wafer (silicon-in-silicon hybrid). The first-level metal is deposited and patterned; then the holes are chemically etched through the backside of the wafer. The chips are stuffed face down in the holes, and the entire structure is placed on an optically flat surface. Then the dice are glued down by polyimide or epoxy binding. A layer of polyimide is spun on the wafer, and a final layer of metal is deposited and patterned to interconnect the dice. A good alignment between the dice and the wafer is required if the same mask is to be used to define the second-level metal in every wafer. If the alignment is not satisfactory, electron beam exposure techniques may be used to compensate the individual misalignments of the dice in the wafer. The alignment problem of this final wiring layer and the cost and low throughput of an e-beam exposure process (if required) are the major drawbacks of this packaging technology. Wire bonding may also be used as a simpler replacement for the second-level metal that connects the dice to global wiring, but wire bonding cannot support as large an I/O count as integrated wires.

In silicon-on-silicon hybrids, the package substrate and chips are both silicon. Accordingly thermal expansion mismatches and resulting stresses are avoided. Of course, the thermal resistance between the silicon dice and substrate must be small; otherwise they may have a large temperature differential. Then although they have the same coefficient of expansion, they will expand differently. If wire bonding or TAB is used, the backside of the dice is in direct contact with the silicon substrate, and the heat flow from the chips to silicon substrate is much better than if the substrate were ceramic or plastic. Additional metal or silicon carbide (SiC) heat sinks can be attached to the substrate to improve heat removal. As an alternative, micro cooling channels

TABLE 3.4 Typical Wiring Pitches in IC and
Packaging Technologies

Technology	Wiring Pitch
Integrated circuit	4 μm
Silicon-on-silicon hybrid substrate	50 μm
Advanced PC board	150 μm

can be etched on the backside of the silicon substrate, and water or air can be circulated through these channels to remove large amounts of heat. This method has been demonstrated to cool heat densities as large as 1 kW/cm^2 [3.49]–[3.51]. (See Section 3.8 for more details on micro cooling channels.)

As described earlier, the advantages of the multichip module over the single-chip carrier are numerous. The multichip module minimizes chip-to-chip spacings and reduces the inductive and capacitive discontinuities between the chips and the substrate. In addition, narrower and shorter wires on the silicon substrate have much less capacitance and inductance than the PC board interconnections. Overall it is estimated that silicon-on-silicon hybrids can reduce the parasitic inductances and capacitances by a factor of five compared to a PC board. The total number of socket or solder contacts to the PC board and the wiring demand on the board are also reduced by placing multiple chips on a single carrier because some of the board wiring is in effect transferred to the chip carrier.

Using customary IC processing techniques, interconnections with 50 μm pitch can be manufactured easily on the silicon substrate. Here pitch is defined as the sum of the wire width and spacing between wires. On the other hand, the most advanced PC board technologies can offer only a 150 μm pitch. A factor of three reduction in wiring pitch causes a ninefold decrease in substrate area. Interconnection densities provided by IC, silicon-on-silicon hybrid, and PC board technologies are compared in Table 3.4.

Well-established IC processing techniques can be directly applied in deposition, photolithography, and etching of the interconnection and insulator layers on the silicon substrate. Existing IC processing equipment and know-how can be put to use. Other packaging and PC board technologies are drastically different from the IC processing and require totally separate fabrication techniques and facilities.

Currently silicon-on-silicon hybrids carry three to ten chips, but it is obvious that as they evolve, the chip count may exceed 100. When their density, speed, and thermal advantages are taken into consideration, silicon hybrids emerge as an attractive alternative to ceramic multichip carriers. The major challenge in silicon-on-silicon packaging is the second-level package that carries the silicon substrate and forms the electrical connections between the silicon substrate and the PC board.

3.6.4   Wafer-Scale Integration

In wafer-scale integration (WSI), "chips" manufactured on a wafer are not diced but instead are interconnected by additional levels of global wafer-scale interconnections [3.31]–[3.41]. Ideally WSI gives the best performance because it eliminates all intermediate levels of packaging by integrating an entire system on a wafer. The performance is superior because WSI offers a dense interconnection network, and shorter wires improve capacitive and inductive loadings and speed up the system. Because the pins (which generally limit the number of connections among chips) are eliminated, increased levels of communication become possible among the subsections of a computer (execution unit, caches, memory management unit, and I/O controllers). This opens up new possibilities for system partitioning in conventional hardware architectures and encourages new and less traditional organizations, such as a massively parallel computer fabricated on a single wafer. WSI systems are not only smaller and faster, but they are also more reliable. Wafer-scale circuits are more reliable because they dissipate less power, and integrated interconnections are more reliable than multilevels of packaging with many soldered wires, pins, cables, and various support attachments. Since the entire structure is integrated on a wafer, thermal mismatches and resulting stresses are avoided. In both wire-bonded and flip-chip/solder-bump-connected chips, die-to-package bonding is a failure mechanism. The fact that WSI eliminates this reliability concern is significant.

So far, manufacturing problems have been responsible for obstructing the commercialization of WSI, and the technological challenge of obtaining acceptable yields at such high levels of integration remains to be conquered. Various redundancy schemes can be used to overcome the yield problem. For example, three copies of logic can be fabricated, and a voting circuit can be used to select the value given by at least two of the circuits (majority voting). Three copies of the voting circuitry are also required to account for the defects in them. The main problem with using triple redundancy is the overhead generated by three copies of the logic plus the voting circuitry. The circuits with triple redundancy typically occupy three and a half times the area required by regular implementation of the same function. In addition, any defects that form shorts between the power and ground lines are not covered by this scheme. Unless a fixing mechanism exists, any wafer with even a single defect that shorts the power and ground has to be discarded. Special care must be taken in layout to keep the power and ground lines far apart in order to minimize the occurrence of shorts. Large-area defects such as catastrophic parametric failures, ion implantations with improper doses, or large-area photolithography defects can affect the yield of the wafer if the area covers all three copies of a subcircuit.

Fuses and laser beams can be used to program and fix the connectivity of the defective wafers. An important component in implementing redundancy is the link technology, which must have a low impedance during the on-state and

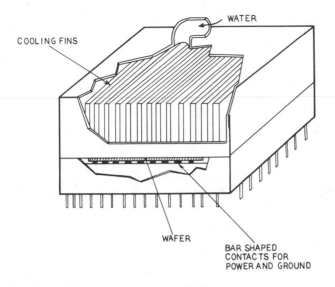

FIGURE 3.25   Wafer-scale module of Trilogy

a very large impedance during the off-state. Fuses can be blown to disconnect failing parts, but their on-impedance is not low enough for all applications (they are especially hard to apply to power lines). Laser beams require a more complicated technology and equipment set-up, but they provide better link removal and adding capability. Laser beams can be used to evaporate and remove material to delete connections, and they can be used also to add connections by removing the dielectric between two normally isolated interconnections and thereby forming new vias [3.40]. Laser beams can also repair opens by laser-enhanced deposition of polysilicon or other conductors. Currently lasers are used routinely to open connections in multimegabit RAMs to remove a defective row and to modify the address decoder so that a spare row is selected instead.

Reconfigurable parallel computers and systolic arrays are architectures that are particularly suitable for wafer-scale integration [3.35],[3.41]. Here many identical processors are used that can be substituted for each other, so perfect yield is not required. The defective processors can be disconnected by laser beams, fuses, or electrical switches. The array can even reprogram itself during every power-up if neighboring processors can test each other and disconnect the faulty cells using electrical switches (self-test and self-configuration) [3.34].

A project at Trilogy Systems Corporation was targeted at building a large mainframe computer with WSI parts (1983) [3.36]. In this WSI mainframe computer, high-speed ECL technology is used to achieve performance in excess of 32 MIPS. The computer contains 40 WSI modules, and every module is water cooled and plugs into a PC board via a 1089-pin connector. The power is not distributed by the PC board; instead it is connected directly to lugs on top of the module that are in contact with the wafer (Fig. 3.25). This simplifies the board design tremendously. The wafer is die bonded to a molybdenum cooling substrate. Molybdenum is selected because it has a thermal expansion coefficient similar to that of silicon and it is also a good thermal conductor (see Table 3.2). The thermal resistance of the package is 0.052°C/W, and as a result a WSI module that dissipates 1000 W experiences a temperature increase of only 52°C. The entire module is filled with helium and hermetically sealed. A combination of triple redundancy and programmable fuses is used to improve the yield of the wafers. The wafer measures 6.2 cm on a side and is divided into 24 one-square-cm zones and eight half-zones to accommodate clock and signal distribution requirements and to simplify the logic design. Each square-cm zone contains 16 power/ground connections and 48 redundant signal connections, and each zone can dissipate up to 50 W of power. The main challenges faced were wafer yield, heat dissipation, package connections, unavailability of computer-aided design (CAD) tools, testing, clock skew, power distribution, resistive losses at the interconnections, and capacitive and inductive couplings between wires. Technical objectives of the project were met, but the product was too expensive to be economically feasible. In other places, more recent work in WSI emphasizes CMOS rather than bipolar.

3.7   Thermal Considerations in Packaging

Fast and highly integrated circuits can generate large amounts of heat. Although a typical ECL gate dissipates less than 10 milliwatts, 10,000 of these gates integrated on a chip can bring the total power consumption easily up to 10–20 W. In CMOS circuits a good portion of the power is dissipated by the output drivers. For example, a CMOS chip that operates at 50 MHz with 3.4 V power supply and has 100 output buffers, each driving 70 pF of capacitance, dissipates 1 W of power only at its output circuits.[1]

The heat generated by the chips must be removed efficiently because virtually all failure mechanisms are enhanced by increased temperatures. Electromigration, oxide breakdown, and hot-electron injection effects worsen, and

---

[1] The generic power dissipation expression is $P = \frac{1}{2} f_c V_{DD}^2 C$. Here, f_c is the clock frequency, V_{DD} is the power-supply level, and C is the capacitive load. The factor $\frac{1}{2}$ accounts for the fact that the capacitance needs to be charged and then discharged to dissipate a total of CV^2 energy. This equation assumes that the node is switching every cycle. When applied to output buffers, the power equation becomes $P = \frac{1}{2} \frac{1}{2} N_D f_c V_{DD}^2 C$, where N_D is the number of drivers and the additional factor of $\frac{1}{2}$ is included, assuming that a buffer will switch every other cycle. Thus, $P = 0.5 \times 0.5 \times 100 \times 50 \times 10^6 \times 3.4^2 \times 70 \times 10^{-12} = 1$ W.

leakage currents increase in reverse-biased junctions and turned-off MOS transistors. At high temperatures, corrosion mechanisms accelerate, and stresses are generated at the material interfaces because of different expansion coefficients; as a result, solder and wire bond failures occur. In addition, CMOS switching speed degrades as the temperature increases. This results in operational failures at elevated temperatures because signals do not reach their targets on time. To prevent reliability and performance degradation, the temperature of semiconductor chips has to be kept in a certain range. Typical temperature ranges for commercial devices are 0 to 70°C during operation and −55 to 125°C during storage. For military parts, operation and storage temperature ranges can be as much as −55 to 125°C and −65 to 150°C, respectively.

To meet the temperature requirements, heat must be removed rapidly from the semiconductor devices. Let us look at the heat transfer mechanism. *Conduction, convection,* or *radiation* can be responsible for the heat transfer. Convection, however, is basically conduction of heat in a moving fluid and can be considered a special case of conduction. When the amount of energy radiated from a chip is computed, it is found that the maximum radiated power with a junction temperature of 120°C is $P \leq \sigma_{rad}T_j^4 = 0.14$ W/cm^2, which is too small a heat flux to meet power dissipation requirements of ICs [3.50]. Thus, in IC packages, conduction is the only mechanism of practical significance, and the package heat transfer mechanism is mainly described by the heat conduction equation

$$J_Q = -\kappa \nabla T \tag{3.1}$$

where J_Q is the heat flux, κ is the thermal conductivity of the material, and ∇T is the thermal gradient.

The generic heat conduction equation expressed in Eq. 3.1 can be used to define the thermal resistivity of a package as

$$\theta = \frac{(T_j - T_a)}{P}. \tag{3.2}$$

This is obtained from Eq. 3.1 by simplifing to a single dimension and assuming a linear temperature gradient. It is based on the fact that the temperature differential between the circuit and its ambient is inversely proportional to the heat removal capability of the package. Here θ is defined as the thermal resistance of the package, T_j and T_a are the junction and ambient temperatures, and P is the total power dissipation. Power dissipation of the chips rises with increased integration density, die size, and circuit speed. The ambient temperature, on the other hand, is usually room or chilled-water temperature. As a result, as chip power consumption increases, the thermal resistance of the packages has to be reduced to keep the junction temperatures within an acceptable range. In low-cost applications, such as personal computers, workstations, and minicomputers, the chip power dissipation is relatively low, the

TABLE 3.5   Thermal Conductivities of Various
Materials (at room temperature unless noted otherwise)

Material	Thermal Conductivity (κ) W/cm-°K
Metals	
Silver	4.3
Copper	4.0
Aluminum	2.3
Molybdenum	1.4
Brass	1.1
Steel	0.5
Lead	0.4
Semiconductors	
Silicon	1.5
Germanium	0.7
Gallium arsenide	0.5
Insulators	
Silicon carbide (SiC)	2.2
Beryllia (BeO) (2.8 g/cc)	2.1
Beryllia (BeO) (1.8 g/cc)	0.6
Alumina (Al_2O_3) (3.8 g/cc)	0.3
Alumina (Al_2O_3) (3.5 g/cc)	0.2
Silicon dioxide (SiO_2)	0.01
High-κ molding plastic	0.02
Low-κ molding plastic	0.005
Polyimide	0.004
Epoxy glass (PC board)	0.003
Liquids	
Water	0.006
Liquid nitrogen (at 77°K)	0.001
Liquid helium (at 2°K)	0.0001
Gases	
Hydrogen	0.001
Helium	0.001
Oxygen	0.0002
Air	0.0002

chip packaging is simple, and the ambient temperature is room temperature. In high-end systems, chips dissipate more heat; accordingly the package is more complicated in order to provide a low thermal resistance. In addition, the ambient temperature is lowered by using chilled water to increase the temperature differential.

Table 3.5 lists thermal conductivities of various materials [3.42]. (A substance with a large thermal conductivity is a good heat conductor.) Metals as a group are better heat conductors than nonmetals, and gases are poor heat conductors. From the kinetic theory, the thermal conductivity κ can be obtained as

$$\kappa = \frac{1}{3}C_Q vl, \tag{3.3}$$

where C_Q is the heat capacity per volume, v is the average particle velocity, and l is the mean free path of a particle between collisions.

Two principal mechanisms are responsible for transport of heat in solids: drift motion of the electrons and the directional cooperative vibration of interacting lattice ions (phonons). In metals and alloys, heat is conducted via electrons, which are light, fast, and have long mean-free paths; as a result metals are good electrical and thermal conductors. In dielectrics, the heat is conducted via phonons, which have comparable mean-free paths to electrons, are slower than electrons, and have larger (lattice) heat capacity than electrons. Overall the heat conductivities of insulators and semiconductors are somewhat smaller but still comparable to metals. Crystalline insulators and semiconductors have a more regular lattice structure and as a result are better thermal conductors than their noncrystalline counterparts. Liquids and gases are poor heat conductors because of the weaker couplings between their molecules. In gases, heat is conducted by the molecules themselves, and as a result gases with lighter and faster molecules (e.g., helium) are better heat conductors than heavier gases (e.g., air). Fluids can be forced to circulate to improve their heat removal efficiency; water can be circulated through pipes, or air can be blown by fans.

3.8   Thermal Properties of Various Packages

In low-cost DIPs, dice are wire bonded to the lead frame and encapsulated in plastic molding. As illustrated in Fig. 3.26, a heat spreader can be placed under the die to improve the heat transfer [3.43]. There are three major heat flow paths between the die and the ambient: through the die bond that glues the chip to the heat spreader, through the wire bonds and lead frame between the chip pads and the package pins, and through the plastic molding surrounding the chip (Fig. 3.26). Thermal conductivities of the plastic, lead frame, and heat spreader, as well as the shape of the lead frame and the heat spreader, are the factors that determine the thermal resistance of the package. With proper design, thermal resistance of a 40-pin DIP can be as low as 38°C/W during natural convection and 25°C/W with forced convection of air. As a result a DIP can dissipate 2 W and 3 W of power with natural and forced air convection and still keep the temperature differential between the junction and the ambient below 75°C [3.43].

Ceramic chip carriers and pin grid arrays are not molded in plastic; instead the die is attached on an alumina (Al_2O_3) or beryllia (BeO) substrate and hermetically sealed with an air cavity above it. A die can be mounted on a ceramic substrate with the air cavity facing up or down (Fig. 3.27). If the die is mounted facing toward the PC board, a heat sink can be placed on its backside to improve the heat flow (Fig. 3.27(b)). Unlike in a DIP where the die is in contact with plastic molding, in ceramic chip carriers and pin grid arrays, the primary heat flow path from the chip to the package is through the backside of the chip and the die bond; as a result, a void-free and thin die bond is essential.

Figure 3.28 plots the thermal resistance of a 68-terminal ceramic chip carrier as a function of air velocity (assuming that air is blown with a fan). The top curve is alumina substrate with no heat sink, the middle curve is the same substrate with a heat sink, and the bottom curve is beryllia substrate with a heat sink. The heat sink has a surface area of 14 cm^2. As can be seen in this plot, thermal resistance of the package can be reduced to below 10°C/W when a beryllia substrate and a heat sink are used.

Figure 3.29 illustrates an air cavity-down PGA with its attached heat

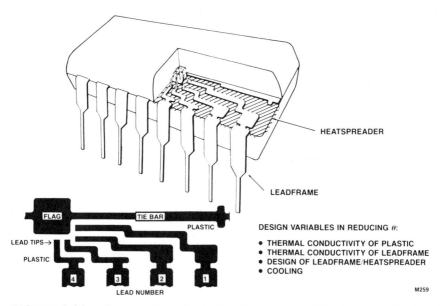

FIGURE 3.26   Cross-section of a dual-in-line package. A heat spreader is placed under the lead frame to improve the heat removal rate [3.43] (© 1985 IEEE)

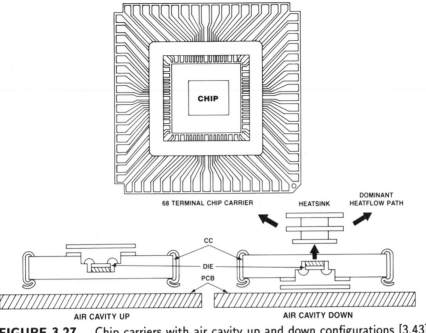

FIGURE 3.27   Chip carriers with air cavity up and down configurations [3.43]
(© 1985 IEEE)

sink. As indicated in the figure, the primary heat-flow path is via the die bond, ceramic substrate, and heat sink. If a high-thermal-conductivity beryllia plug is substituted for alumina in the area immediately below the chip, the thermal resistance of the package can be reduced. Since the multilayer signal distribution lines and brazed pins are still on alumina, this package combines the relatively low cost of Al_2O_3 with the good thermal conductivity of BeO.

Figure 3.30 plots the thermal resistance of packages as a function of air flow rate with variations on the structure illustrated in Fig. 3.29. The top curve is the thermal resistance of a PGA without a heat sink, and the one below it is the thermal resistance of the same package with an attached heat sink. The third curve is the alumina substrate with a beryllia or CuW plug and a heat sink. The heat sink is made of an aluminum alloy and has an effective area of 128 cm^2 [3.43].

Various heat sink structures are illustrated in Fig. 3.31. The structure and the material of the heat sink are very important in determining the thermal resistance of the package. The efficiency of the heat transfer at the interface between the die and the heat sink is also very important. The thermal resistance of an electronic package has six components: the spreading resistance between

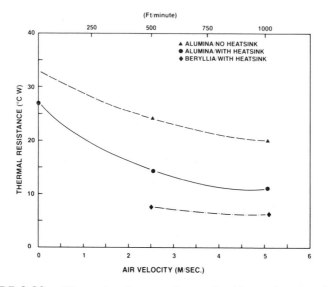

FIGURE 3.28 Thermal resistance of ceramic chip carriers as a function of air flow rate [3.43] (© 1985 IEEE)

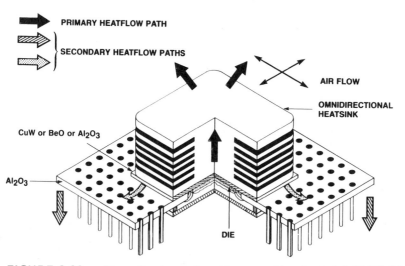

FIGURE 3.29 Air cavity-down pin grid array with a heat sink [3.43] (© 1985 IEEE)

the individual devices and the semiconductor chip substrate θ_{SPREAD}, bulk resistance due to heat conduction through the semiconductor chip θ_{BULK}, the interface resistance between the die and the heat sink θ_{INTERF}, the spreading and bulk thermal resistances of the heat sink itself θ_{SINK}, the convective thermal resistance between the heat sink and the coolant fluid θ_{CONV}, and the caloric thermal resistance due to heating of the coolant as it absorbs the heat θ_{CAL} [3.50]. The total thermal resistance can be modeled as the sum of these individual resistances, analogous to electrical resistors connected in series.

If any one of these resistances is much larger than the rest, it dominates the overall resistance. In VLSI circuits θ_{BULK} is small because silicon is a good heat conductor and the semiconductor die is relatively thin (500 μm). At high levels of integration, a large number of devices that individually dissipate small amounts of power are placed on the chip, and the heat is distributed evenly. No

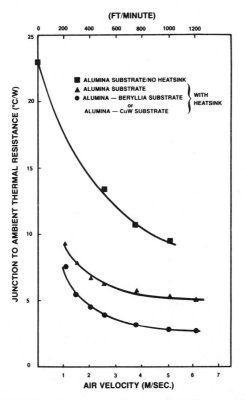

FIGURE 3.30   Thermal resistances of PGAs as a function of air flow rate [3.43] (© 1985 IEEE)

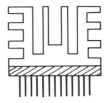

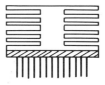

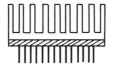

FIGURE 3.31 Various heat sink structures. The material and shape of the heat sink are very important in determining the efficiency and thermal resistance of the package.

hot spots exist, and as a result the θ_{SPREAD} is small. The thermal resistance of the heat sink θ_{SINK} can be minimized by selecting a good heat conductor and designing the shape to reduce spreading resistance from the interface to the fins. The caloric thermal resistance θ_{CAL} can be minimized by forcing a strong flow of a fluid with a large heat capacity because $\theta_{CAL} = 1/C_Q f$ where C_Q is the volumetric heat capacity and f is the volume flow rate. For example, water has a heat capacity of 4.18 J/C-cm^3, and with a flow rate of 10 cm^3/sec, it adds a caloric thermal resistance of only 0.024 C/W [3.50]. In VLSI circuits the convective and interface thermal resistances dominate the overall thermal resistance of the entire structure. To minimize θ_{CONV} the shape of the heat sink is optimized to allow a smooth fluid flow and to maximize the surface area between the fluid and heat sink. The thermal resistance of the interface between the die and the heat sink θ_{INTERF} can be reduced by using a thin and void-free die bond or by eliminating the interface and fabricating the heat sink directly on the silicon substrate [3.49]–[3.50].

Dice also can be flip-mounted on the substrate via collapsible solder balls or pressure contact bumps [3.48]. Then the major thermal path from the die to the substrate is through the solder balls and contact bumps. If a large

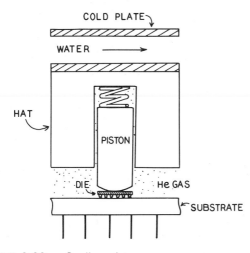

FIGURE 3.32   Cooling pistons

amount of power dissipation is required, a thermal path from the backside of the die has to be established.

Figure 3.32 shows a scheme that can remove large amounts of heat from densely packed flip-mounted chips, and was first used in the IBM 3081 processor unit (1981) described in Section 3.6.2. Here spring-loaded pistons press against the backside of the dice [3.15]. The heat is transferred from chips to the pistons, which in turn conduct it to a cold plate cooled by circulating water. The cavity is also filled with helium, which has a higher thermal conductivity than air because He molecules are lighter than air molecules (see Table 3.5). The total resistance of the package is 11°C/W per chip site. As a result the chips can dissipate up to 4 W of power and raise their junction temperatures only 44°C above the temperature of the cooling water [3.13]. Because the chips are flip mounted and the piston diameter is same as the chip size, dice can be brick walled very close to each other. To dissipate the same amount of power with air-cooled heat sinks would require a large-area heat sink and increase the footprint size per chip. Since chip-to-chip distance directly affects the performance, placing the dice close together and still being able to remove large amounts of power is very important.

In the NEC SX supercomputer (1985), metal studs are placed on FTCs to remove the heat generated by the chips [3.18] (described in Section 3.6.2). There is one stud for every FTC, and these studs are placed in holes machined in a heat transfer block, which in turn is attached to a water-cooled cold plate (Fig. 3.33). After the entire structure is fitted together, the studs are attached to the FTCs by a thermal compound. The total resistance between a chip

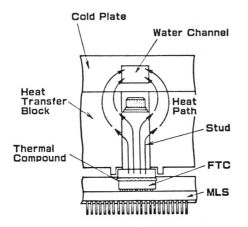

FIGURE 3.33   Cooling studs [3.18] (© 1985 IEEE)

and the cooling water is 4.5°C/W. By this method, a module with 36 chips can dissipate up to 250 W of power and still have junction temperatures only 31°C above the cooling water temperature.

A structure that brings the chips in direct contact with the coolant fluid is illustrated in Fig. 3.34. Here a jet of chilled water hits directly on the backside of the die, and a flexible bellow structure provides conformance to the chip surface. Using this concept, a thermal resistance of 2–3°C/W for 7.5 × 7.5 mm chips has been achieved [3.53]. As a result a chip can dissipate 75 W of power and maintain a junction temperature only 25°C above the cooling water temperature.

The efficiency of the heat transfer can be tremendously improved by fabricating the heat sink directly on the semiconductor wafer itself (Fig. 3.35). This eliminates the interface between the die and the heat spreader and improves the thermal resistance of the package. In a study conducted at Stanford University, microscopic channels 50 μm wide and 400 μm deep were mechanically cut on the backside of a silicon wafer and were closed by a cover plate to confine a fluid flow forced through them [3.49]–[3.50]. These microscopic laminar flow heat exchangers are shown to provide heat removal rates better than 0.1 W/°C-cm^2. Using this method, heat fluxes as large as 1 kW/cm^2 can be removed while keeping the junction and ambient temperature differential below 100°C. In an independent study conducted at Motorola, similar microchannels were mechanically cut on a 3.8×3.8 cm silicon substrate, and by forcing water flows of 12 and 63 cm^3/s, thermal resistances of 0.03 and 0.02°C/W were obtained [3.43]. This corresponds to heat removal rates of 2.4 and 3.6 W/°C-cm^2. When air was used as a coolant, thermal resistances of 1.0–0.7°C/W were obtained, which correspond to a heat removal rate of 0.07–0.1 W/°C-cm^2.

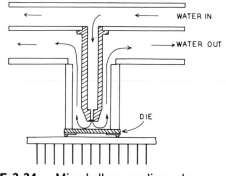

FIGURE 3.34   Microbellows cooling scheme

These microchannels can be fabricated on the chips themselves, or by using a silicon-on-silicon hybrid technology the chips can be mounted on substrates with micro cooling channels. The existence of an interface between the chip and the substrate reduces the efficiency of the heat removal mechanism in a silicon-on-silicon hybrid, but since the substrate can remove so much heat,

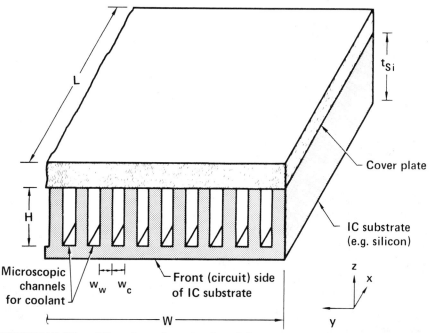

FIGURE 3.35   Microchannel cooling [3.50] (Figure courtesy of David B. Tuckerman)

the overall result most likely will be superior to other alternatives. Long-term reliability of this cooling scheme needs to be carefully studied. A source of concern is the possible clogging of some of the microchannels by accumulation of the particles in the cooling water. This may give rise to local hot spots, which will accelerate a number of device and packaging failure mechanisms.

3.9 Superconductors

Certain materials, when cooled to low temperatures, attain zero dc electrical resistance, a phenomenon called superconductivity [3.54]. Superconducting materials can be used to build very fast switching devices, as well as lossless[2] interconnections on the chips and packages. Wires with zero resistance provide perfect transmission lines with a high bandwidth and a very low attenuation and dispersion.

Electrons have identical charges and they electrostatically repel each other. There is also a net attractive force between the electrons near the Fermi surface. The motion of the positive lattice ions can overscreen the electrostatic repulsion, and pairs of electrons can then attract each other. The existence of the other $(N - 2)$ orbit electrons is essential for the electron pair to attract each other because, through the Pauli exclusion principle, the rest of the electrons restrict the wave vectors the electron pair can attain and forces them in a bound state in spite of the weakness of their attraction. This pairing of electrons enhanced by the interactions of the electrons with the vibrating positive lattice ions (electron-phonon interaction) puts the material in a highly ordered superconducting state. Actually this interaction between the electrons is always present, but at high temperatures thermal energy overwhelms the attraction between the electrons. At very low temperatures, the pairing of electrons forms a very coherent state where the electron pair is in harmony with the lattice and can travel long distances without scattering. As a result the dc resistivity of the material appears to be zero. A common charactersitic of all superconducting material is an energy gap Δ_0 at T=0°K centered about the Fermi energy, where no one-electron levels are allowed.

The transition to superconducting state appears sharply at a *critical temperature* T_c. Above this temperature the material behaves regularly, and once cooled below T_c, it becomes superconducting. In the past, the materials had to be cooled to around 20°K before they exibited superconducting properties. Recently, using compounds of yttrium/barium/copper oxide, and lanthanum/strontium/copper oxide, researchers have demonstrated superconductivity at 77°K and as high as at 90°K. There are even reports of room-

---

[2]The ac resistance does not diminish. Thus there is some loss, which vanishes as the temperature approaches 0°K. The important fact is that superconducting transmission lines are largely free of dispersion and can carry high frequency signals and attain high bandwidths.

temperature superconductivity exhibited by a blend of copper, barium, and other oxides. These new materials have energy gaps Δ much larger than previously observed superconductors since $\Delta \propto kT$. Achieving superconductivity at liquid nitrogen temperature (77°K) makes superconducting materials economically more attractive because liquid nitrogen is easier to handle and cheaper than liquid hellium, which was required to obtain temperatures around 20°K. Bipolar device performance degrades with reduced temperatures, but CMOS transistors get faster as they are cooled down to liquid nitrogen temperature. Silicon CMOS and GaAs HEMT devices operate best at liquid nitrogen temperatures. Superconductors therefore are promising for cryogenic CMOS and GaAs IC interconnections and their packaging. Ordinary metals, such as aluminum and copper, also have liquid nitrogen temperature resistivities that are only one-sixth of their room temperature resistivities. Consequently, the resistivities of ordinary metals are sufficiently low at liquid nitrogen temperature to make superconductors unnecessary for most electronic packaging applications. If superconductivity can be achieved at room temperature and above, then they become attractive for electronic packages.

Besides cooling the specimen below the critical temperature, a number of other conditions must be satisfied to sustain a superconducting state. Superconductivity is destroyed if the material is subject to a large magnetic field (critical field) or if the current density exceeds a critical current. The magnitude of the critical current is dependent on the geometry of the specimen and the magnetic field intensity generated at its surface by this current. Recently current densities as high as 100,000 A/cm^2 have been observed using superconducting materials at 77°K. At high frequencies superconductivity again disappears. Skin effect also becomes a problem because, at high frequencies, skin effect forces the current away from the bulk where the material is superconducting to the outer surface of the conductor where it is not superconducting. Since electronic applications that will benefit from superconducting wires operate at very high frequencies, this is a dilemma that must be resolved.

To use superconducting compounds on the chips, these materials also have to be produced in thin film form. Recently researchers obtained thin film superconductors at 77°K. This makes it possible to use these materials in Josephson junction devices or as zero-dispersion interconnection material in ICs and packages. Thick film wires can be used on package substrates. Brittleness of the ceramic superconductors is a major concern. The reliability of the circuits may suffer when brittle ceramic wires are temperature cycled between room temperature and 77°K.

Evaporation, sputtering, chemical vapor deposition, and laser beam deposition can be used to deposit thin films of superconductors. In one technique, using a combination of evaporation and sputtering known as coevaporation, just the right proportions of yttrium, barium, and copper are deposited on a substrate at 500°C. The specimen is then heated in oxygen to 900°C in order to form a superconducting ceramic. These high processing temperatures make superconductor interconnections unfeasible for conventional ICs because the devices fabricated before the wires would lose their structure at such high temperatures. Lower temperature processes compatible with conventional IC processes must be developed before superconducting interconnections can be integrated with circuits. For packaging applications, high-temperature processing poses no problems. Another concern is the loss of oxygen in the superconductor when heated at vacuum or in an atmosphere that lacks oxygen. This is a common IC processing step (for example, to anneal materials). If the specimen is briefly exposed to an oxygen plasma at a high temperature, however, the supeconductivity is regained.

High contact resistance is another problem that needs to be addressed. This not only adds a series resistance that undermines the advantage of using a superconducting wire but also generates heat and destroys the superconducting properties of the surrounding materials. New techniques that protect the contact surface have been developed. They minimize the air exposure during processing that causes surface degradation. The surface is sputter etched to remove any degraded material and then immediately sputter coated with silver or gold. Using these techniques contacts have been obtained with a surface resistivity of 10 $\mu\Omega$-cm^2 at liquid nitrogen temperature. This is 1000 to 10,000 times lower than the previous methods that use indium solders, silver paint, or silver epoxy.

The substrate under the superconductor is also very important. If, for example, silicon dioxide is used as the substrate, the critical temperature degrades and becomes less than 77°K. The most sucessful substrate, strontium titanate, has a large dielectric constant, which results in large parasitic capacitances and long transmission line delays. The transmission line propagation speed v is expressed as $v = c_0/\sqrt{\epsilon_r}$, where c_0 is the speed of light at vacuum and ϵ_r is the relative permitivity of the medium.

In summary, superconductors have great promise, but many material and processing problems must be solved before they can be applied to integrated circuits and electronic packages. In addition, for superconductors to be useful as IC interconnections, the critical temperature may need to be as high as room temperature. This is required for bipolar chips because bipolar performance degrades with reduced temperature. In addition, at liquid nitrogen temperatures common metals such as copper and aluminum attain conductivities that are a factor of six better than their room-temperature values. Such good resistivities are sufficiently good for many CMOS applications and can make the use of more exotic superconductors unnecessary.

3.10   Optical Interconnections

Optical interconnections offer a fast, reliable, easy-to-route, and noise-free alternative to electrical wires [3.55]–[3.59]. Already many fiber optic local area networks (LAN) with data transmission rates of tens of megabits per second are used in computer-to-computer communication, and AT&T's ESS-5 digital switching computer utilizes optical fibers at the backplane level to connect its boards (processor-to-processor interconnections). The lower levels of communication hierarchy (module-to-module connections at the board level, chip-to-chip connections at the module level, and gate-to-gate connections at the chip level) are currently all electrical; however, research on optical alternatives at all levels is showing promise. The question that remains is how far down the hierarchy optical wires can advance and still be more attractive than their electrical counterparts.

3.10.1   Properties of Optical Interconnections

Optical interconnections typically have higher bandwidth, lower dispersion, and lower attenuation than electrical wires, and they can support larger fan-outs.

A major advantage of optical interconnections is that they do not generate noise because they lack mutual coupling effects such as capacitive and inductive couplings seen in electrical wires. This is why optical signals can cross through each other without any coupling or loss of information. For the same reason, optical signals do not require any ground or reference planes to confine their energy and to avoid cross-couplings.

Optical interconnections behave a lot like perfectly matched transmission lines. The propagation speed in a transmission line with a small permeability and permittivity ($\mu_r \approx 1$, $\epsilon_r \approx 2\text{--}4$) is the same as in optical fibers or wave guides. To achieve transmission line speeds, however, electrical wires must be driven by small impedance drivers, and the end of the wire must be terminated. Optical wires always behave as if they were driven by a small impedance source, and the optical signal reflections from line ends can be more easily controlled. In addition, optical wires exibit less attenuation and dispersion.

Still, the advantages of optical wires for ICs are not clear, and optimal technology depends on the length of the line. Optical wires do act like perfect "transmission lines," but they may also require just as much or even more power than a 50 Ω matched transmission line driver. If the electrical line is driven by a relatively high impedance driver and is in an RC mode, however, optical wires have a speed advantage because their delay is dominated by the parasitic capacitances associated with the detector, whereas the electrical system must charge up the line capacitance, as well as the capacitance of the receiver circuit. As a result compared to the electrical lines where interconnection capacitance is dominant and the line driver has a high impedance, the optical method has a clear advantage. In large fan-out situations where there are many loads on a line, optical and electrical methods are fundamentally in a similar position; the photon flux must split just like the electrons to feed multiple receivers. Optical drivers, however, are claimed to do a better job in driving fan-outs as large as 100 [3.58].

Optical signals are more flexible than electrical ones. They can cross through each other and do not require a nearby ground plane. They can even travel in free space without requiring any "wires" and can be reflected and focused by mirrors, lenses, and holograms. This provides tremendous freedom in routing optical signals. In addition, the "connection" can be modified by reprogramming the interconnection patterns. This can be achieved by rewriting a hologram (described in the next section) on a photorefractive material using two laser beams. Some photorefractive materials such as lithium niobate and bismuth silicon oxide can hold the hologram for seconds, days, weeks, or indefinetely, and some others, such as gallium arsenide, can store the pattern for only nanoseconds [3.57]. As a result, it is possible to change the connectivity of a computer at almost every machine cycle or instruction. This opens up new possibilities for future computer architectures, such as highly parallel processors with dynamically programmable interconnectivity.

3.10.2   Types of Optical Interconnections

There are two basic types of optical connections: index-guided and free-space interconnections [3.55]–[3.56].

Index-guided interconnections are optical fibers and integrated optical wave guides. Here the light wave is confined in a fiber or wave guide that has a different index of refractivity than its surroundings, and, as such, the light is guided from the source to the receiver. Optical fibers are well studied. The wave guides can be formed, for example, by sputtering materials with a different refractive index on a substrate (such as sputtering glass on SiO_2) or by selectively implanting impurities into transparent materials such as SiO_2 to change their refractive index.

In the free-space approach, there are no fibers or wave guides. Optical signals are broadcast and can be reflected and focused by mirrors, lenses, or holograms. Figure 3.36 illustrates a concept developed for processor-to-processor communication in the fifth-generation Dialog.H computer of Japan's Ministry of International Trade and Industry [3.57]. Here multiple processors with their local memories are arranged around a convex mirror, and a light beam emitted from a processor is reflected by the mirror, to be received simultaneously by the rest of the processors.

In chip-to-chip communication, an unfocused free-space approach can be wasteful of photons because the signal is broadcast on the entire chip (Fig. 3.37). This requires a large amount of energy to be cast on the surface to ensure that a sufficient amount is captured by the sensors. The total amount of power can be prohibitively large. Also this limits the number of signal sources to one (or to a few, if detectors sensitive to different regions of

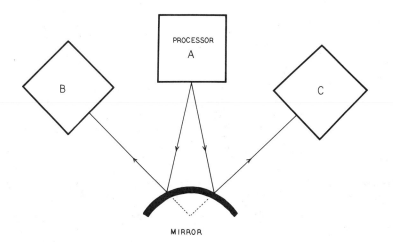

FIGURE 3.36   Interprocessor communication via optical beams reflected from a convex mirror

the spectrum are employed), and as a result it is useful for distributing only global signals such as clocks. In addition, the light falling on the other portions of the die may disturb circuit operation. For example, in CMOS chips, light generates minority carriers in the substrate and can induce latch-up.

Holograms are the most complicated but also the most efficient way of routing optical energy because they can take multiple light sources and focus their energy to selected spots. Holograms of three-dimensional objects are prepared by illuminating a photosensitive film with a reference laser beam and reflecting another beam from the object. Two beams form an interference pattern on the film to record the three-dimensional nature of the object, which can be recreated by illuminating the developed film with light. Physical principles of the holograms for optical interconnections are similar, but the information stored on the hologram is where the light beams need to be focused instead of the three-dimensional nature of an object. The focusing is accomplished again via interference of light waveforms. Figure 3.38(a) shows how an external light source can be focused on multiple spots on a substrate using a hologram, and in Fig. 3.38(b) the source as well as detectors are on the same substrate and the hologram is of reflective type.

One can build efficient silicon detectors, but Si is not suitable as a laser or LED material because it is an indirect bandgap semiconductor. Modulators can be fabricated on the silicon chips that can modulate a uniform light source cast on the chip. Even better, GaAs (a direct bandgap material) can be selectively deposited on Si using molecular beam epitaxy (MBE) or liquid phase epitaxy techniques, and light sources can be built right on the chip itself on

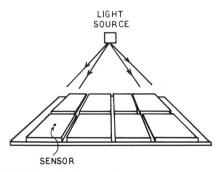

FIGURE 3.37 Unfocused free-space interconnections. Here a global signal such as a system clock is broadcast on a module populated by dice. The detectors on the chips sense the optical pulses and convert them to electrical signals.

these GaAs islands. If GaAs light sources are not integrated on silicon, they can be wire bonded next to the logic chips in a hybrid fashion. This would be less efficient and less reliable but easier to realize in the near term.

3.10.3 Challenges Associated with Optical Interconnections

Tight alignment requirements are some of the biggest difficulties associated with optoelectronic components. To obtain a tight packing density, light sources and wave guides must be aligned to within 1 μm. Free-space interconnections, such as holograms, require precise alignment techniques—for example, an on-chip alignment monitor and a feedback system. This, however, can account for misalignments between only two components of the system and cannot easily compensate multiple independent perturbations. In addition, variations in the thermal expansion of the components must be kept under control to avoid generating misalignments. Techniques borrowed from IC technologies can be used for building alignment fixtures. For example, V-grooves etched in silicon can be used to align multiple fibers with integrated detectors fabricated on a substrate [3.58]. Complete integration, however, is the ultimate solution for good, lasting alignment. In order for the optical wires and components to achieve the same levels of miniaturization as VLSI devices, they have to be manufactured with the same type of integration technology and on the same substrate as electronic parts.

Size incompatibility with ICs (which is related to alignment requirements) is another problem with optoelectronic devices, fiber optics, and wave guides. Although potentially optoelectronic components can be made as small as electronic ones, currently they are an order of magnitude larger.

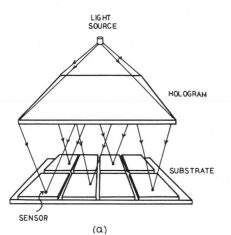

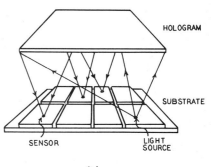

FIGURE 3.38   Free-space interconnections using holograms. (a) The light source is external to the substrate, and the hologram is of transparent type. (b) Both the light sources and sensors are on the substrate, and the hologram is of reflective type.

Optoelectronic components also dissipate large amounts of power. Assuming that logic and memory technologies are electrical, the transmitted energy will have to change its form at least twice (once from electrical to optical to drive the light source and a second time from optical to electrical at the detecting end). According to the laws of entropy, any energy conversion process has an efficiency of less than one, and sometimes, due to practical constraints, this efficiency can be much less than one. For example, refraction efficiencies of holograms are typically less than 20 percent, and laser efficiencies for devices with uncoated mirrors are less than 30 percent. In addition, the minimum power dissipation of a solid-state laser is approximately 1 mW because of the

1 V potential difference needed to turn on the diode and 1 mA current fundamentally required for lasing [3.55]. This is still much less than the 120 mW required by a 75 Ω driver (smallest wire characteristic impedance that can be practically achieved) across a 3 V supply ($3V \times 3V/75\Omega = 120$ mW).

Currently optoelectronic sources and sensors require a lot more than the 1 mW fundamental lower bound. According to one study, the power requirements for a 1 Gbit/sec link between two chips separated by a few centimeters are the same whether optical or electrical interconnections are used. If the distance is longer, optical wires become more attractive. If the distance is shorter (for example, between two nearby chips or gate-to-gate wires on the chip), electrical lines are superior to optical interconnections [3.56]. As a result optical wires do not have any clear speed or power advantage over electrical wires in on-chip and chip-to-chip interconnections if the chips are only a few centimeters apart. If optical wires are used on chips and small modules, it will be to harness their routability and lack of noise couplings. At higher levels of the packaging and communication hierarchy such as module-to-module, board-to-board, and computer-to-computer interconnections, optical wires become very attractive.

3.11   Summary

Packaging has always played an important role in determining the performance, cost, and reliability of high-speed systems, such as supercomputers, mainframes, and military electronics. As the performance of microprocessors, application-specific ICs (ASICs), and other custom CMOS designs improves, packaging will also become important in entry-level systems.

With today's miniaturized devices and high levels of integration, the speed of the on-chip circuitry is so fast that a significant portion of the total delay in a processing unit comes from the time required for a signal to travel from one chip to another. In order to minimize this delay, chips must be placed close together. In addition, VLSI circuits with large numbers of gates require many signal and power connections, which can be satisfied only by a dense interconnection network. These wires should exibit good transmission line behavior and should be shielded to minimize the couplings between the lines and to ensure signal integrity. High-speed and densely packed circuits also generate a large amount of heat that has to be removed efficiently. It is obvious that chip-to-chip delays, noise, and large I/O count requirements will pose significant challenges to the performance increase of computers. Packaging, in fact, may become the most important technology that will determine the leadership in computer and semiconductor industries.

As IC technology improves, entry-level systems are facing packaging problems that once were significant only in mainframes and supercomputers. Currently in the low- and middle-end systems, dual-in-line packages (DIP) and pin grid arrays (PGA) are very common, and surface-mounted chip carriers are becoming more popular because they offer larger pin counts, smaller footprint sizes, and reduced parasitic inductance and capacitance. Multichip packages for midsize computers with ceramic and silicon substrates are also being developed. With the proliferation of tape-automated bonding (TAB) and the availability of bare dice on tapes from the IC vendors, direct die mounting on PC boards or insulated metal substrates will be more common. This chip-on-board technology may eliminate the chip carrier as we know it and become the future standard package in low-end systems.

High-end computers traditionally employed multilayer ceramic substrates and sophisticated cooling schemes. Recent developments in this field include thin film copper lines and polyimide interlayer dielectric, all integrated on a ceramic substrate. Another alternative substrate is the ever-resourceful silicon. Here thin film copper or aluminum interconnections are fabricated on a wafer, and separately processed and tested dice are mounted on this silicon substrate via TAB or flip-chip techniques. This minimizes chip-to-chip spacings and reduces the inductive and capacitive discontinuities between the dice and the substrate. In addition, narrower and shorter wires on the silicon substrate have much less capacitance and inductance than other package interconnection technologies. Using a silicon substrate also eliminates the thermal expansion mismatches and may improve the heat removal efficiency. Silicon-on-silicon packaging may eventually replace ceramics in high-speed systems. Wafer-scale integration is still the most elegant scheme, but yield problems prevent it from becoming a practical alternative.

The recent developments in superconductivity promise superconducting wires at liquid nitrogen temperature and maybe even at room temperature. This may have a significant impact on electronic packaging if processing and material problems associated with superconductors are solved. Interconnection resistance is a major concern in ICs because the wires on the chips are narrow and thin. So far this has not been a problem in packages where the lines are much wider and thicker than the on-chip interconnections, but it may not be too long before package wires are scaled down to dimensions where resistance may become significant. Issues such as skin effect and current density limits and brittleness of ceramic superconductors must be resolved before the superconductors are used in practical applications.

Optical wires are another promising alternative interconnection technology. They offer fast, reliable, easy-to-route, and noise-free lines. Already many fiber-optic LANs are used in computer-to-computer communication, and some systems employ optical fibers at the backplane level, but the benefits of optical wires are still undetermined for the lower levels of the packaging hierarchy. Tight alignment requirements, size incompatibility with ICs, and high power consumption limit the applicability of optoelectronic wires.

4

INTERCONNECTION CAPACITANCE

With increasing chip dimensions, the parasitic capacitance of on-chip wires dominates gate capacitances, and as a result the speed improvement expected from simple device scaling does not apply to global nets. Because the critical paths that determine the clock frequency of an IC consist of such nets, faster local circuits do not necessarily enhance the overall performance of ICs.

The capacitance of chip-to-chip wires is an order of magnitude larger than on-chip interconnections, and this makes board-level communication even more challenging than chip-level signal transmission. To improve overall system performance, sizes of the devices that drive long on-chip lines and chip-to-chip wires are increased. This in turn increases total chip current and power consumption. As a result in high-performance CMOS microprocessors, a good portion of the power is dissipated by the off-chip drivers. In DRAMs, all the power is consumed in driving and sensing long column and row lines and in off-chip drivers. A wafer-scale integrated (WSI) circuit or a multichip module with dense thin film interconnections may partially resolve this problem by reducing the system-level wire lengths and the capacitive loading associated with them. Nevertheless on-chip and chip-to-chip global interconnections will still dictate the overall system performance and consume a major portion of the available power.

One way to reduce both the power dissipation and delay at high-capacitance nodes is to limit the voltage swing at these nets. Some of the CMOS circuits described in this chapter attempt to improve the power delay product of the nets with long lines by precharging the interconnection to the high-gain region of the receiver and reducing the voltage swing at these nodes.

In this chapter, first SRAM and DRAM sense amplifiers are described that reduce the memory access time by minimizing the delay of driving high-capacitance bit lines. Later various CMOS logic families and driver/receiver circuits for logic applications are presented and compared. Finally BiCMOS circuits are described that employ densely packed, low-power CMOS devices to implement memory and logic functions and fast bipolar transistors to drive large-capacitance nodes and to implement sense amplifiers. Bipolar transistors can also be used to build high-speed ECL drivers and receivers for chip-to-chip communication.

4.1   Scaling of Interconnection Capacitance

As minimum feature size is scaled down and chip size increases, wiring capacitance becomes more important than device capacitances. Transistor input capacitance decreases because gate, gate-drain overlap, and source/drain diffusion capacitances are reduced when feature sizes are miniaturized. Wiring capacitance increases because interconnection capacitance per unit length remains approximately constant and global wire lengths increase as chips get larger. It should be pointed out that if an existing chip is scaled down, wire lengths are reduced. In practice, however, as the technology is scaled down, designers want to put more and more circuits on a chip, and the die size increases rather than decreases. Consequently, global wire lengths get longer from one generation of chips to the next.

In a simplified form, the 50 percent delay[1] from the first to the second gate in Fig. 4.1 can be expressed as

$$T_{50\%} = R_{tr}(C_{int} + C_{gate}),\qquad(4.1)$$

where R_{tr} is the effective on-resistance of the driving gate, C_{int} is interconnection capacitance, and C_{gate} is the input capacitance of the receiving gate.

The scaling properties of the delay components for a critical path that

---

[1]Fifty percent delay, $T_{50\%}$, is defined as the delay from the time when input potential reaches the midpoint between V_{DD} and ground to the time when the output reaches the same midpoint.

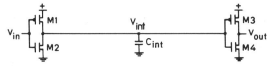

FIGURE 4.1   Gate delay. The gate delay is determined by the on-resistance of the driver R_{tr} and capacitances of the wire and the receiving gate C_{int}; C_{gate}.

extends across the chip, to a first order, are as follows:

$$R_{tr} \approx \frac{1}{\frac{W}{L}\mu C_{gox}(V_{DD} - V_T)} \qquad \propto 1$$

$$C_{gate} = \epsilon_{ox}\frac{W_n L_n + W_p L_p}{t_{gox}} \qquad \propto 1/S \qquad (4.2)$$

$$C_{int} = \epsilon_{ox}\frac{W_{int} l_{int}}{t_{ox}} \qquad \propto S_C$$

Here W and L are the width and length of a transistor, μ and C_{gox} are the surface mobility of the carriers and the gate oxide capacitance per unit area, V_{DD} and V_T are the power supply and threshold of a transistor, t_{gox} and t_{ox} are the gate and field oxide thicknesses, and l_{int} and W_{int} are the length and width of an interconnection; S and S_C are the scaling factors for the minimum feature and chip sizes, both greater than 1.

The on-resistance of a transistor remains approximately constant as feature size is reduced. This results from a number of interacting factors. First, both L and W are scaled down by $1/S$, and the W/L ratio remains constant. Second, as gate oxide thickness is reduced, C_{gox} increases. As a result more carriers can be generated at the channel region with a given gate potential; this improves the on-resistance of the transistor. Third, the reduction of the voltage levels V_{DD} and V_T cancels the effects of increased C_{gox}. More carriers can be generated with a smaller gate potential, but the available gate potential is also smaller; current drive capability of the device remains constant. As described in Chapter 2, the power-supply level must be reduced to avoid oxide and junction breakdowns. In any case, if the potential levels are not reduced, carrier velocity saturation cancels the on-resistance improvement suggested by simple MOS theory (see Chapter 2). As a result the on-resistance of a transistor remains approximately constant as devices are scaled down.

The gate input capacitance C_{gate} improves because both W and L are reduced by $1/S$. This reduces the gate area and capacitance by $1/S^2$. The gate oxide is thinned down by $1/S$, which cancels one of the $1/S$ improvement factors, and the net reduction in C_{gate} is $1/S$. The wiring capacitance increases by S_C because the interconnection capacitance per unit length is approximately constant and global wires are longer in larger chips.

Delay resulting from interconnection capacitance becomes significant when the chip dimensions are increased such that

$$C_{int} = C_{gate}$$

$$C_{int}l_{int} = \epsilon_{ox}\frac{W_n L_n + W_p L_p}{t_{gox}}, \qquad (4.3)$$

where W and L are the transistor width and length, and subscripts n and p refer to the nMOS and pMOS devices in a CMOS inverter. Here C_{int} is the wiring capacitance per unit length. Assuming a large-size inverter in 2 μm CMOS technology: $W_n = 10$ μm, $W_p = 20$ μm, $L_n = L_p = 2$ μm, $t_{gox} = 400$ Å, and $C_{int} = 2$ pF/cm, the wire length for which interconnection capacitance becomes comparable to gate capacitance is obtained as

$$C_{int}l_{int} = \epsilon_{ox}\frac{W_n L_n + W_p L_p}{t_{gox}}$$

$$2.0\ pF/cm \times l_{int} = 0.35\ pF/cm \times \frac{10\mu m \times 2\mu m + 20\mu m \times 2\mu m}{0.04\mu m} \qquad (4.4)$$

$$l_{int} = 260\ \mu m = 0.26\ mm.$$

By today's standards, this is not a long wire. This calculation basically illustrates that the capacitance of a 0.26 mm wire is equal to the input capacitance of a large CMOS inverter ($W_n/L_n = 10\mu m/2\mu m$, $W_p/L_p = 20\mu m/2\mu m$). The critical interconnection length would be even smaller for a small size inverter. Accordingly the critical paths of large logic chips are usually dominated by wiring rather than device capacitances because the die sizes of state-of-the-art chips are 1×1 cm^2 or larger [4.1]. The above example assumes a fan-out of 1, but buses in data paths have large fan-outs. When many transistors are connected to a single bus, the sum of the drain capacitances of the driver transistors and the gate capacitances of the receiver circuits may constitute a major portion of the total load. Still, even with a fan-out of 10, the critical interconnection length in the above example is 2.6 mm—only one-quarter of a side of a large die.

The interconnection capacitance has three components: area component (also referred to as parallel plate capacitance component), fringing field component, and wire-to-wire capacitance component. To improve the delay resulting from wiring capacitance, dielectric thickness can be increased; however, this approach is not effective when the insulator thickness becomes comparable to the interconnection width and thickness because of the fringing fields [4.2]. As demonstrated in Fig. 4.2, capacitance of a single wire can be modeled by a parallel plate capacitor with a width approximately equal to W_{int} and a cylindrical wire with a diameter equal to H_{int} [4.3]. A more accurate version of this model for $W_{int} > H_{int}/2$ assumes an effective interconnection width

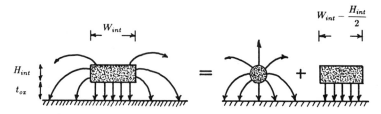

FIGURE 4.2 Modeling of the contribution of fringing fields to interconnection capacitance. Total wire capacitance can be thought of having two components: a parallel plate capacitance determined by the perpendicular field lines between the wire and the ground plane and a fringing field component, which can be approximated by the capacitance of a cylindrical wire with a diameter equal to interconnection thickness [4.3]. Reprinted by permission of Addison-Wesley Publishing Co.

reduced by $H_{int}/2$ to account for some second-order effects [4.4]. This yields the interconnection capacitance per unit length C_{int} as

$$C_{int} = \epsilon_{ox} \left\{ \frac{W_{int}}{t_{ox}} - \frac{H_{int}}{2t_{ox}} + \frac{2\pi}{\ln\left[1 + \frac{2t_{ox}}{H_{int}}\left(1 + \sqrt{1 + \frac{H_{int}}{t_{ox}}}\right)\right]} \right\}. \quad (4.5)$$

Figure 4.3 plots C_{int} as a function of W_{int}/t_{ox} for two H_{int}/t_{ox} ratios. As can be seen, C_{int} stops decreasing when $W_{int} \approx t_{ox} \approx H_{int}$ and approaches an asymptote of 1 pF/cm due to fringing fields.

The third component is the capacitance between neighboring lines. The fringing field and wire-to-wire capacitances are becoming significant in today's chips. Combined together, they are usually larger than the parallel plate component. To improve packing density and at the same time maintain a small interconnection RC constant, it is desirable to keep the conductor and oxide thicknesses constant and to reduce the wire width and spacing. As shown in Fig. 4.4, this increases the capacitance between the lines, and the total interconnection capacitance reaches a minimum of approximately 2 pF/cm when $W_{int} = W_{sp} \approx H_{int} = t_{ox}$ [4.2]–[4.5]. In a multilevel-interconnection environment, the capacitance to the lines in other levels must also be included in the calculations.

The parasitic capacitance between interconnections has two undesirable effects: it degrades the switching speed and causes cross talk between neighboring signal lines, which can result in faulty operation if not controlled properly. Two- and three-dimensional modeling can be used to obtain accurate values for wiring capacitance. They have been reported extensively in the literature [4.4]–[4.10].

SINGLE RUNNER CAPACITANCE

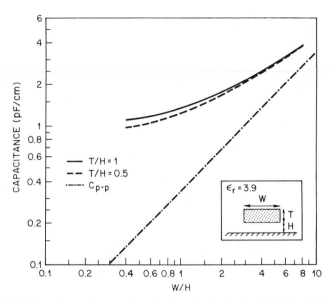

FIGURE 4.3   Capacitance of a single wire including fringing field effects. The actual capacitance, which is the sum of parallel plate and fringing field capacitances, is compared with the parallel plate approximation as a function of W_{int}/t_{ox} for two H_{int}/t_{ox} ratios. Wire capacitance decreases as its width is reduced with respect to its thickness. This continues until the wire width is approximately equal to insulator thickness; then the capacitance levels off at approximately 1 pF/cm [4.2]. Here, W, T, and H are same as W_{int}, H_{int} and t_{ox}. (© 1983 IEEE)

4.2   Circuit Techniques for High-Capacitance Nodes

As described in the previous section, the fringing fields and capacitance between neighboring lines will establish a lower bound around 2 pF/cm for the interconnection capacitance in any scaling scheme. Consequently current state-of-the-art logic chips with large die sizes can have on-chip large fan-out signal nodes with 10 pF capacitance, and clock lines on a large chip can have as much as 100 pF of total capacitance. In a WSI circuit built on a 6-inch wafer, a single interconnection can easily have more than 10 pF of capacitance. The driver size for these nodes must be increased to achieve satisfactory performance. Increasing the driver size, however, has the following adverse effects:

• Large drivers may require excessive chip area.

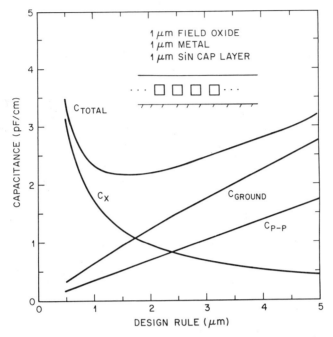

FIGURE 4.4   Interconnection capacitance including the capacitance between neighboring lines. Here, $H_{int} = t_{ox} = 1\ \mu m$ and $W_{int} = W_{sp}$ is indicated as "design rule." The total capacitance (C_{TOTAL}) and its two subcomponents, capacitance to ground plane (C_{GROUND}) and between conductors (C_X), are plotted as a function of W_{int}. The parallel plate approximation C_{P-P} is also shown as a reference point. When the wire width (W_{int}) is much greater than insulator thickness (H_{int}), wire capacitance to ground is larger than the wire-to-wire capacitance. When W_{int} is smaller than H_{int}, capacitance between the wires dominates. The cross-over point is at $W_{int}/H_{int} = 1.75$ where the total capacitance reaches a minimum at 2 pF/cm [4.2]. (© 1983 IEEE)

- It is difficult to drive a large buffer because it has a high input capacitance. The total delay of switching the buffer and driving the line should be optimized by using cascaded drivers which will be described later in this chapter (Fig. 4.29) and again in the next chapter. Even with cascaded drivers, as the driver size grows, at one point, the total delay starts to increase rather than decrease.

Despite these second-order effects, increasing the driver size (sometimes referred to as powering up the circuit) is the most common technique to reduce the delay of the critical paths in CMOS logic chips and to decrease chip-to-chip transmission delays. Increasing the driver size is popular because it is simple

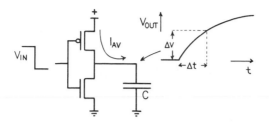

FIGURE 4.5   Delay associated with a capacitive load. Delay is proportional to the capacitive load C and the potential swing at the output ΔV and inversely proportional to the average driver current I_{av}.

and straightforward. There are other alternatives. For example, memory circuits take advantage of their regular structure to implement sense amplifiers. Some of the circuits described in this chapter are designed to reduce the delays at high-capacitance nodes in logic chips without dissipating excessive power or requiring large areas. These alternative techniques require careful circuit design because they reduce the voltage swings and employ some form of a sense amplifier that reduces the delay but compromises the noise margins. The circuits are no longer in the simple "1" "0" digital realm, and the analog effects need to be carefully considered to ensure proper circuit operation. This is difficult to manage on a 1 million transistor chip, which is usually under schedule pressures. That is why increasing the driver size is the most commonly used method except in very regular structures like memory circuits.[2]

As illustrated in Fig. 4.5, the delay associated with a capacitive load C_L can be expressed as

$$T = \frac{C_L \Delta V}{I_{av}}, \tag{4.6}$$

where ΔV is the voltage swing required for a logic state change and I_{av} is the average drive current. Power consumption is expressed as

$$P = f_D V_{DD} I_{av}, \tag{4.7}$$

where V_{DD} is the power-supply level, and f_D is the duty factor, which is a measure of the portion of the time I_{av} is active. As a result the power-delay

---

[2]This is also why it is harder to do custom bipolar ECL design compared to custom CMOS design. Most ECL designs are gate arrays partly because it is easier to solve all the analog circuit and noise problems once, generate a set of design rules, and use the same gate array chip multiple times. The other reasons are the high power dissipation of ECL circuits (which prohibits packing many circuits on a chip because it cannot be cooled), less sensitivity to capacitive loading and large fan-in/fan-out in ECL circuits (CMOS requires custom layout with short wires more than ECL does), and the relatively lower volumes of the ECL parts (which makes ECL custom design too expensive for most applications).

product becomes

$$P \times T = f_D C_L V_{DD} \Delta V. \tag{4.8}$$

The duty factor f_D is approximately 1 for bipolar ECL, 0.5 for nMOS, and as low as 0.01 for CMOS at low clock frequencies or when only a small portion of the circuit is active at a given time. Duty factor is 1 for ECL gates because either one or the other branch of the differential amplifier is conducting (see Figs. 2.48 and 2.49). Enhancement/depletion nMOS gates conduct static current only when one of the pull-down transistor paths is turned on and the gate output is low. When the output is high, there is no path to ground and no steady-state current flows (see Fig. 2.6). As a result the duty factor of nMOS circuits is 0.5 because there is 50 percent chance that the output of a logic gate will be a 1. Duty factor is low in CMOS circuits because a gate delay is usually a small fraction of the cycle time (2–10 percent), and CMOS gates conduct current only during switching. There is no dc current flow. Once the output is charged high or low, the current flow stops. As a result CMOS gates conduct current only in short bursts, and their duty factor can be as low as 0.01.

The power-supply level V_{DD} is established by the available device technology and system-level integration considerations.[3] Interconnection capacitance per unit length C_{int} is determined by the electrical characteristics of wires. Device and process engineers strive to optimize these parameters. From a circuit and chip designer's point of view, the first step in minimizing interconnection delays is to select a chip logic partition, data flow, floor plan, and circuit layout such that wire lengths are kept as short as possible. If many levels of logic are required to implement a function, it is essential to keep the gates close together; global chip interconnections must be avoided. In multichip CPUs, the number of chip crossings in a given path must be minimized. It is best to confine a critical path in a single chip. This is usually not possible because of architectural and functional requirements. The number of chip crossings in multichip paths must be limited to one or at the most two. Similar to on-chip interconnections, card-level wire lengths must be kept short by placing the chips next to each other if they are in critical paths.

After all the effort is put into minimizing wire lengths and chip crossings, the only factors a circuit designer can improve are f_D and ΔV. In CMOS circuits, f_D depends on the ratio of the propagation delay to clock frequency, the portion of the circuits that are active, peak-to-peak voltage swings on large-capacitance nodes V_{PP}, and static power dissipation of the circuits. The efficiency factor E_f represents the portion of the current actually used to charge or discharge the load capacitance. Even in static CMOS, some of the

---

[3]It is highly desirable to have all the chips in a given system work with the same power-supply level because it is expensive to have multiple power supplies. It is also much harder to distribute multiple power-supply levels on a card. This also requires an expensive card with more layers.

current is wasted because of the power-to-ground path that exists when the pMOS and nMOS devices are switching. Almost any circuit solution that increases speed also degrades E_f because, when biased at their high-gain region, circuits conduct static current. This is unavoidable because if the transistors are turned off, their inputs need to be charged or discharged to put them in a conducting state where they can be used to amplify signals. To minimize the time delay associated with turning on the transistors, the devices can be kept in a conducting state, which implies that the circuit dissipates static power.

4.3   Random Access Memory (RAM)

In the next few sections, various circuit techniques will be described that are used in memory chips. They show how voltage swings and noise margins are traded off against speed and illustrate some of the concepts described in the previous section. Random access memories (RAMs) are used to store data in digital systems. Data can be written in and read from a RAM. Figure 4.6 illustrates the architecture of a typical RAM. The chip is organized to provide

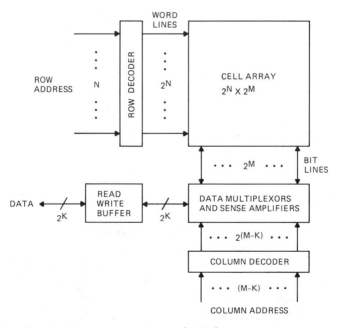

FIGURE 4.6   Random access memory (RAM) organization

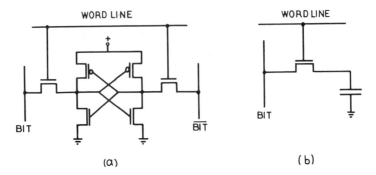

FIGURE 4.7 Memory cells: (a) six-transistor CMOS static memory cell, (b) single-transistor dynamic memory cell formed by a pass gate and a capacitor

a large amount of storage capacity on a small die area. Most of the chip is occupied by a memory array, which consists of densely packed memory cells. Each memory cell can store one bit of data. Surrounding the cell array are the row and column decoders, which determine the cells that are being addressed. Data multiplexors select the column that is being addressed, and sense amplifiers are used to improve the access time of the chip.

There are two kinds of RAMs: static (SRAM) and dynamic (DRAM). As shown in Fig. 4.7(a), SRAMs use a static latch as a memory cell. Consequently the data stay intact as long as power is supplied to the chip. DRAMs store the information on a capacitor (Fig. 4.7 (b)). The data in a DRAM cell must be refreshed periodically because the charge on the capacitor slowly leaks through the pass transistor (see Section 2.4.7). The charge is also lost due to electron-hole pair recombination at the surface and depletion regions near the cell capacitor. Substrate noise, power-supply noise, and adjacent cell noise also contribute to the charge loss. Because of their dynamic nature, DRAM cells are also more prone to soft errors. Soft errors result from the recombination of the charge stored on the cell capacitor with the electron-hole pairs generated by an alpha particle penetrating inside the silicon substrate (see Fig. 1.5). Consequently SRAMs are faster, more reliable, and easier to use. Because of their large cell sizes, the storage capacity of SRAMs is typically a factor of four smaller than the DRAMs of the same time period. In addition, to achieve high speeds, SRAMs consume more power. Static RAMs also cost more per bit than dynamic RAMs. DRAMs, on the other hand, have a smaller cell size and can offer a larger storage capacity on a single chip. DRAMs are also more prone to soft errors caused by alpha particles. They are sensitive to noise generated on the chip and noise at the power lines. DRAMs usually require additional logic to support memory refreshing, and they are not as fast as SRAMs.

TABLE 4.1   State of the Art in Experimental Memory
Chips (1989)

Memory Type	Technology	Storage Capacity	Access Time
Static RAMs	CMOS	4 Mbit	25 ns
	CMOS	1 Mbit	9 ns
	BiCMOS	1 Mbit	8 ns
	ECL	64 Kbit	5 ns
	GaAs	4 Kbit	1 ns
Dynamic RAMs	CMOS	16 Mbit	45 ns
	BiCMOS	1 Mbit	30 ns

Static RAMs are used in high-speed applications, such as cache memories and high-speed buffers in minicomputers and mainframes or very fast main memory in supercomputers. There are many varieties of static RAMs that satisfy different niche applications (by-four, by-eight, by-nine, tag-compare, etc.). The main advantage of SRAMs is their high speed. Currently, 256 kilobit CMOS SRAMs with 25 nsec, 256 kilobit BiCMOS SRAMs with 15 nsec, 64 kilobit BiCMOS SRAMs with 12 nsec, 4 kilobit ECL SRAMs with 5 nsec, and 1 kilobit ECL SRAMs with 2 nsec access times are commercially available. CMOS is trying to enter the low-density, high-speed arena with 6 nsec 4 kilobit CMOS SRAMs. In research laboratories, 4 megabit CMOS SRAMs with 25 nsec, 1 megabit BiCMOS SRAMs with 8 nsec, 64 kilobit ECL SRAMs with 5 nsec, and 4 kilobit GaAs SRAMs with 1 nsec access times have been built. The power dissipation of an SRAM varies from 0.1 W in CMOS to 2–5 W in ECL technologies. Dynamic RAMs, on the other hand, target low-cost, high-volume markets. The main advantages of DRAMs are high density, low power dissipation, and low cost per bit of storage. Currently 1 megabit DRAMs with 70–100 nsec access times are commercially available, 4 megabit DRAMs are at the sampling stages and are avilable in limited volumes, and DRAMs with storage capacities up to 16 megabits with access times of 45 nsec have been built in research laboratories [4.11]–[4.13]. Typical power dissipation of a DRAM is 100–500 mW in the active state and 0.25–10 mW at standby. The state of the art in experimental SRAM and DRAM technologies is summarized in Table 4.1.

Memory chips usually quadruple in capacity from one generation to the next, and it takes about three years to bring up the next generation. The

TABLE 4.2   Commercial DRAM
Technology Trends

Year in Production	Memory Capacity	Access Time
1975	4 Kbit	250 ns
1978	16 Kbit	180 ns
1981	64 Kbit	130 ns
1984	256 Kbit	100 ns
1987	1 Mbit	80 ns
1990	4 Mbit	65 ns
1993	16 Mbit	50 ns

access time is reduced approximately by a a factor of 1.3 between generations. Table 4.2 summarizes the evolution of commercial DRAM chips.

4.4   Static RAM Circuits

In this and the following two sections, circuit techniques will be described that minimize the delay of driving large-capacitance nodes in memory chips. Figure 4.8 shows the organization of an SRAM chip, and Fig. 4.9 illustrates a simplified circuit diagram for the read and write paths of an SRAM. The bit lines are precharged to a high potential and equalized. The chip then waits for a transition in any one of its address, write enable (WE), and chip select (CS) pins. When a transition is detected in one of these pins, an internal clock is generated that controls the timings of the chip.[4] First, the precharge clock is turned off, and the address pin inputs are fed into address buffers that drive the inputs of row and column decoders, also known as X and Y decoders. The address is decoded using NAND gates. To minimize the number of NAND gates, some of the address bits can be predecoded and fed into a NOR gate (secondary decoder) along with the output of the first NAND gate (primary decoder). In Fig. 4.9, one address bit is predecoded. This NOR gate serves as

---

[4]It is possible to design an SRAM with no internally generated clocks. The bit lines do not need to be precharged; they can be tied to V_{DD} via clamp circuits. The sense amplifiers also do not need to be clocked; the chip select pin input can be used to enable the sense amplifiers. Early SRAMs were designed using such techniques, but they dissipated too much power per bit. Internal clock generation by address transition detection circuitry reduces the power dissipation and makes relatively low-power 1 Mbit CMOS SRAMs possible.

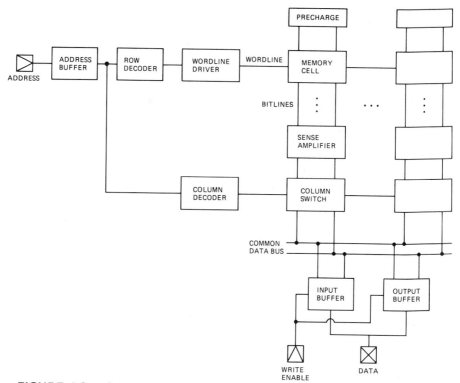

FIGURE 4.8   Organization of an SRAM chip

the second stage of the decoder and as the word line driver. This two-stage decoding is required to reduce the number terms in the row decoder in order to match the pitch of the decoder with the cell pitch. Depending on which word line is driven high, one of the many rows is connected to BIT and $\overline{BIT}$, and the cells in this selected row start driving the bit lines. After a row of cells is selected, the voltages on the bit lines move very slowly because the total capacitance of a bit line (the wire capacitance plus the diffusion capacitance of the drains of the pass transistors tied to the bit line) is large. Usually the cell size is determined by the contact vias and wiring (power, ground, word line, bit line). Within this cell area, devices are made as large as possible. Once the cell area is optimized, the transistors in the memory cell cannot be made stronger to improve the access time because then the cell size would grow, and both the wire and diffusion capacitance on the bit line would increase by an amount equal to the rise in the current drive capability of the cell. The net effect is an increase in the die size with no speed enhancement.

To minimize the delays at the high-capacitance bit lines, sense amplifiers

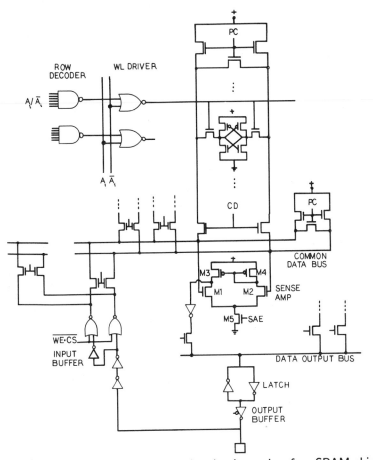

FIGURE 4.9 Circuit diagram of the read and write paths of an SRAM chip

that can detect a potential swing equal to a small fraction of the full power-supply range are employed. Most DRAM and SRAM chips use a differential sense amplifier with a good common mode noise-rejection ratio. After the memory cell establishes a small potential difference between the bit lines, the sense amplifier is activated, and the difference between the bit lines is rapidly amplified to a full swing and fed to the output buffer. An important issue is the position of the column multiplexors. If the column selectors are placed after the sense amplifiers, more amplifiers are required, and this increases power dissipation and area. In order to match the pitch of the sense amplifier with bit line pitch, the sense amplifier must be very simple. If the column multiplexors are placed before the sense amplifiers, the inputs to the amplifier will be slower and the access time will degrade because a larger differential signal is required due to pass transistors. On the positive side, a better sense

amplifier can be used because more area and power can be allocated per sense amplifier since there are fewer of them. As a compromise, some column multiplexing may be done both before and after the sense amplifier, as shown in Fig. 4.9. Here multiple bit lines are connected to the common data bus and feed the same sense amplifier. The column decoder will select only one of these bit lines and connect it to the sense amplifier. The common data bus is also precharged to the same value as the bit lines. In high-speed SRAMs there is usually one sense amplifier per column and, to improve the access time, a column is usually divided into two halves, with the sense amplifier placed between them. The output of the sense amplifier is tied to the data output bus by a pass transistor, which is driven again by the column decoder. Finally the data output bus is latched and fed to the output pin by a buffer. Input data come through the same pin and go through the input buffer. The sense amplifier is bypassed during a write operation, and the write buffer directly feeds the memory cell selected by the column and row decoders.

In the circuit shown in Fig. 4.9, a source-coupled pair with an active load is used as the sense amplifier. The sense circuit amplifies the signal and performs double- to single-ended conversion. The pMOS transistor pair M3 and M4 have equal gate-to-source potentials; if the transistors are identical, to a first degree, the currents through them are also equal. Appropriately this configuration is called a *current mirror*. When switched on, the nMOS transistor M5 acts as a current source, and the gain of the amplifier is provided by the differential pair M1 and M2. At the bias point when the two inputs of the amplifier are set equal by the precharge circuitry, the current gain of the amplifier $G_m = dI_{out}/dV_{in}$ is equal to the current gain g_m of M1 and M2:

$$G_m = g_{mM1}.$$

The voltage gain A_v of the amplifier is obtained as

$$A_v = g_m r_o,$$

where r_o is the parallel combination of the output resistances of the pMOS current mirror M3 and the nMOS gain transistor M1:

$$r_o = r_{oM1} \| r_{oM3}.$$

In a memory sense amplifier, there are other considerations besides the gain: the output swing, output current, and power dissipation.

To obtain a high gain, M1, M2, M3, and M4 should be in saturation at the precharged state. This has two advantages. First, during saturation, the increase in the transistor current I_D is larger for per unit increment in the gate-to-source potential V_{GS}. In other words, the gain $g_m = dI_D/dV_{GS}$ of M1 and M2 is larger when they are saturated. Second, the output resistance $r_o = dV_{DS}/dI_D$ is larger at saturation; this is desirable for M1,M2, M3, and

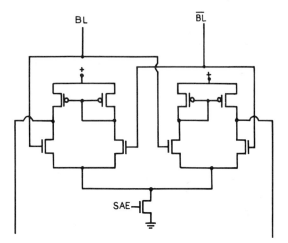

FIGURE 4.10   Two single-ended current-mirror-loaded amplifiers connected together. This amplifier provides a symmetric differential output with large swings in both polarities.

M4 because the gain is proportional to output resistance. These can be observed from the I-V characteristics of nMOS transistors in Figs. 2.2 and 2.11. Transistor gain is proportional to the distance between the I-V curves, and the output resistance is inversely proportional to the slope of an I-V curve (the flatter the curve, the larger the output resistance). As the bit lines are driven away from the bias point, it becomes difficult to keep all transistors in saturation, so the gain of the amplifier decreases and the output potential saturates. This limits the magnitude of the output voltage swing. It is important to keep the output potential swing and output current large in order to minimize the overall sensing delay. The delay is inversely proportional to the bias current of the amplifier, but the area and power dissipation also increase with the bias current.

Some of the major advantages of this amplifier are its large differential-mode gain and small common-mode gain. This means that if the inputs to M1 and M2 move in the opposite direction, the output varies greatly (large differential gain), but if the inputs to M1 and M2 move up or down together, the output does not change (small common-mode gain). The circuit has a good common-mode rejection ratio. This is advantageous in memory circuits with differential bit lines because if the bit lines are laid out next to each other, any noise will couple to both of them simultaneously, and it will have little effect on the operation of the circuit. As a result, the current mirror load sense amplifier has a very high-speed data transmission rate and high immunity to noise. On the negative side, it has a small voltage amplification and a small output potential swing. The minimum potential swing that can be sensed by

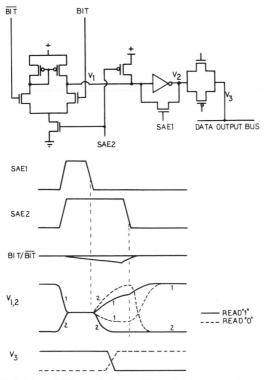

FIGURE 4.11   Differential amplifier with a second stage precharged to $V_{DD}/2$. The feedback loop around the inverter precharges the circuit to its high-gain region when enabled by SAE1. The dashed line represents a "1" being read, and the solid line a "0" being read.

this amplifier is mainly limited by the threshold mismatches between M1 and M2 because the differential pair amplifies the mismatch between the threshold potentials of M1 and M2 (approximately 20 mV between adjacent devices) just like the potential difference between their gate terminals.

Figure 4.10 shows two single-ended active load differential amplifiers connected together symmetrically. This provides a double-ended output signal that can be fed to another stage of a similar amplifier, to a differential-input latch, or to an output buffer with a differential input. The simple current mirror loaded amplifier in Fig. 4.9 cannot provide a good differential signal because the side of the amplifier that has the pMOS transistor M4 with its gate tied to its drain does not have a large potential swing. The circuit in Fig. 4.10, however, is perfectly symmetric and provides large swings in both outputs [4.14]–[4.15].

Figure 4.11 shows another modification to the active load differential am-

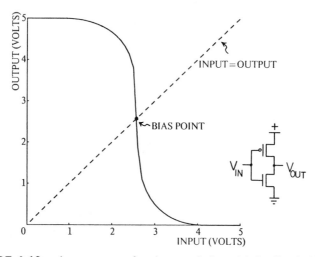

FIGURE 4.12   Inverter transfer characteristics with feedback. When the input and output of an inverter are shorted together, the circuit autobiases itself at the transition region where $V_{IN} = V_{OUT} \approx V_{DD}/2$. This bias point is also the highest-gain region of the inverter—the portion of the transfer curve with the steepest slope in both directions.

plifier. Here the second stage, which is a simple inverter, has a feedback path that connects its input to its output during the precharge period. When the feedback path is closed, as illustrated in Fig. 4.12, the differential amplifier and the inverter are autobiased in their high-gain region (biased at their unity gain point). When the feedback path is broken, both the differential amplifier and the inverter start from the equlibrium voltage. This guarantees a rapid response time because the equilibrium voltage is the high-gain region of the inverter [4.16]–[4.20].

In the idle state, SAE1 and SAE2 are low. As a result the differential amplifier is turned off, and node 1 is pulled high by the pMOS transistor that is clocked by SAE2. In this state, neither the differential amplifier nor the inverter dissipates any static power. When the chip is accessed, SAE1 and SAE2 go high. Then the input and output of the inverter are shorted together, the pMOS pull-up that ties V_{DD} to the input of the inverter is turned off, and the differential amplifier is activated. This precharges the nodes 1 and 2 to the high-gain region of the inverter and amplifier. After a memory cell is selected and a potential difference is established between the bit lines, SAE1 goes low, the feedback loop around the inverter is broken, and the difference between BIT and $\overline{BIT}$ is rapidly sensed. Then the output pass gate is turned on, and the result is driven to the data output bus.

At the end of the read cycle, SAE2 goes low, and the differential amplifier

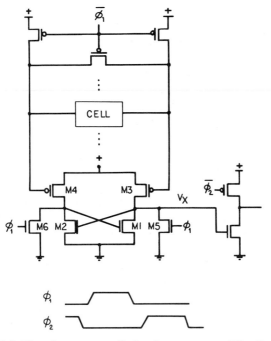

FIGURE 4.13   Low power-dissipation sense amplifier for static arrays

and inverter are again forced to the power-down state. The threshold level of the inverter is designed to be equal to the output of the differential amplifier when the bit lines are precharged. As a result device parameter mismatches are a source of concern. In a similar sense amplifier designed in 1.3 μm CMOS technology, the increase in sensing time has been shown to be less than 2 ns, with a threshold mismatch of 0.2 V between nMOS and pMOS transistors and a transconductance ratio variation of 20 percent [4.17].

A sturdy amplifier with a low static power dissipation is shown in Fig. 4.13 [4.21]. During ϕ_1, the bit lines are precharged to V_{DD} by pMOS pull-ups, and the sense amplifier output nodes are precharged to 0 V by the nMOS pull-down transistors M5 and M6. Later the word line is driven, and the sense amplifier is enabled during ϕ_2. Since the bit lines are precharged to V_{DD}, the response time of this sense amplifier is not very fast because one of the bit lines has to be discharged one $|V_{TP}|$ below V_{DD} before one of the pMOS pull-ups in the amplifier starts conducting. This makes the timing of ϕ_1 and ϕ_2 clocks less critical because the sense amplifier is guaranteed not to respond until a difference of $|V_{TP}|$ is established between BIT and $\overline{BIT}$. Unlike the previous circuits, if this sense amplifier is enabled too early, the result will be less catastrophic. An additional factor that makes this amplifier slow is the fact that amplification is provided by pMOS transistors, which have less gain than nMOS transistors.

Once the pMOS pull-up tied to the falling bit line starts conducting, the output node connected to that pMOS will be pulled up. Meanwhile the cross-coupled nMOS pair will keep the node that is being charged up isolated from ground while tying the opposite node to ground. If M3 is turned on, V_X will rise and activate the nMOS output device to pull down the data bus that was precharged high during ϕ_2.

This sense amplifier is slow, but its main advantage is the simplicity of its clocking requirements and the lack of any static power dissipation [4.21]. All the current is used to charge the output; no supply-to-ground path is turned on during the operation of the circuit. If one of the pMOS pull-ups is conducting, the corresponding nMOS pull-down is turned off, and vice-versa. Consequently this amplifier is more suitable for small static arrays that may be embedded in logic chips. Such arrays will usually have memory cells with larger current drive capabilities relative to high-density SRAM cells. This results in a smaller storage capacity, but the clocking simplicity and circuit ruggedness may be attractive enough to make this amplifier preferable in a logic chip.

To achieve a reasonable speed, the capacitance at the intermediate node (V_X) of this amplifier should be small. The amplification is provided by pMOS pull-ups, and the cross-coupled nMOS pull-down transistors must be kept small because they conduct hardly any current. Circuit speed may be improved by precharging the bit lines to ($V_{DD} - V_{TN}$) via nMOS precharge devices, but this would degrade the noise margins and introduce static power dissipation. In addition, it would require a more complicated clocking scheme that disables the cross-coupled nMOS pull-down section of the sense amplifier for a short period after the word line goes high. This would have been required in order to provide time for the memory cell to establish a potential difference between the bit lines connected to the gates of the pMOS pull-ups before the amplifier is turned on. The sizing of the cross-coupled nMOS pull-downs also would become important because both pMOS pull-ups may be conducting simultaneously—one conducting more than the other.

Another sense amplifier also suitable for logic-oriented arrays and register files is shown in Fig. 4.14. Both rails of the data bus are precharged to V_{DD} during PC. When the PC is turned off and one of the word lines is activated, the selected static memory cell slowly pulls one of the bit lines low. After a sufficient potential difference is established between the bit lines, SAE goes high, and the sense amplifier is activated. Then the cross-coupled nMOS transistors start pulling the bit lines down. The bit line that has a lower potential than the other bit line gets pulled down faster, and this potential difference then rapidly gets amplified. The cross-coupled nMOS pair will be analyzed in more detail in the section on dynamic RAMs because this technique is used widely in DRAMs. The pMOS pull-up transistors with grounded gates hold

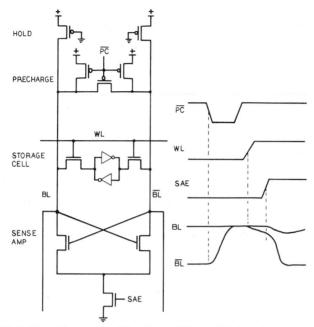

FIGURE 4.14   Sense amplifier for a differential data bus

the bit line with the high value against any charge leakage. These pMOS pull-ups should not be too strong for three reasons. One, there is no need to make them large since their function is just to compensate for leakage. Two, if they are too strong, they will slow down the drive of the memory cell. Three, if the pMOS load transistor is too large, the low value of the bit line can become too large because this low value is determined by the ratio of the pMOS pull-up hold device and the nMOS pull-down chain of the sense amplifier.

4.5 Dynamic RAM Circuits

Dynamic RAMs differ from SRAMs because DRAMs are optimized for low power dissipation and high density. For example, DRAMs usually do not have separate row and column address pins; their address pins are time multiplexed to minimize the pin count and packaging and board costs. If the chip organization shown in Fig. 4.6 described a typical DRAM, M would be approximately equal to N, and K would be 1. As illustrated by the timing diagram in Fig. 4.15, row address is received during row address strobe (RAS) and decoded. Later during column address strobe (CAS), column address

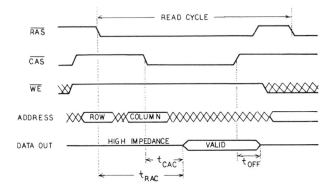

FIGURE 4.15   DRAM timing diagrams for a read cycle. Here, t_{CAC} is column access time, t_{RAC} is row access time, and t_{OFF} is the time required to disable the output buffer. Two read operations must be separated by a minimum time period defined as *read cycle*. This is a simple DRAM read cycle. Static column, page mode, and nibble mode operations have more complicated timing sequences where larger chunks of consecutive data can be read or written to memory without having to send the same row address multiple times. RAS is activated only once, and CAS is toggled to do multiple reads and writes.

comes in through the same pins, and it is decoded while the cells drive the bit lines. The decoded column address will be ready to drive the column switches after the sense amplifier is activated and the bit lines are amplified. References [4.22]–[4.25] present a very good four-part review of DRAMs from a user's perspective (internal architecture, timing diagrams, various access modes, bank-oriented design, interleaving, DRAM controller design, and DRAM board layout).

Figure 4.16 shows the read path of a DRAM circuit. When the precharge clock PC is high, the bit lines are precharged to V_{BL}. The chip access starts when RAS is activated and triggers a host of internal clocks. DRAM chips have many more internal clocks than SRAMs; the signals shown in Fig. 4.16 are only a small subset. First, the precharge clock is turned off, and then the selected word line (WL) is driven. At the same time, a dummy word line $(DWL0$ or $DWL1)$ is selected. This dummy word line is the one that is connected to the bit line opposite the selected memory cell. During precharge, dummy cells are charged to a dummy cell reference potential V_{DC}. The scheme shown in Fig. 4.16 is only one of many possible implementations of one basic idea; the bit lines are precharged to a value between V_{DD} and ground, and a memory cell and the dummy cell opposite it are connected to BL and $\overline{BL}$. In this way, the value stored in the memory cell is compared to a reference value (V_{DC}) stored in the dummy cell. Depending on the data stored in the

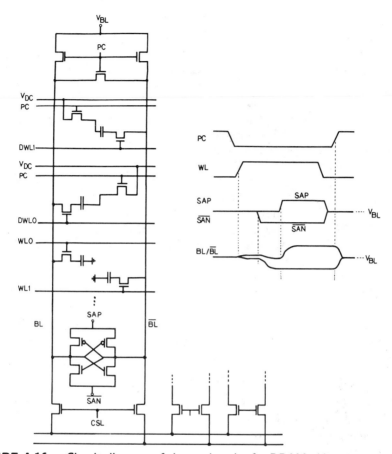

FIGURE 4.16   Circuit diagram of the read path of a DRAM chip

cell, a positive or negative potential difference is established between BL and $\overline{BL}$. The magnitude of this potential difference is a function of the ratio of the storage cell capacitance and the bit line capacitance. To improve the noise immunity of a DRAM, the two bit lines (BL and $\overline{BL}$) are laid out parallel to each other. Any outside noise will couple to both lines. This structure is referred to as *folded bit lines*. Depending on the bit line precharge level, dummy cells may not be necessary. Even then dummy cells are implemented to minimize the noise by keeping the circuit symmetric.

After a small potential difference is established between the bit lines, the sense amplifier is activated to amplify it. First, the cross-coupled nMOS pull-down section of the sense amplifier is activated by the signal $\overline{SAN}$, and the bit line with the lower potential is pulled down even lower. Then the pMOS pull-up section is activated by SAP, and the bit lines are driven to V_{DD} and V_{SS} to restore the data back to the memory cell. (Here V_{SS} is same as ground or 0 V.) The cross-coupled nMOS pull-down pair should be designed such that the bit line with the higher potential stays approximately at the same level while the other bit line is rapidly pulled down to ground. This can be achieved by making the current source that feeds the $\overline{SAN}$ clock increase slowly at the beginning and quickly after the signal starts being latched. If the pull-down current increases too rapidly, the potential difference between the bit lines will not have a chance to grow, and both will collapse to ground. A similar problem also exists with the magnitude of the pull-down current source. It should be large in order to minimize the delay. On the other hand, if it is too large, when the sense amplifier is enabled, both bit lines will be pulled down to ground. Usually one carefully designed current source feeds a large group of amplifiers by signals SAP and $\overline{SAN}$.

This amplifier structure is essential for DRAMs because, unlike in SRAM circuits, the DRAM read operation is destructive. That is why it is necessary for the sense amplifier to drive the bit lines to a full swing in order to restore the data in the memory cell. The sense amplifiers described in the SRAM section cannot be used for DRAMs because SRAM amplifiers are capacitively coupled to the bit lines, and they drive the output but not the bit lines. If such amplifiers were used in DRAMs, the data could not be written back to the memory cell. In addition, the DRAM sense amplifier has to fit the bit line pitch because a sense amplifier is required for every column in DRAMs. Every cell selected by the word line needs its data to be refreshed by a sense amplifier at the end of a read cycle because the read operation is destructive in DRAMs.

If it is desired to write back a logical "1" equal to full V_{DD}, the word line must be boosted at least one V_{TN} above V_{DD} to compensate for the threshold drop across the nMOS pass gate that connects the cell to the bit line. Again, this is not necessary in SRAMs because their memory cells are regenerative.

Near the end of the read cycle, the column address strobed by CAS is decoded, and the column select signal CSL ties the addressed column to data output bus. When RAS is deactivated, word line is turned off. The current sources that feed the sense amplifier clocks SAP and $\overline{SAN}$ are also turned off and SAP and $\overline{SAN}$ go back to V_{BL}, and the precharge clock PC goes high. The DRAM chip is ready for another read cycle [4.26]–[4.30].

There are many variations in how the bit lines are precharged and the dummy cells implemented [4.26]–[4.27]. Some of the early DRAM designs used a midpoint reference scheme [4.31]–[4.32]. Many 16 Kbit DRAMs precharged the bit lines to full V_{DD} and used a dummy cell half the area of a regular storage cell and precharged the dummy cell to V_{SS}. In this way the final value of the reference bit line is $V_{DD}C_{BL}/(C_{BL} + 0.5C_{CELL})$. The potential on the reference bit line is compared with V_{DD} if the value stored in the addressed cell is a 1, and to $V_{DD}C_{BL}/(C_{BL} + C_{CELL})$ if the value is a 0. There are many different ways to implement a dummy cell: for example, a reference voltage can be coupled to the cell from its bottom plate and the

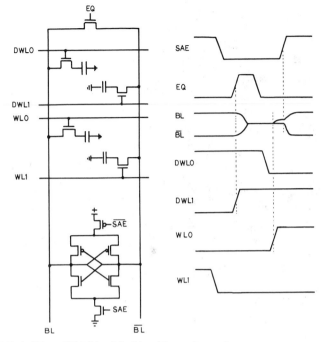

FIGURE 4.17   DRAM with $V_{DD}/2$ sensing scheme

top plate can be permanently connected to the bit line. In this case there is no pass transistor, and the word line is connected to the bottom plate of the dummy storage capacitor. When the dummy word line goes high, a reference charge is injected to the bit line. Starting with the 64 Kbit DRAMs, midpoint sensing again gained popularity. Almost all new CMOS parts use midpoint sensing. Here a full-sized dummy cell is precharged to $V_{DD}/2$ and compared with a V_{DD} or V_{SS} stored in a full-size memory cell. The dummy cell can be precharged to $V_{DD}/2$ either by directly connecting it to the reference value, as shown in Fig. 4.16, or via charge sharing, as shown in Fig. 4.17. Actually if the bit lines are precharged to $V_{DD}/2$, there is no need for a dummy cell. Both bit lines are precharged to $V_{DD}/2$, and when a memory cell is accessed, the bit line is either pulled up or down and away from $V_{DD}/2$; the direction depends on the value stored in the cell. As a result $V_{DD}/2$ on the opposite bit line is compared with $V_{DD}/2 - \Delta$ or $V_{DD}/2 + \Delta$ on the bit line the accessed cell is connected to. A full-sized dummy cell, however, is usually included to keep the bit lines symmetric, to minimize the susceptibility of sense amplifiers to bit-line-precharge-level variations, and to improve speed and reliability [4.37].

Figure 4.17 illustrates a half-V_{DD} DRAM sensing scheme in which the bit lines and the storage capacitors of full-sized dummy cells are precharged to $V_{DD}/2$. As seen in the timing diagram in Fig. 4.17, at the end of a read cycle, one of the bit lines is at V_{DD} and the other is at 0 V. After the sense amplifier is turned off, the equalization clock EQ goes high, and the bit lines are connected. At the same time, both dummy word lines, $DWL0$ and $DWL1$, are forced high, and full-sized dummy storage cells are connected to the bit lines. Since both bit lines have equal capacitances, the bit lines and dummy cells share their charges and reach a value approximately at the midpoint between V_{DD} and ground at the end of the equalization period. This minimizes the power dissipation of the chip and the dI/dt noise at the power lines because no current is drawn from the power supply during the precharge period; half the electrical charge on one bit line is simply transferred to the other bit line. Later the address is decoded, and one of the word lines is selected (for example, $WL0$). Then the dummy cell connected to the same bit line as the selected word line ($DWL0$) is disconnected, and the other dummy cell ($DWL1$) is left tied to the opposite bit line. As a result the value stored in the selected memory cell (V_{DD} or V_{SS}) is compared with $V_{DD}/2$ stored in a full-sized dummy cell [4.33]–[4.35].

Precharging the bit lines and the dummy cells to $V_{DD}/2$ has many advantages. It is fast because building sense amplifiers with a large gain at the midpoint between the power levels is easier than building one with a large gain near V_{DD} or ground. It reduces the dI/dt noise at the power lines and decreases the dynamic power dissipation by a factor of two. It also reduces the current spikes and peak currents, which decreases the IR drops at the power lines and improves electromigration. The main disadvantage of this method is that to store back a high value in the memory cell, the bit line has to be driven from $V_{DD}/2$ all the way up to V_{DD}, which takes a long time and requires a relatively long cycle time. In the alternative sensing schemes where the bit lines are precharged to V_{DD} or ($V_{DD} - V_{TN}$), the high value is already on the bit line and the low value is obtained relatively fast because nMOS pull-downs in the sense amplifier are faster than the pMOS pull-ups. In addition, writing a zero is less critical because the low value is passed to the cell rapidly and with no degradation by the nMOS pass gate. Of course, to store a full V_{DD} high value in the cell, both methods need to boost the word line voltage at least one V_{TN} above V_{DD}. Precharging the bit lines to $V_{DD}/2$ instead of V_{DD} or $V_{DD} - V_{TN}$ also increases the bit line capacitance because the $n+$ junctions of the pass gates are less reverse biased.

Figure 4.18 shows a sense amplifier that is resistively isolated from the bit lines. This improves the access time because the effective capacitances of the n-channel flip-flop nodes are reduced, and the sense amplifier will drive the output much faster. The bit lines will be driven more slowly, and it will take longer for the sense amplifier to restore the value back to the memory cell. This is not as important because the output needs to be fed to the data output bus and then to the output buffers. During that time the bit lines can be driven to V_{DD} and V_{SS} even with the resistive couplings slowing down the path from the sense amplifier to the memory cell [4.36]–[4.37].

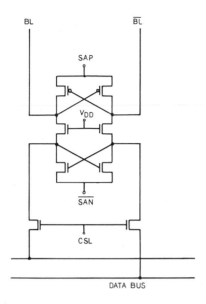

BL $\overline{BL}$

SAP

V_{DD}

$\overline{SAN}$

CSL

DATA BUS

FIGURE 4.18   Resistively isolated sense amplifier. The access time is improved because the bit line capacitance is isolated from the sense amplifier output capacitance.

4.6   ROM and EPROM Circuits

Read-only memory (ROM) and electrically-programmable ROM (EPROM) chips usually do not have double-rail signals available from their memory cells; as a result, they employ single-ended sense amplifiers. There are two basic approaches. The first is to use a differential amplifier and feed the signal from the memory array to one of the inputs of the amplifier. The other input of the amplifier is tied to a reference potential. Figure 4.19 shows such an amplifier. The second approach is to design a truly single-ended amplifier.

Figure 4.20 shows a single-ended amplifier for a ROM or EPROM circuit [4.38]. The signal CE is the chip enable, which is activated to enable the chip. When none of the columns is selected, the node X does not have any path to ground, and the pMOS transistor M3 pulls the node X high. During pull-up, when node X reaches to the logical threshold of the inverter I1, the potential is amplified by I1 and I2, and the negative feedback loop turns M3 off. As a result M3 cannot pull V_X all the way to V_{DD}. The pMOS M2 may pull V_X somewhat higher, but V_X eventually will be clamped by the ratio of M2 and M4. At this point, the output V_{OUT} is 0 V, and the potential at node X is clamped at near $V_{DD}/2$.

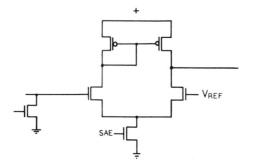

FIGURE 4.19   Differential amplifier used as a single-ended sense amplifier

When the memory is accessed, a row and column are selected, and if the cell is programmed in a conducting state, V_X is pulled down. Because V_X was near the high-gain region of I1, the sense amplifier responds immediately, and the output switches. The pMOS transistor M2 prevents V_X from being pulled all the way to ground and minimizes the recovery time. The low value of V_X will be determined by the ratio of M2, M4, and the memory cell path. The clamp transistors M2 and M4 limit both the high and low values of the potential swing.

The clamp transistors M2 and M4, pMOS pull-up M3, and the inverter I1 dissipate dc power during various stages of the operation. The dc path to ground is turned off at the standby state. When the chip is not selected, $\overline{CE}$ is high; as a result M1 is turned off, and the power to M2 and M3 is

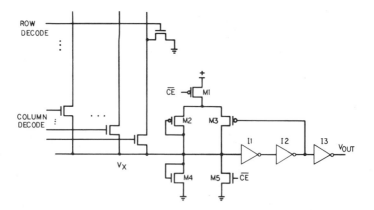

FIGURE 4.20   Single-ended sense amplifier of an EPROM chip

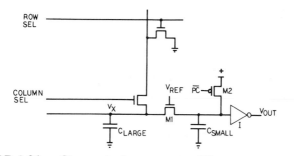

FIGURE 4.21   Charge-sharing sense amplifier

disconnected. In addition, M5 is turned on, and the input to the inverter is grounded. Accordingly the sense amplifier does not dissipate any static power at the standby mode.

A charge-sharing (or charge-balance) sense amplifier is shown in Fig. 4.21. This circuit can be used in ROMs or PLAs in logic chips. The pass transistor M1 isolates the large-capacitance node from the input node of the inverter, which has a small capacitance. The basic principle of this circuit is based on the imbalance between these two capacitors. When the precharge clock PC is activated, C_{SMALL} is charged to V_{DD}. Because of the threshold drop across M1, C_{LARGE} is charged only to $(V_{REF} - V_{TN})$. At this state, the input of the inverter I is at V_{DD}, and V_{OUT} is low. The inverter does not dissipate any dc current. When the potential on C_{LARGE} is pulled down, M1 immediately starts conducting. Because of the size imbalance between C_{LARGE} and C_{SMALL}, the charge stored on C_{SMALL} is rapidly transferred to C_{LARGE}, and input of the inverter is discharged to $V_X \approx (V_{REF} - V_{TN})$. If the reference voltage V_{REF} is low and the logical threshold of I is not very far from $(V_{REF} - V_{TN})$, the inverter output very quickly switches to high. The noise margin of this circuit is small. If charge leakage or coupled noise causes V_X to go down, C_{SMALL} can be erroneously discharged to the level of V_X. Of course, the precharge period should be sufficiently long to charge V_X all the way to $(V_{REF} - V_{TN})$ in order to avoid losing charge after the clock is turned off.

4.7   Logic Circuits with Reduced Potential Swings

Although differential amplifiers used in RAMs can detect very small potential changes, they are hard to apply in logic circuits. Chip area in logic-intensive designs is normally determined by interconnections; as a result the die size of a logic chip may increase by 30 percent or more when double-rail signals are used. In addition, the large area and power dissipation of sense amplifiers may not be acceptable for logic applications.

A major difference between memory and logic circuits is their interconnectivity structure and fan-in and fan-out of their basic building blocks: memory cells and logic gates. Memory cells are organized as an array, and one long interconnection serves as an input and output to many cells. Numerous cells are fed by the same source and feed a single sink. A large sense amplifier is economical in arrays because it is shared by many cells. In logic circuits, however, this is usually not the case. Another major difference is the degree of control of the circuit parameters. Memories are based on regular array structures; the cell and bit line capacitances can be accurately calculated in advance, and the sense circuit can be honed to detect a 500 mV potential change. Design style in logic applications is totally different. There are too many gates in a logic chip to treat every path as an analog circuit. In addition, layout and placement of gates and interconnections are not known until the final stages of the chip design. This requires the building blocks of logic circuits to be generic so that they will function with any capacitive load at their inputs and outputs. The capacitive loading will add delay but should not prevent the circuit from performing its basic function. Of course, array-based building blocks in logic chips, such as register files, programmable logic arrays (PLA), and on-chip RAM and ROM structures, are exceptions. Internal circuitry of these array-based structures can be carefully optimized, and they can be modified and used as many times as necessary. This is why regular array structures are becoming increasingly popular in logic chips. Other advantages of arrays are their high density and quick customization.

Some of the single-ended sense amplifiers used in memory chips may be applied to logic gates. The differences between logic and memory circuits, however, should be carefully considered. In memory circuits, the driver is the cell, which has to be very small to minimize the chip area and the capacitive load on the word lines and bit lines. A sense amplifier is shared by many memory cells and therefore can be made larger. Even then one needs to be careful that the sense amplifier pitch does not cause the bit line pitch to grow because sense amplifiers need to fit at the bottom of bit lines. As a result the cell (driver) is optimized for minimal area, and the sense amplifier (receiver) is optimized for minimal delay and power consumption. In logic circuits, on the other hand, driver and receiver circuits play equal roles in determining the area and capacitance. Consequently the driver is the traditional delay control element in logic circuits because it is easier to design and adjust. Usually there is no separate receiver circuit; the signal is directly fed to the receiving logic gate. To reduce the delay, the driver size can be increased until the delay of turning on the driver becomes comparable to the propagation delay from the driver to the "receiver." Since there are many more "receivers" in a logic subsystem than in a memory chip, the power limitations are more stringent for sense amplifiers in a logic environment.

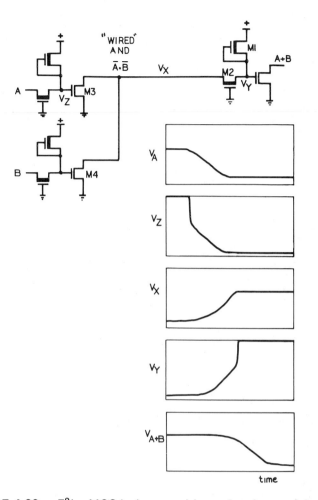

FIGURE 4.22   F^{2}L nMOS logic gate with a reduced potential swing

Some nMOS logic circuits reduce the potential swings at high-capacitance nodes in order to minimize gate delays [4.39]–[4.40]. An F^{2}L gate is shown in Fig. 4.22. Here the node Y is pulled up to V_{DD} by M1, a depletion-mode nMOS device. If both nMOS pull-downs M3 and M4 are off, the node X will be also pulled up by M1. The potential at node X cannot go all the way to V_{DD} because the depletion-mode nMOS pass transistor M2 turns off when V_X reaches $|V_{TD}|$. This is similar to the charge-sharing amplifier in Fig. 4.21, except $V_{REF} = 0\ V$, V_{TD} is negative, and the circuit is static. In this gate, the small-capacitance internal node Y has a full swing, but the external nodes with large interconnection capacitances have limited swings. This reduces the delay significantly. Similar to I^{2}L bipolar circuits, these nMOS gates have one input and multiple outputs. The outputs from two gates can be tied together to obtain a "wired" AND gate. Consequently these circuits are referred to as F^{2}L gates.

If the pass transistor M2 is eliminated, the gate formed by transistors M1, M3, and M4 has the same structure as a regular E/D nMOS NOR gate. There are two differences between a regular E/D nMOS gate and an F^{2}L gate. First, transistors in an F^{2}L gate are configured differently. The pull-ups and pull-downs are far apart, so the internal node of the regular E/D nMOS gate becomes the external node, and the external node becomes the internal node. Second, because of the pass gate M2, the potential swing on the external net (which corresponds to the internal net of a regular gate) is limited. Just like the E/D nMOS gate, F^{2}L gate dissipates static power when a pull-down transistor is on.

The low values of V_X and V_Y are determined by the resistive network formed by M1, M2, and M3. Because there is a potential drop across the depletion-mode pass device M2, the low value at node Y is proportional to $(R_{M2} + R_{M3})/(R_{M1} + R_{M2} + R_{M3})$. Device sizes must be selected to ensure that this is low enough to turn off the output device that generates $A + B$. Degradation of the low potential at node Y and the reduced swings at node X affect the noise margins negatively.

4.8   Differential CMOS Logic Circuits

Figure 4.23 shows a differential CMOS version of the F^{2}L circuit. At first sight, the circuits look different, but the underlying principles are similar. The nMOS depletion load M1 is replaced by a pMOS pull-up M1, and the depletion-mode pass transistor with a grounded gate (M2) is replaced by an enhancement nMOS device (M2) with its gate connected to a reference potential V_{REF}. Because of the pass transistor M2, the high value of the potential at the long interconnection is limited to $(V_{REF} - V_{TN})$. If V_{REF} is set to $(0.5V_{DD} + V_{TN})$, the high level at the large-capacitance global nodes will be clamped at $V_{DD}/2$. Just as in the previous circuit, the potential swings at the internal nodes cover the full range from V_{DD} to V_{SS}. This CMOS circuit structure is referred to as *differential split-level (DSL) logic*. Using a 2 μm CMOS process, complex DSL logic gates with 0.8 ns gate delays have been fabricated [4.41].

Compared to an F^{2}L gate, cross-coupled pMOS pull-ups in DSL reduce the power dissipation and improve the logic zero level at the interconnection and internal nodes. The power dissipation and logic zero level improve because the pMOS transistor corresponding to the side that is being pulled down has a potential of $V_{REF} - V_{TN}$ at its gate, so it is turned on only lightly and

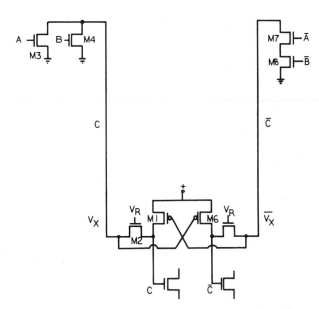

FIGURE 4.23   A CMOS differential split-level logic (DSL) gate

does not conduct a large current. If M3 or M4 is on, V_X will be pulled down toward the ground. At the same time, either M7 or M8 will be off, and $\overline{V_X}$ will be pulled high to $(V_{REF} - V_{TN})$. Accordingly V_X is grounded, and the pMOS pull-up M1 is turned on lightly. At the opposite side, $\overline{V_X}$ has no path to ground, and the other pMOS pull-up M6 is turned on strongly and charges $\overline{V_X}$ high. The logic zero level of the external wires is about 100 mV, and the internal node is at 300 mV. Later, when both A and B become zero, M7 and M8 are turned on, and M3 and M4 are turned off. The node $\overline{V_X}$ is pulled down rapidly from $(V_{REF} - V_{TN})$ to 100 mV, and pMOS pull-up M1 turns on strongly. The internal node C that is tied to V_X is quickly pulled up to V_{DD} by M1, and the wire with V_X is charged to $(V_{REF} - V_{TN})$.

Because of the overhead introduced by four transistors in the pull-up network, DSL gates are more suitable for complicated gates where this overhead can be shared. Also it is not desirable to keep the "true" and "complement" legs of the logic tree completely separate. In complex gates, there is more opportunity to merge the two legs in order to share the common functions and reduce the transistor count.

Just as F^2L gates are a modification of common E/D nMOS gates, DSL gates are a variation of a CMOS logic family called *cascode voltage switch logic* (CVSL) [4.42]. A CVSL gate is shown in Fig. 4.24. It can be seen that a DSL gate is derived from a CVSL gate the same way F^2L is derived from E/D nMOS logic. First, by changing the placement of the transistors, internal nodes of a CVSL gate are transferred into external nodes in DSL. Then, by

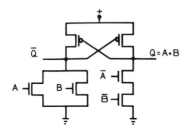

FIGURE 4.24   Cascode voltage switch logic (CVSL)

placing a clamping pass transistor next to the pull-up, the high value of the potential on the external node of DSL is limited.

The cross-coupled pMOS pull-ups in CVSL provide amplification and function as a latch. Once the latch is set, there is no dc path between V_{DD} and V_{SS}. Unlike DSL, the gate of the pMOS is charged all the way to V_{DD} and not to $(V_{REF} - V_{TN})$. Consequently the pMOS pull-up is completely off.

The two legs of the gate are complements of each other and can have common paths to ground in more complicated gates. This saves transistors. Because the outputs drive only nMOS transistors, the loading at the outputs is about one-third smaller than the static CMOS gates that require complementary pMOS and nMOS pairs. The major disadvantage of CVSL is the wiring complexity due to differential signals.

A clocked version of CVSL is shown in Fig. 4.25. The operation of this circuit is similar to domino logic described in Chapter 2. The outputs Q and $\overline{Q}$ are precharged to zero when ϕ is low. When ϕ turns on, all the inputs are initially zero, and none of the nMOS transistors in the combinational pull-

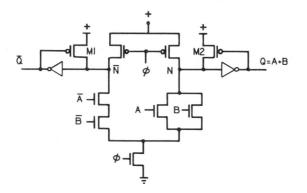

FIGURE 4.25   Clocked cascode voltage switch logic (CVSL) gate

down network can conduct. After a short time, outputs of the preceding logic stages propagate, similar to falling dominoes, and some of the inputs go high. Then either N or $\overline{N}$ is pulled down by the nMOS combinational network. The pMOS transistors M1 and M2 hold the nodes N and $\overline{N}$ high prior to switching. They reduce charge-sharing noise within the tree and improve the noise margins with minimal performance penalty. After the gate switches, the internal node that has a high value continues to be held high.

Domino logic provides only uncomplemented logic such as AND and OR, while clocked CVSL provides both types of logic (NAND and NOR as well as AND and OR) at the expense of differential signals. Just like DSL and domino logic, CVSL is also more suitable for complex logic gates because of the overhead introduced by the pull-up network. This is especially true for clocked CVSL, which requires nine transistors for clocking. In complex gates, this overhead is shared by a bigger function, and the true and complement sections of the pull-down tree can be merged together to reduce the transistor count. In a static CMOS gate, the true and complement pull-up and pull-down sections cannot be merged; a static CMOS gate with n inputs requires 2n transistors. Because of these savings, a complex CVSL gate usually requires fewer transistors than a fully static CMOS gate but more than a domino gate. CVSL logic is claimed to be four times as fast as static CMOS logic. The major drawback of CVSL is the differential signals and wiring complexity. CVSL may require fewer transistors, but it also requires both true and complemented inputs. Static CMOS requires half as many inputs because it feeds the same inputs to both nMOS and pMOS transistors; inversion is provided by the complementary nature of the pMOS transistors.

Figure 4.26 shows various implementations of a NOR function in various MOS logic families. The transistors labeled M1 and M2 serve the same function in all gates. The evolution of MOS logic gates can be observed in the clockwise order. First, the enhancement load in the early E/E nMOS gate is replaced by a depletion transistor, and E/D nMOS gate with improved noise speed and margins is formed. Push-pull nMOS attempts to minimize the power dissipation by tying inverted versions of the input signals to depletion pull-ups. Because of the overhead associated with double-rail signals, however, truly low-power push-pull gates become practical only with the advent of CMOS. The static CMOS gate does not require complements of the inputs because the inversion is provided by pMOS devices. A static CMOS gate has a large transistor count and a high input capacitance. In addition, the pull-up time is usually long because pMOS devices are about half as fast as nMOS devices. Domino logic addresses all three drawbacks of static CMOS by precharging the output and using only nMOS devices to implement the logic. Noise problems and unavailability of inverted logic limit the applicability of domino logic. Clocked CVSL provides inverted logic but requires double-rail signals. Static CVSL has smaller input capacitance than static CMOS and may require fewer transistors in complex gates. It also requires

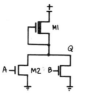
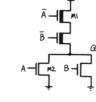
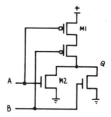

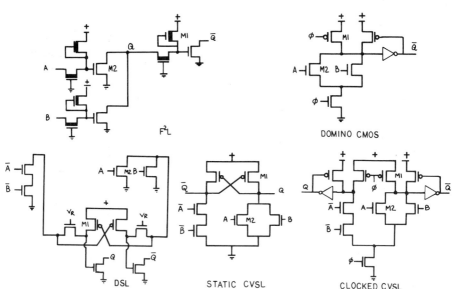

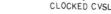

FIGURE 4.26 MOS logic families

double-rail signals. F²L and DSL are modifications of E/D nMOS and static CVSL. They separate the pull-up and pull-down portions of a gate and place a clamping pass transistor in between. This reduces the potential swings at the high-capacitance nodes and improves the speed.

Figure 4.27 illustrates another differential driving scheme. During ϕ_1, both interconnections are precharged to one threshold below V_{DD} through nMOS pull-ups. Later, during ϕ_2, depending on the value of V_{IN}, one of the interconnections is pulled down, and a differential amplifier rapidly senses this potential difference. An output stage restores the signal levels to 0 and V_{DD}. The potential on the interconnection can be clamped to avoid its being discharged to 0 V. If the interconnections and circuits are laid out symmetrically, they will be affected by the outside noise sources equally, and, because of the high common-mode signal-rejection ratio of the differential sense amplifier,

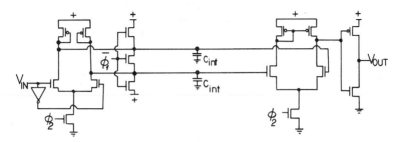

FIGURE 4.27   Differential driver/receiver circuit

the circuit will operate reliably. The smallest potential difference that can be sensed is determined by how closely the threshold voltages of the transistors in the circuit are matched. Double-rail interconnections increase the area requirement, but its superior speed and reliability may make this circuit attractive for certain applications.

4.9   Drivers and Receivers for Logic Circuits

In this section experimental CMOS driver and receiver circuits are described that reduce the delay at high-capacitance nodes by precharging them to the high-gain region of the receiver. These circuits limit the potential swing by adjusting the driver-transistor size or using clamping circuits [4.49]–[4.50]. Circuit operation of driver/receiver circuits will be described, and their propagation delays, power dissipations, and areas will be compared with more traditional circuits such as inverters and simple precharged buses.

Because these circuits improve the power-delay product by decreasing ΔV and V_{PP}, noise margins are degraded. As a result the noise sources must be reduced along with ΔV and V_{PP} to ensure reliable circuit operation. The major noise sources in VLSI circuits are capacitive, inductive, and resistive couplings between various sections of the circuit, power-level fluctuations resulting from simultaneous-switching noise, and deviations from the ideal design caused by parameter variations and mismatches.

Since voltage swings and current surges are decreased and noise generation resulting from couplings is proportional to these swings and surges, on-chip noise generation is reduced along with the noise margins. Switching noise in the power lines is lower because of a decrease in the dynamic component of the signals and an increase in their dc component (see Chapter 7). Noise resulting from device-parameter variations and mismatches is minimal in the circuits developed in this section because of their self-biasing nature. The only noise component that remains unchanged is off-chip generated noise. Consequently

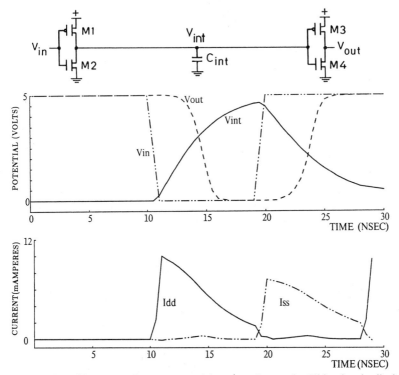

FIGURE 4.28   Two inverters as driver/receiver pair. This circuit dissipates no static power and has wide noise margins. The delay and dynamic power dissipation, however, are large because of the large potential swing on the interconnection. The transistor dimensions used in simulation are M1: 50/2 μm, M2: 20/2 μm, M3: 10/2 μm, M4: 4/2 μm, $C_{int} = 10$ pF.

improved shielding against off-chip noise may be needed to ensure reliable operation of reduced potential swing circuits.

4.9.1   Pair of Inverters

A pair of inverters is the most straightforward driver/receiver configuration (Fig. 4.28). It is simple, has wide noise margins, and does not dissipate any static power. It is the most commonly used CMOS circuit configuration. It will be used as a comparison base for other circuits. This approach is not very fast because the potential on the interconnection is a slowly rising waveform that must be driven between V_{SS} and V_{DD}. In addition, the receiving inverter does not respond until its input reaches approximately $V_{DD}/2$. The circuit speed can be improved only by increasing the driver size. The currents from

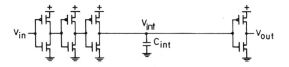

FIGURE 4.29 Cascaded drivers. The driver is a chain of inverters that increase in size until the last device is strong enough to drive the load. If the interconnection was driven directly by a large inverter, which in turn was driven by a small device, the turn-on delay of the large inverter could become larger than the delay of charging or discharging the wire.

the power supply (I_{DD}) and to ground (I_{SS}) are shown in Fig. 4.28. The lack of static power dissipation is demonstrated by the nonoverlap between I_{DD} and I_{SS}. No current is wasted; it is all used to charge or discharge the output node. The large potential swing, however, is the major factor that degrades the switching speed and power consumption.

Figure 4.29 is a slight modification of this circuit. In this configuration, the driver is a chain of inverters that increase in size until the last device is large enough to drive the load. This is necessary because if the load is driven by a large inverter, which in turn is driven by a small device, the turn-on delay of the large inverter may dominate overall delay [4.51]–[4.54]. (See also Chapter 5.) This circuit minimizes the total delay, but the problems associated with the inverter pair remain. The waveforms for cascaded drivers are not any faster than the waveforms shown in Fig. 4.28 because the simulation in Fig. 4.28 assumes a fairly fast rise time (1 ns) as an input to a large driver. This, in effect, fullfills the function of the cascaded drivers. It also provides a more favorable set of assumptions for the two inverters with respect to the following precharged circuits, which require smaller transistors and are easier to drive.

4.9.2 Precharge to V_{DD}

A precharged driving scheme is shown in Fig. 4.30. When ϕ_1 is high, the interconnection is precharged to V_{DD} through M1. During this period, M2 is forced to the nonconducting state, irrespective of V_{IN}, via the NOR gate that gates V_{IN} with $\overline{\phi_2}$. This prevents M2 from discharging V_{int} during precharge. Once ϕ_1 goes low, the input V_{IN} must be valid, and, during ϕ_2, V_{IN} passes through the NOR gate, and the interconnection may or may not be discharged via M2, depending on the value of V_{IN}. Circuit operation is similar to domino logic described in Chapter 2 [4.43]. It is essential that V_{IN} has obtained its final value before ϕ_2 goes high because if V_{IN} is inadvertently high and assumes its correct value *after* ϕ_2 is activated, the interconnect will be incorrectly

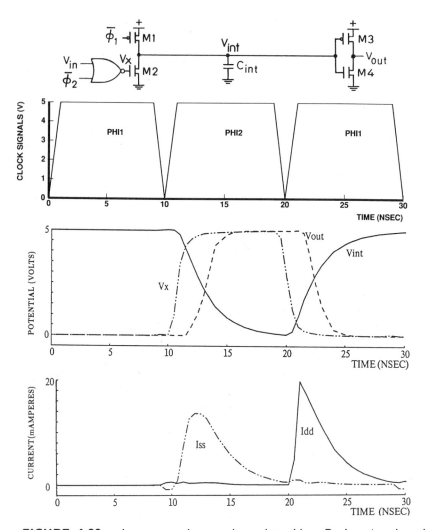

FIGURE 4.30   Interconnection precharged to V_{DD}. During ϕ_1, the wire is precharged to V_{DD} via M1, while the M2 is forced to the nonconducting state. During ϕ_2, the wire may or may not be discharged by M2 depending on the input V_{IN}. M1: 100/2 μm, M2: 50/2 μm, M3: 10/2 μm, M4: 4/2 μm, C_{int} = 10 pF.

discharged. Unlike static circuits, the lost charge is not gained back when V_{IN} goes low.

Multisource bus designs based on this method are area efficient because one large pull-up (precharge transistor M1) is shared by many pull-downs. This circuit, however, has the same speed and power problems as the two-inverters scheme. As mentioned above, it is also sensitive to glitches and sub-ject to charge-sharing mechanisms. If, after the precharge period, V_X inad-vertently goes one threshold above ground, M2 may discharge the precharged interconnection.

4.9.3   Precharge to $V_{DD}/2$

A configuration in which the interconnection is precharged to the high-gain region of the receiver circuit ($V_{IN} = V_{OUT} \approx V_{DD}/2$) is shown in Fig. 4.31. This circuit uses an inverter with a feedback path between its output and input. When the input and output of an inverter are tied together, the inverter is automatically biased at the high-gain region (Fig. 4.12). Similar techniques are widely used in the comparator circuits of A/D converters [4.55]–[4.57]. They are also applied to RAM sense amplifiers, as shown in Fig. 4.11.

During ϕ_1, the input is disabled via the NAND and NOR gates that feed the driver transistors; both M1 and M2 are turned off. At the same time, the input and output of the receiver (inverter formed by M3 and M4) are connected[5] through M5, and as a result the bus is precharged to $V_{DD}/2$. Later, during ϕ_2, the feedback loop around the inverter is broken by turning M5 off, and the input is enabled through the NAND and NOR gates that feed the driver transistors. Depending on the value of the input, either M1 or M2 will turn on, and the driver will tip the interconnection potential to either side of $V_{DD}/2$. Then the receiver will respond immediately because the wire was precharged to the transition point of the receiving inverter.

The main difference between this approach and the previous two is that, in the previous circuits, the driver has to bring the interconnection potential from V_{SS} or V_{DD} to the threshold of the receiver ($V_{TH} \approx V_{DD}/2$) before the receiver responds. In contrast, in the precharge to $V_{DD}/2$ method, the interconnection is already precharged to the high-sensitivity region of the receiver in Fig. 4.31, and the inverter does not require much potential change at its input to detect state changes.

At very high clock frequencies, this configuration may also conserve power because, unlike the previous circuits, the potential on the interconnection will be confined around $V_{DD}/2$ and will not be charged and discharged between

---

[5]Because of the large interconnection capacitance, the inverter with the feedback loop cannot oscillate. Even with a small interconnection capacitance, the circuit will not oscillate because the CMOS inverter is not fast enough. For an inverter to oscillate when its input and output are connected, the delay through the inverter must be comparable to the delay of the wire that ties the inverter input and output together so that, at a given time, the inverter input and output will be complements of each other. This 180° phase difference is essential for oscillation. Oscillations can be also obtained if the "inverter" has multiple "internal" states. A good example of this is a ring oscillator built by an odd number of inverters connected to form a ring. By adding more inverters, the effective delay of the "wire" is increased that ties the input and output of a given inverter in the chain. At least three or five CMOS inverters are required to make a ring oscillator that will work.

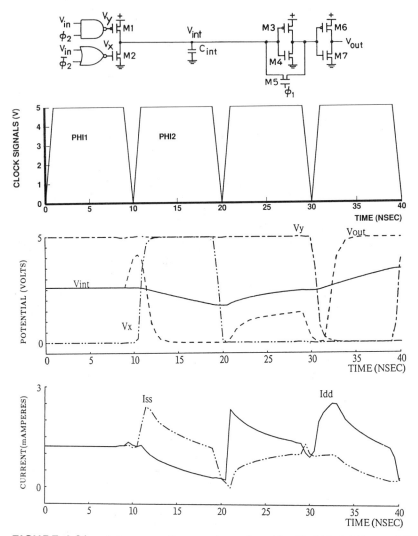

FIGURE 4.31   Interconnection precharged to $V_{DD}/2$. M1: 10/2 μm, M2: 4/2 μm, M3: 40/2 μm, M4: 16/2 μm, M5: 34/2 μm, M6: 10/2 μm, M7: 4/2 μm, $C_{int} = 10$ pF.

V_{DD} and ground. The magnitude of the fluctuation around $V_{DD}/2$ and power consumption will be determined by the noise margin requirements. The power advantage of this circuit with respect to two inverters diminishes if the propagation delay is much smaller than the cycle time or if the circuit switches infrequently. Two inverters in Fig. 4.28 stop conducting once the potential

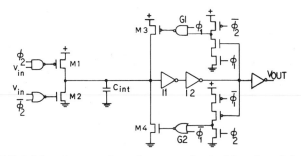

FIGURE 4.32   Improved precharge method with reduced dc power dissipation. This $V_{DD}/2$ precharge with fast feedback has no static power dissipation.

at the output of the driver has reached its final value. Precharge to $V_{DD}/2$ circuit dissipates power every cycle even if the input value does not change. In addition, if the input to the circuit changes seldom, two inverters consume power only when the input switches. In the precharge to $V_{DD}/2$ scheme, however, the power dissipation is the same irrespective of the ratio of propagation delay to cycle time or how often the input switches because there is a dc current flow during the precharge period and the interconnection is precharged at every cycle even if the input does not change.

The supply-to-ground current that flows when the inverter is biased at $V_{DD}/2$ can be eliminated by a fast feedback loop that turns off the large precharge transistors when the potential on the interconnection reaches $V_{DD}/2$ (Fig. 4.32). During ϕ_2 (evaluation period), the input of G1 is precharged low and the input of G2 is precharged high, and M3 and M4 are disabled by ϕ_1 and $\overline{\phi_1}$ that feed G1 and G2. Later, during ϕ_1 (precharge period), M3 and M4 are enabled. M1 and M2 are disabled; the potential on the interconnection is amplified by inverters I1 and I2, and the input of either G1 or G2 is immediately discharged. This turns off the corresponding pull-up (M3) or pull-down (M4). The one that remains on (M3 or M4) charges or discharges the interconnection toward $V_{DD}/2$, and when the interconnection potential reaches $V_{DD}/2$, the corresponding gate is turned off and precharge is completed. At this point, both M3 and M4 are off, and there is no dc power dissipation. The reason for precharging the inputs of G1 and G2 is to avoid oscillations because the direct feedback loop in this circuit forms a three-stage ring oscillator. The precharged gates do not oscillate because, after they are turned off, there is no way to turn them on until the next clock cycle. During ϕ_2 (evaluation period), M3 and M4 are disabled and, depending on the input, M1 or M2 tips the interconnection potential to either side of $V_{DD}/2$ as in the previous precharge to $V_{DD}/2$ circuit.

There are two limitations in these circuits. First, if M1 or M2 is much faster than the nominal design value because of parameter variations (such as threshold potential or channel-length variations), the interconnection may be completely discharged to ground or pulled all the way up to V_{DD}. This will not reduce the speed of the circuit but will increase power dissipation. Second, it is difficult to build reliable large fan-out circuits using this technique because matching the threshold potentials of receivers located at two ends of the chip (or at two separate chips if this technique is applied for chip-to-chip communication) is not easy. Consequently this technique is not suitable for off-chip drivers. To improve the matching among the receiver circuits on the same chip, all should have identical layouts and should be oriented in the same direction.

4.9.4   Static Sense Amplifier

A "single-sided" and static version of the improved $V_{DD}/2$ circuit in Fig. 4.32 is presented in Fig. 4.33. It is "single sided" because only one of the feedback and driver paths is kept. On close inspection, it can be seen that the static sense amplifier is the same as the portion of the improved $V_{DD}/2$ circuit in Fig. 4.32 consisting of the transistors M2 and M3 and inverters I1 and I2. This static amplifier does not require any clocking; consequently the transistors used for gating the signals with various clocks are not necessary.

In the static sense amplifier, when V_{IN} is low, M1 pulls up the interconnection until it reaches a value somewhat larger than $V_{DD}/2$. At this point, V_{INT} is amplified by inverters I1 and I2, and V_X becomes equal to V_{DD} and turns M1 off. Because of this negative feedback path, M1 cannot charge the interconnection all the way up to V_{DD}. Later, when M2 turns on, it discharges the interconnection until V_{INT} drops slightly below $V_{DD}/2$. Then M1 also turns on, and M1 and M2 form a voltage divider that clamps the interconnection potential between 0 V and $V_{DD}/2$. As a result the large-capacitance interconnection node is not charged and discharged between V_{SS} and V_{DD}; instead it is confined around $V_{DD}/2$. This is a very fast circuit, but it dissipates dc power when both M1 and M2 are on.

The noise immunity is good because there are no precharged nodes and the amplifier is regenerative. If noise coupled to node V_{INT} forces the circuit to a faulty state, it will pull itself back to the correct state. There are no precharged nodes that are susceptible to charge leakage. The interconnection floats when the pull-up transistor M1 is turned off, but if the circuit is forced to a wrong state by noise coupled to the interconnection, M1 will turn on and pull the wire to the correct state and then turn itself off again. This is not true for precharged circuits that leave the interconnection floating, such as domino logic or precharge to V_{DD}. The precharge to $V_{DD}/2$ does not have this problem because during the evaluation period the wire is pulled either high or low depending on the input. This is unlike precharge to V_{DD} in which the output is driven only if its value is low. In addition, the static sense amplifier lends itself well to large fan-in circuits and is therefore a good candidate for

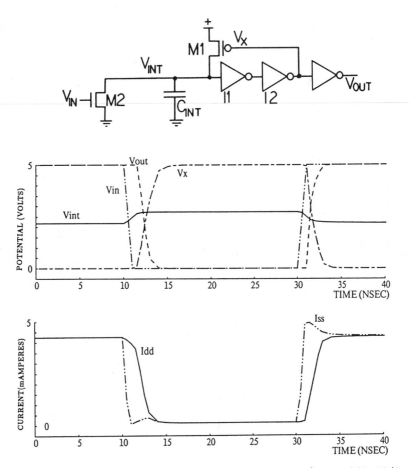

FIGURE 4.33   Static sense amplifier with feedback. M1: 26/2 μm, M2: 16/2 μm, all other nMOS: 2/2 μm and pMOS: 5/2 μm, $C_{int} = 10$ pF.

high-speed PLA circuits. On the negative side, the matching required between the pMOS pull-up M1 and the nMOS pull-down M2 increases the challenge in making this circuit work. It is hard to match pMOS and nMOS transistors because the device parameters of nMOS and pMOS transistors are determined by different process steps. It is much easier to match transistors of the same type. A modified version of this circuit that requires matching only between nMOS transistors is shown in Fig. 4.34.

The sense amplifier in Fig. 4.34 eliminates one of the inverters and replaces that inversion function by exchanging the pMOS transistor pull-up M1 with an nMOS transistor. In the resulting circuit, both M1 and M2 are nMOS; consequently they are easier to match. The swing on node V_X is limited because

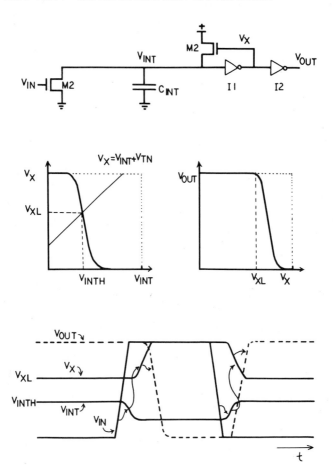

FIGURE 4.34   Static sense amplifier with improved matching

V_X has to be one threshold above V_{INT} for the nMOS pull-up M1 to be in a conducting state. As can be observed from the V_X-V_{INT} transfer curve, the *low* value of V_X is above $V_{DD}/2$, and the *high* value of V_{INT} is below $V_{DD}/2$. To make this circuit work, the inverter I1 should have a low switching point (a strong nMOS pull-down and a weak pMOS pull-up) so that V_{INT} does not have to go too high to switch it. In contrast, I2 should have a high switching point (a weak nMOS pull-down and a strong pMOS pull-up) because V_X does not go very low. Of course, if I1 has a very low switching point, it will be susceptible to noise. One way to improve the potential swing range at the input of I2 is by adding an amplifying-level shifter in inverter I1, as shown in Fig. 4.35. The pMOS transistor M3 acts like a cascode and enhances the potential swing at node A.

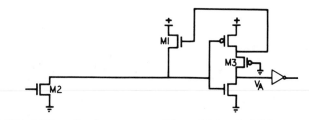

FIGURE 4.35   Static sense amplifier with level shifting

4.9.5 Clocked Sense Amplifier

The sense amplifier in Fig. 4.36 is the clocked version of the previous static sense amplifier. Two improvements are introduced by the clocked amplifier. One, the ratio between M1 and M2 is not important. Two, there is no continuous current flow. The interconnection potential is pulled all the way down to ground, and therefore charging it to a high value will require a longer time. Because the more important delay in many applications is the evaluation (pull-down) period, the longer precharge may be acceptable. Also after ϕ_1 turns off, the interconnection becomes a floating node, and if its potential changes as a result of charge leakage or noise, it will not be restored. Because the interconnection is precharged very close to the high-gain region of the receiver, this becomes a major reliability concern. The previous static circuit does not have this problem.

4.9.6 $|V_{TP}|$ Sense Amplifier

The receiver of the circuit in Fig. 4.37 is faster than the standard inverter that acts as the receiver for the circuit in which the wire is precharged to V_{DD} (Fig. 4.30). The speed improvement is a result of the difference between the thresholds of the two receivers. The new receiver is activated when the interconnection potential goes one threshold ($|V_{TP}|$) below V_{DD}. The simple static inverter, on the other hand, does not respond until V_{INT} is discharged down to approximately $V_{DD}/2$. Power dissipation is the same as in any dynamic precharge type of circuit. Of course, a smaller receiver threshold also means smaller noise margins. If the precharged interconnection leaks such that its electrical potential is reduced by $|V_{TP}|$, the circuit will fail. Or if V_{DD} of the receiving gate bounces up and goes one $|V_{TP}|$ above V_{int}, it will cause failure. The slower circuit, which uses a simple static inverter as its receiver, has wider noise margins and does not fail until the interconnection potential leaks to approximately $V_{DD}/2$.

   If the pull-up time is critical or further power reduction is desired, M2 can be sized such that it never pulls down the interconnection potential all

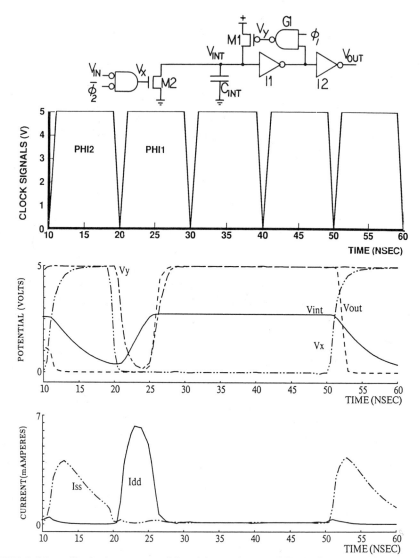

FIGURE 4.36   Clocked sense amplifier. M1: 40/2 μm, M2: 20/2 μm, all other nMOS: 2/2 μm and pMOS: 5/2 μm, C_{int} = 10 pF.

the way to 0 V. This is too dependent on device parameters and hard to achieve because of process variations. An alternative technique is clamping the interconnection potential to a value between $V_{DD}/2$ and 0 V, which is more reliable but also dissipates a greater amount of power. The clamping

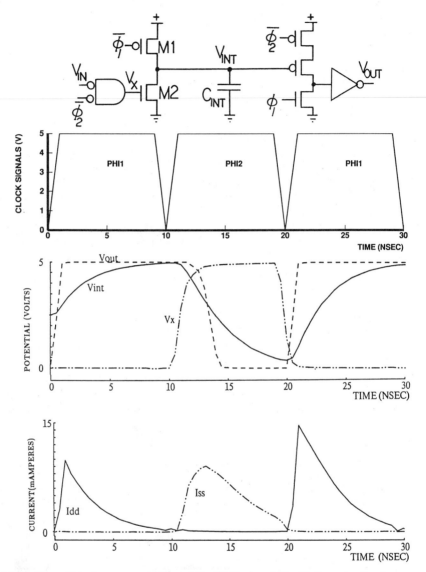

FIGURE 4.37 $|V_{TP}|$ sense amplifier. M1: 75/2 μm, M2: 30/2 μm, all other nMOS: 2/2 μm and pMOS: 4/2 μm, $C_{int} = 10$ pF.

technique is illustrated in Fig. 4.38 and can be applied also to the circuits in Figs. 4.36 and 4.37.

Table 4.3 shows a quantitative comparison of the described receiver/driver circuits for logic applications. The waveforms in all the figures and the infor-

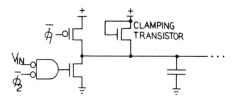

FIGURE 4.38   Voltage clamping for large-capacitance nodes

mation in the table were obtained using the SPICE circuit simulation program [4.58]. Table 4.4 summarizes the 2 μm CMOS parameters used in the simulations [4.59].

4.10   BiCMOS Circuits

By integrating both bipolar and CMOS devices on the same die, advantages of bipolar and CMOS can be combined to obtain superior digital, analog, and power ICs. A typical BiCMOS technology is described in Chapter 2; in this section, various digital BiCMOS circuit applications will be presented.

Bipolar devices switch fast, can drive large currents, and have excellent matching properties. CMOS transistors dissipate very little power, offer a high functional density, and have high yields. In a digital BiCMOS IC, most of the

TABLE 4.3   Comparison of the Interconnection Driving Methods

$(V_{DD} = 5\ V,\ C_{int} = 10\ \text{pF},\ C_L^\dagger = 0.1\ \text{pF, area} \approx 6\sum W \times L)$

Circuit Type	$\Delta t^{\dagger\dagger}$ (nsec)	Worst Case Power (mW)	Power Delay Product (pJ)	Noise Margin (V)	ΔI_{max} (mA)	Area (1000 μm^2)
Two inverters	6.5	12.5	81	2	10	1
Precharge to V_{DD}	4	12.5	50	0.85	20	2.3
V_{TP} sense amplifier	3	12.5	38	0.85	15	1.6
Precharge to $V_{DD}/2$	3	5	15	1	2	1.6
Static sense amplifier	3	20	60	2	5	0.8
Clocked sense amplifier	2.5	6	15	0.4	7	1.1

† C_L is the load at the output of the receiver.
†† Δt is defined between the 10% and 90% points of V_{IN} and V_{OUT} waveforms.

TABLE 4.4   SPICE Parameters for the Stanford 2 μm CMOS Process

Parameter	SPICE Symbol	nMOS	pMOS
Zero-bias threshold	VTO (V)	0.85	-0.85
Transconductance parameter	KP $(\mu A/V^2)$	53	21
Bulk threshold parameter	GAMMA $(\sqrt{V})$	0.175	0.715
Surface potential	PHI (V)	0.56	0.7
Channel-length modulation	LAMBDA $(1/V)$	0.03	0.066
Bulk junction potential	PB (V)	0.89	1.0
Gate-source overlap capacitance	CGSO $(fF/\mu m)$	0.24	0.34
Gate-drain overlap capacitance	CGDO $(fF/\mu m)$	0.24	0.34
Gate-bulk overlap capacitance	CGBO $(fF/\mu m)$	0.2	0.2
Junction bottom capacitance	CJ $(fF/\mu m^2)$	0.13	0.23
Junction sidewall capacitance	CJSW $(fF/\mu m)$	0.65	1.5
Substrate doping	NSUB $(1/cm^3)$	7×10^{14}	1.1×10^{16}
Gate-oxide thickness	TOX (A)	400	400
Junction depth	XJ (μm)	0.35	0.50
Lateral diffusion	LD (μm)	0.28	0.40
Surface mobility	UO $(cm^2/V\text{-}s)$	650	240

function can be implemented in CMOS (memory cells, logic gates, registers, multiplexors, etc.), and bipolar devices can be used sparingly in critical paths to enhance performance (word line drivers, sense amplifiers, buffers to drive on-chip global wires, critical logic paths such as carry-propagate, clock drivers on microprocessors, and chip-to-chip drivers and receivers).

BiCMOS is also very suitable for chips that integrate analog and digital functions on the same chip. Most of the analog portion of the circuit can be implemented in bipolar, because bipolar transistors make perfectly matched differential pairs and low drift operational amplifiers, provide excellent reference circuits, and can be used to obtain high current output buffers. CMOS devices can be used to implement dense logic and memory sub-blocks, and to provide input stages with high input impedance.

One of the first successful applications of BiCMOS technology is high-speed static RAMs with ECL-compatible I/O levels. In a BiCMOS SRAM, the memory cells that make up the core of the chip are MOS, and the peripheral circuitry is BiCMOS or pure bipolar. Input buffers are bipolar, decoders and drivers are BiCMOS, sense amplifiers are bipolar, and output buffer is bipolar. Table 4.5 illustrates the performance enhancements brought by BiCMOS [4.62]–[4.63], and Figs. 4.39 and 4.40 illustrate the input buffer and the circuit diagram of the read path of a BiCMOS SRAM [4.61].

TABLE 4.5 Comparison of CMOS and BiCMOS SRAMs (1 μm design rules)

Delay Component	CMOS (TTL-level I/O)	BiCMOS (ECL-level I/O)
Input buffer	5 ns	2 ns
Address decoder and word line driver	5 ns	2 ns
Memory cell driving the bit line	3 ns	1 ns
Sense amplifier	5 ns	1 ns
Output buffer	8 ns	2 ns
Total (Read access time)	26 ns	8 ns

There are many reasons why BiCMOS SRAMs are faster than CMOS SRAMs. Having small swing ECL-level ($V_H = -0.9\ V$ and $V_L = -1.7\ V$)[6] I/O signals rather than TTL ($V_H = 2\ V$ and $V_L = 0.8\ V$) or CMOS ($V_H = 5\ V$ and $V_L = 0\ V$) reduces the delays at input and output buffers. BiCMOS word line drivers are more than twice as fast as CMOS word line drivers. Bipolar sense amplifiers switch faster than CMOS amplifiers. In addition, because bipolar devices are matched very well, sense amplifiers can be activated with a smaller potential difference between the bit lines, and the potential swing on the common data bus can be limited to a range as small as 300 mV. This improves the delay of the cell driving the bit line and sense amplifier delay. Of course, the power dissipation increases, but it is possible to build a 64 kbit BiCMOS SRAM with 1.3 μm design rules that has ECL-level I/O signals, and a 7 ns read access time, and dissipates only 350 mW [4.62].

BiCMOS has great promise for application-specific, semicustom, and full custom logic as well. The most straightforward application is in gate arrays; CMOS gates can be replaced by faster BiCMOS gates. In all logic chips, off-chip drivers can be replaced by fast ECL drivers, and chip-to-chip signals can be converted to ECL levels. The main drawback is the conversion delay from CMOS to ECL levels at the off-chip driver and the conversion delay from ECL to CMOS at the receiving chip. The sum of the conversion delays swamps out the benefits of ECL-level I/O signals. The conversion delay is not present at the read path of a BiCMOS SRAM because the internal memory cells do not have to drive a full swing as logic circuits do.

The more subtle applications are in custom and semicustom designs where bipolar devices can be substituted for CMOS in the critical paths of the chip.

---

[6]In bipolar circuits, it is customary to refer to power levels in the negative direction from 0 V—for example, $V_{CC} = 0$ V and $-V_{EE} = -5.2$ V. In CMOS circuits, power levels are referenced in the positive direction, such as $V_{DD} = 5$ V and $V_{SS} = 0$ V.

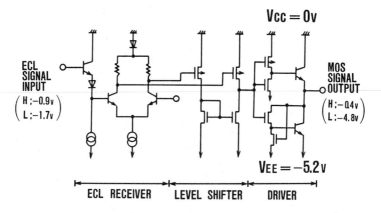

FIGURE 4.39 BiCMOS input buffer circuit that converts ECL-level signals to CMOS levels [4.61] (© 1986 IEEE)

This can be done either manually, such as redesigning portions of the carry propagation circuit of an ALU using BiCMOS, or automatically by a computer aided design tool that identifies the critical paths and, following a rule-based algorithm, replaces some of the CMOS gates with BiCMOS ones.

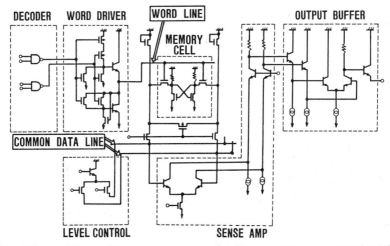

FIGURE 4.40 Circuit diagram of the read path of a high-speed BiCMOS SRAM [4.61] (© 1986 IEEE)

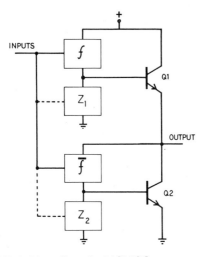

FIGURE 4.41   Generic BiCMOS gate

A generic BiCMOS gate is shown in Fig. 4.41. When f is a 1, Q1 conducts; and when $\overline{f}$ is a 1, Q2 is turned on. Because Q1 and Q2 are never turned on simultaneously, there is never a dc path between V_{DD} and V_{SS}, and the power dissipation is low. When Q1 or Q2 is turned off, the base charge is removed via Z_1 or Z_2. This speeds up the gate by reducing the time period when both Q1 and Q2 are on.

Five BiCMOS implementations of a NAND gate are shown in Figs. 4.42 to 4.46. The implementation shown in Fig. 4.42 uses a CMOS NAND gate to turn on the bipolar pull-up Q1. In the state where either one of the inputs is 0, gate output is a 1 and the base of Q1 is connected to V_{DD}. Then Q1 is turned on and charges the output to $(V_{DD} - V_{BEon})$, where V_{BEon} is the forward bias potential drop across the base-emitter junction. At the same time, N5 is on, and the base of Q2 is grounded. At the state where both inputs are 1, the gate output is 0 and the base of Q1 is grounded via N3, N4, and N5 or the base of Q2. Then the base of Q1 is clamped at either V_{BEon} or V_{TN} (whichever is smaller), and Q1 is turned off but is ready to turn on. Concurrently the base of the bipolar pull-down Q2 is connected to the output of the gate via N1 and N2 and conducts until the output is discharged to V_{BEon}. There is no dc power dissipation, and $V_{OH} = V_{DD} - V_{BEon}$ and $V_{OL} = V_{BEon}$. It is important that both Q1 and Q2 have paths that bleed their base charge and turn them off quickly. It is also important that their bases are not discharged to 0 V and the output of the BiCMOS gate never goes all the way to V_{DD} or V_{SS}. This prevents bipolar transistors from saturating and makes them easier to turn on. It also reduces the possibility of triggering latch-up.

The NAND gate in Fig. 4.43 is similar to the one in Fig. 4.42, but the

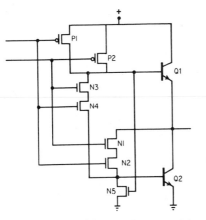

FIGURE 4.42   BiCMOS NAND gate with an *n*MOS chain that turns off the bipolar push-pull pair

base of the bipolar pull-up is driven by a regular CMOS NAND gate. When Q1 is off, its base is fully grounded. It may take a longer time to turn on Q1 compared to the previous circuit. The gate of N5 is also controlled by the BiCMOS gate output rather than the CMOS NAND gate. When N1 and N2 starts conducting, N5 is also on. As a result both Q2 and the *n*MOS chain N1, N2, and N5 discharge the output; the low value of V_{OUT} goes down to V_{BEon} or V_{TN}, whichever is smaller.

In the NAND gate in Fig. 4.44, N5 is replaced by a resistor. As a result the output is pulled all the way down to 0 V. It is necessary that $R_1 > R_{N1} + R_{N2}$; otherwise R_1 may rob the base current of Q_2, and Q_2 may never turn on. To improve the performance of this circuit, its output may be clamped with a clamping pull-up in order to avoid its being discharged all the way to 0 V.

The NAND gate in Fig. 4.45 has the fewest number of transistors; it uses two resistors to discharge the base currents of Q1 and Q2. Because of the resistive paths to V_{DD} and V_{SS}, the output of this BiCMOS gate has a full swing from V_{DD} to 0 V. When Q1 and Q2 turn off, the paths through R1 and R2 continue charging or discharging the output. The role of the resistor R_2 is to generate a positive potential across the base-emitter junction of Q1 to allow it to turn on. Without R_2, Q1 would never turn on. (It is interesting to note the similarity between the resistors in this circuit and the "undesirable" parasitic resistors in CMOS latch-up.)

Figure. 4.46 shows yet another NAND implementation in BiCMOS. Here R_2 is eliminated, but the bases of Q1 and Q2 are tied to the output via a diode D1. There is no need for R_2; the output cannot rob the base current of Q1 because of the diode. The direction of the diode allows the charge from

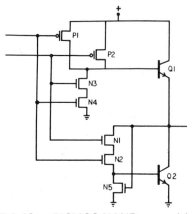

FIGURE 4.43 BiCMOS NAND gate with two isolated nMOS chains that turn off the bipolar output stage

the output to go to the base of Q2 but prevents the base current of Q1 from being sucked away by the output node. The output potential is clamped at V_{BEon} and does not get discharged to ground, again because of D1.

Figure 4.47 shows a comparison of typical BiCMOS and CMOS gates designed in 1 μm technology. For small output loads, a simple CMOS gate is faster because it does not have an overhead of internal MOS transistors and bipolar output buffer that a BiCMOS gate has. When the gate is heav-

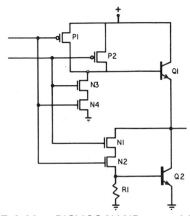

FIGURE 4.44 BiCMOS NAND gate with a resistive path that bleeds the base charge of the bipolar pull-down device

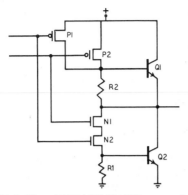

FIGURE 4.45   BiCMOS NAND gate with resistive paths that remove the base charge from the bipolar output pair

ily loaded, either by large fan-out or a long interconnection, BiCMOS gate becomes faster because of its small output impedance. The crossover point is roughly 0.1 pF—about 0.5 mm of wire. This figure illustrates the fundamental difference between the CMOS and BiCMOS gates, but it is also sometimes used to unfairly compare the two technologies. One would not drive a 3 pF wire with such a small CMOS gate; one would buffer up the output using a cascaded driver and bring down the CMOS delay near the BiCMOS delay. With the addition of a buffer, the base delay (delay with 0 pF load) of the CMOS gate will increase above BiCMOS, and CMOS will form a line almost

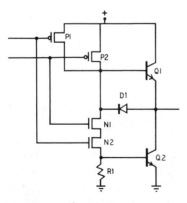

FIGURE 4.46   BiCMOS NAND gate with a diode that directionally discharges the ouput node

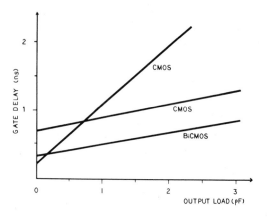

FIGURE 4.47   Comparison of CMOS and BiCMOS gate delays

parallel to BiCMOS and slightly above it. Consequently, one would select the small CMOS gate with a small base delay and a large slope at small output loads and switch to the buffered gate at large loads. This is easier to do at custom circuits but becomes harder at gate arrays. BiCMOS covers a larger output load range and simplifies gate array implementations.

4.11   Summary

As feature sizes are scaled down, interconnection capacitance per unit length remains approximately constant at 2 pF/cm. At the same time, die size increases, and the capacitance of on-chip wires becomes larger than the gate capacitances and dominates propagation delays.

In memory circuits, sense amplifiers and creative precharging schemes are used to reduce the delays associated with large-capacitance word lines, bit lines, and data output buses. SRAMs usually employ a differential pair of nMOS transistors coupled to bit lines via their gate capacitances to sense a small potential difference between the bit lines. This amplifier is very fast but also has a small gain and output swing. DRAMs make use of a cross-coupled nMOS pair tied to bit lines through their drains. In this way, the bit lines are directly tied to the output of the sense amplifier, and the memory cell is refreshed as the sense amplifier drives the data output bus. This is essential because the read operation destroys the data dynamically stored in the cell. ROMs and EPROMs use single-ended amplifiers.

Various nMOS and CMOS logic families, such as F^2L and DSL, limit the potential swing at the large-capacitance nodes to reduce gate propagation delays. Domino logic and clocked CVSL gates precharge the interconnections to V_{DD} and use fast nMOS transistors to discharge them selectively.

Driver and receiver circuits can be constructed that reduce the voltage swings at the high-capacitance long interconnections and precharge them to the high-gain region of the receiver circuit. These circuits, however, are sensitive to noise and variation in device parameters. Care must be taken to reduce the noise generation and to satisfy transistor parameter matching requirements.

BiCMOS circuits combine the advantages of bipolar and CMOS technologies. CMOS transistors are used to implement dense memory and logic building blocks with low power dissipation, and bipolar devices are used selectively in the critical paths and I/O circuitry to improve the performance. BiCMOS is becoming established as the technology of choice for high-speed SRAMs. Combining bipolar and CMOS devices, logic gates can be designed that implement the logic using CMOS and employ a push-pull bipolar pair to drive the output. These gates have no dc power dissipation and provide bipolar speeds. They can be used in gate arrays, signal processors, and microprocessors.

Depending on the speed, power, area, noise margin, and functional constraints of a specific application, a combination of the above circuits can be used to optimize the overall performance of an integrated circuit. The methods described in this chapter can be applied to reduce the delay of large-capacitance on-chip loads, as well as chip-to-chip interconnections.

5

INTERCONNECTION
RESISTANCE

Once considered to be electrically negligible, interconnections are becoming a major concern in high-performance integrated circuits because the *capacitance* and *resistance* of wires increase rapidly as chip size grows larger and minimum feature size is reduced. In the previous chapter, the effects of interconnection capacitance on the critical paths of memory and logic chips were analyzed, and numerous circuit techniques were presented that reduce these capacitive delays. In this chapter, wiring resistance and the resulting distributed RC delays will be analyzed, and various methods to reduce them will be investigated.

First, scaling of local and global interconnections is reviewed and consequences of distributed RC delays are described. Then delay equations for distributed RC lines are developed, and a model for interconnection delay is described that includes the effects of scaling transistor, interconnection, and chip dimensions. Based on this model, the delays of aluminum, tungsten silicide, and polysilicon lines are compared, and wiring delays in future VLSI circuits are projected. In addition, effects of increased line resistance on clock skew and IR voltage drops along the power lines are investigated.

Multilevel interconnections with wider and thicker lines at the upper levels are suggested as a technology solution. These-low resistance lines can be used for global communication and for power distribution, and densely packed wires at the lower levels can be used for local communication. Repeaters are

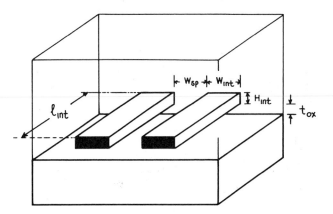

FIGURE 5.1   Interconnection parameters

investigated as a potential circuit method for reducing the distributed RC delay. When the line is broken into smaller buffered subsections, the parts are isolated from each other, and the distributed delays are reduced significantly. When the optimal cross-sectional dimensions and repeater configurations obtained from the model are employed, the distributed RC delays can improve by more than an order of magnitude.

5.1   Scaling of Interconnection Resistance

As device dimensions are miniaturized, interconnection dimensions illustrated in Fig. 5.1 must also be reduced to take full advantage of the scaling process.

A straightforward approach to satisfying the processing and layout constraints is to scale all cross-sectional interconnection dimensions (W_{int}, W_{sp}, H_{int}, t_{ox}) by the same factor as used for transistors (*ideal scaling*).[1] The devices and interconnections then require the same relative accuracy from lithography, pattern etching, and material deposition technologies, and the aspect ratios of interconnections and steps do not change as sizes are reduced. The effects of ideal scaling on local and long-distance interconnections are

---

[1] Throughout the book, the term *ideal scaling* is used to refer to simple scaling of transistors and interconnections where all horizontal and vertical dimensions are reduced by the same factor. No special measures are taken to minimize the adverse effects. Simple scaling is not always "ideal." In Chapter 2, it was shown that when transistors are scaled down following simple scaling rules, a number of performance and reliability problems surface, and the scaling approach needs to be modified. In this and later chapters, it will be shown that the same is true for interconnections, and improved schemes will be explored.

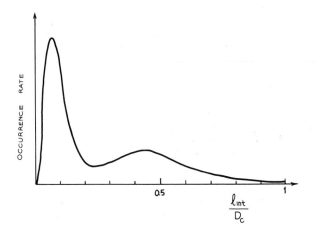

FIGURE 5.2   A typical distribution for interconnection lengths. The distribution has two peaks: one around $0.1D_c$ (representing local interconnections) and another around $0.5D_c$ (representing global interconnections). Here D_c is the length of a chip. For nonsquare dice, D_c can be approximated as $\sqrt{A_c}$; where A_c is the chip area.

listed in Tables 5.1 and 5.2 [5.1]–[5.5]. Although gate delay decreases by $1/S$ in ideal scaling, the response time of local interconnections remains the same. Even more troublesome, the delay time of global interconnections (such as those extending from corner to corner on a die) increases by $S^2 S_C^2$, where S_C is the scaling factor for chip size and accounts for the increase in die size from one generation of ICs to the next. Figure 5.2 illustrates the distribution of interconnection lengths in a logic chip [5.6]–[5.7]. A typical global interconnection has a length of $\sqrt{A_c}/2$ where A_c is the chip area. Rising wire resistance not only degrades the signal propagation delays but also makes designing a good power distribution network very difficult. As wiring resistance increases, IR voltage drops are formed along the power lines that degrade the noise margins of the circuits. Additionally, growing current density increases the electromigration failure rates.

Deviation from ideal scaling can be advantageous. Scaling interconnections and insulator thickness by factors smaller than S will lower R_{int} and C_{int}. Alternative approaches to minimizing propagation delays are presented in Tables 5.1 and 5.2. In *quasi-ideal scaling* of local interconnections, the horizontal dimensions are scaled by $1/S$ (as are transistors) to improve overall packing density by the factor S. On the other hand, to maintain a small RC time constant, the vertical dimensions are reduced only by $1/\sqrt{S}$, and as a result delay decreases by $1/\sqrt{S}$. In *constant-R scaling*, all cross-sectional dimensions are reduced only by $1/\sqrt{S}$, and consequently propagation delay is lowered by $1/S$. This degrades the packing density because the pitch of local

TABLE 5.1   Scaling of Local Interconnections

Parameter	Ideal Scaling	Quasi-Ideal Scaling	Constant-R Scaling	Generalized Scaling
Thickness (H_{int})	$1/S$	$1/\sqrt{S}$	$1/\sqrt{S}$	$1/S_H$
Width (W_{int})	$1/S$	$1/S$	$1/\sqrt{S}$	$1/S_W$
Separation (W_{sp})	$1/S$	$1/S$	$1/\sqrt{S}$	$1/S_{sp}$
Insulator thickness (t_{ox})	$1/S$	$1/\sqrt{S}$	$1/\sqrt{S}$	$1/S_{ox}$
Length (l_{loc})	$1/S$	$1/S$	$\approx 1/S$	$1/S$
Resistance (R_{int})	S	$\sqrt{S}$	1	$S_W S_H/S$
Capacitance to substrate	$1/S$	$1/S^{3/2}$	$\approx 1/S$	S_{ox}/SS_W
Capacitance between lines	$1/S$	$1/\sqrt{S}$	$\approx 1/S$	S_{sp}/SS_H
RC delay (T)	1	$\approx 1/\sqrt{S}$	$\approx 1/S$	$\approx S_W S_H/S^2$
Voltage drop (IR)	1	$1/\sqrt{S}$	$1/S$	$S_W S_H/S^2$
Current density (J)	S	$\sqrt{S}$	1	$S_W S_H/S$

S: Scaling factor for device dimensions.

wires is not reduced as much as transistor dimensions. As a result it is not very attractive. In terms of performance, local interconnections do not pose a serious problem because of their short lengths. Their share of the gate delay may be increasing, but since it is a small portion of the total delay, it is not very significant. The density considerations will be the major driving factor

TABLE 5.2   Scaling of Global Interconnections

Parameter	Ideal Scaling	Constant Dimension	Constant Delay	Generalized Scaling
Thickness (H_{int})	$1/S$	1	S_C	$1/S_H$
Width (W_{int})	$1/S$	1	S_C	$1/S_W$
Separation (W_{sp})	$1/S$	1	S_C	$1/S_{sp}$
Insulator thickness (t_{ox})	$1/S$	1	S_C	$1/S_{ox}$
Length (l_{int})	S_C	S_C	S_C	S_C
Resistance (R_{int})	$S^2 S_C$	S_C	$1/S_C$	$S_W S_H S_C$
Capacitance (C_{int})	S_C	S_C	S_C	$\approx S_C$
RC delay (T)	$S^2 S_C^2$	S_C^2	1	$S_W S_H S_C^2$

S: Scaling factor for device dimensions.
S_C: Scaling factor for chip size.

behind local interconnection scaling for a few more generations of CMOS.

Global interconnections are more difficult because their RC constitutes a nonnegligible portion of the total delay and also because of the additional burden introduced by the increasing chip size. In *constant-dimension scaling* of global interconnections, all cross-sectional dimensions are held constant, and propagation delay rises by S_C^2 because of the growth in chip size. In *constant-delay scaling*, cross-sectional interconnection dimensions are increased such that the improvement in the RC delay per unit length cancels the effect of larger chip size, and total delay remains constant. Scaling schemes in which horizontal interconnection dimensions are kept constant or increased, such as constant-R, constant-dimension, and constant-delay scalings, require multiple levels of wires to avoid an excessive increase in chip size. (This will be discussed in Section 5.4.1.)

As described in Chapter 4, when the ratios H_{int}/W_{int} and t_{ox}/W_{int} are increased, the two-dimensional fringing fields and capacitances between neighboring lines become important, and after a certain point larger aspect ratios yield no additional advantage. As a result the capacitance per unit length of the wires in a multilevel interconnection scheme approaches a lower limit of 1–2 pF/cm with SiO$_2$ as the dielectric material; an ultimate limit of 1 pF/cm is projected with improved insulators [5.8]. To obtain accurate results, two-dimensional fringing fields and the contributions of neighboring lines must be included in determining the capacitance of the interconnections; neglecting these effects can introduce substantial error. Capacitance formulas used in the following derivations have less than 5 percent error over a wide range of interconnection parameters, and their results are in agreement with computer-aided models of two-dimensional interconnection capacitance [5.8]–[5.18].

In the following calculations, conductivity is assumed to be independent of the cross-sectional dimensions of the wire, and this is a valid assumption if the mean-free path of the carriers is much smaller than interconnection thickness and width. The resistivity starts increasing when the interconnection width is scaled down to approximately 0.1 μm [5.19]–[5.20].

5.2   Distributed RC Lines

The responses of distributed and lumped RC networks to a unit step potential input are shown in Fig. 5.3. The response of a lumped RC network is

$$V_{OUT}(t) = 1 - e^{-t/RC}. \tag{5.1}$$

The response of the distributed network is harder to calculate. The frequency

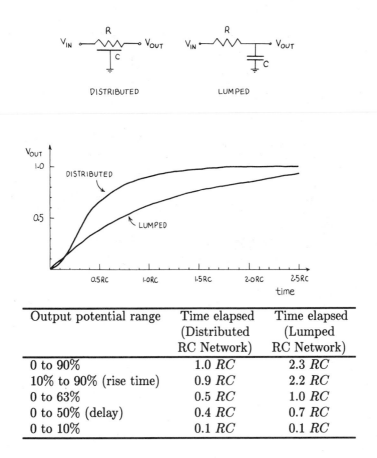

Output potential range	Time elapsed (Distributed RC Network)	Time elapsed (Lumped RC Network)
0 to 90%	1.0 RC	2.3 RC
10% to 90% (rise time)	0.9 RC	2.2 RC
0 to 63%	0.5 RC	1.0 RC
0 to 50% (delay)	0.4 RC	0.7 RC
0 to 10%	0.1 RC	0.1 RC

FIGURE 5.3   Step response of distributed and lumped RC networks. A potential step is applied at V_{IN}, and the resulting V_{OUT} is plotted. The time delays between commonly used reference points in the output potential are also tabulated.

domain solution is given by

$$V_{OUT}(s) = \frac{1}{s \cosh \sqrt{sRC}}. \qquad (5.2)$$

There is no closed-form time-domain transfer for this function, but solutions valid for $t \ll RC$ and $t \gg RC$ can be obtained by expanding the hyperbolic function

$$\cosh(x) = \frac{e^x + e^{-x}}{2}$$

$$= 1 + \frac{x^2}{2!} + \frac{x^4}{4!} + \frac{x^6}{6!} + \cdots \qquad (5.3)$$

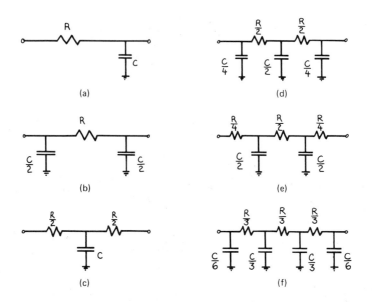

FIGURE 5.4   Lumped circuit approximations for distributed RC lines: **(a)** L model, **(b)** β model, **(c)** T model, **(d)** β2 model, **(e)** T2 model, **(f)** β3 model [5.29]

and using the following approximation

$$
\cosh(x) \approx
\begin{cases}
\dfrac{e^x}{2} & x \gg 1 \\[2ex]
1 + \dfrac{x^2}{2!} + \dfrac{x^4}{4!} & x \ll 1
\end{cases}
\tag{5.4}
$$

The approximation that is valid for $x \gg 1$ can be used to calculate the high-frequency portion of the output—the leading edge. And the approximation valid for $x \ll 1$ can be used for the low-frequency portion of the output—steady-state as the waveform levels off. The resulting time-domain solution is [5.21]

$$
V_{OUT}(t) =
\begin{cases}
erfc\sqrt{\dfrac{RC}{4t}} & t \ll RC \\[2ex]
1 - 1.366 e^{-2.5359 t/RC} + 0.366 e^{-9.4641 t/RC} & t \gg RC
\end{cases}
\tag{5.5}
$$

Using this equation, the values tabulated in Fig. 5.3 are calculated. Various approaches to modeling distributed RC delays of interconnections can be found in references [5.21]–[5.29].

The π and T models of distributed RC lines shown in Fig. 5.4 are suitable for quick calculations. They are also useful for including the effects of inter-

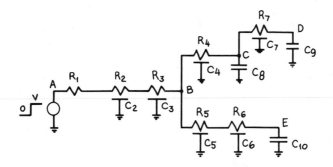

FIGURE 5.5   Typical RC tree for a CMOS VLSI circuit

connection resistance in circuit simulations because most CAD programs do not support distributed RC lines. The accuracy of the calculation increases with the number of sections. In a three-section $\pi 3$ approximation, the relative error is usually less than 3 percent [5.29].

When the RC tree forms branches, 50 percent delays for the branching nodes can be calculated independently and accumulated to obtain the delays at the outputs. This can be illustrated by an example of a typical RC tree shown in Fig. 5.5. The 50 percent delay from A to B is

$$
\begin{aligned}
T_{AB} = \; & R_1(C_2 + C_3 + C_4 + C_5 + C_6 + C_7 + C_8 + C_9 + C_{10}) \\
& + R_2\left(\frac{C_2}{2} + C_3 + C_4 + C_5 + C_6 + C_7 + C_8 + C_9 + C_{10}\right) \\
& + R_3\left(\frac{C_3}{2} + C_4 + C_5 + C_6 + C_7 + C_8 + C_9 + C_{10}\right).
\end{aligned}
\tag{5.6}
$$

The 50 percent delay from B to D is

$$
\begin{aligned}
T_{BD} = \; & R_4\left(\frac{C_4}{2} + C_7 + C_8 + C_9\right) \\
& + R_7\left(\frac{C_7}{2} + C_9\right).
\end{aligned}
\tag{5.7}
$$

The 50 percent delay from B to E is

$$
\begin{aligned}
T_{BE} = \; & R_5\left(\frac{C_5}{2} + C_6 + C_{10}\right) \\
& + R_6\left(\frac{C_6}{2} + C_{10}\right).
\end{aligned}
\tag{5.8}
$$

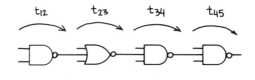

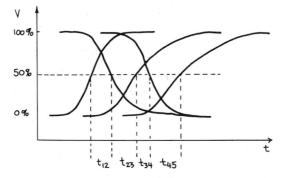

FIGURE 5.6   Delay calculation in logic circuits. Fifty percent delays of succeeding stages can be summed to obtain the overall delay.

The time delays from A to D and E can be obtained as

$$
\begin{aligned}
T_{AD} &= T_{AB} + T_{BD} \\
T_{AE} &= T_{AB} + T_{BE}.
\end{aligned}
\tag{5.9}
$$

5.3   Delay Modeling

In this section, gate delay calculations that take into account distributed RC delays of interconnections will be described. Effects of larger chip dimensions and reduced feature sizes will be evaluated, and projections for future circuit delays will be presented.

In static logic circuits, it is customary to use the delay from 50 percent point of the input waveform to 50 percent point of the output waveform because, as shown in Fig. 5.6, the 50 percent delays of multiple levels of logic that feed each other can be summed to obtain the overall delay.[2] In precharged

---

[2]This assumes that the switching points of all gates are at approximately $V_{DD}/2$. To obtain more accurate results, delay from the switching point of the driving gate to the switching point of the receiving gate should be calculated. In practical applications, such as CAD tools, the detailed approach may require a complicated CAD program and an excessive amount of data to describe the switching points of various subblocks.

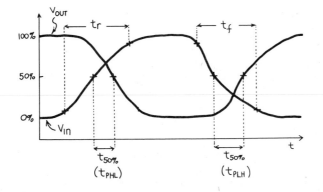

FIGURE 5.7   Definition of rise, fall, and delay times. Rise and fall times of the input are t_r and t_f. The propagation delays $t_{50\%}$ from the input to output are t_{PHL} (output transition from high to low) and t_{PLH} (output transition from low to high).

circuits and in certain critical memory paths, such as fall and rise times of a word line that controls the gate of pass transistors of memory cells, it may be more appropriate to use the 90 percent delay because a transistor switch is not fully turned off until the waveform reaches 90 percent of its final value. The voltage of interest determines the definition of delay. In the following analysis, both 90 percent and 50 percent delays will be calculated for networks with resistive interconnections. Traditionally the time period from the 50 percent point of the input to the 50 percent point of the output is referred to as *delay*. The time from the 10 percent point of a waveform to its 90 percent point is referred to as its *rise time* and from 90 percent to 10 percent as its *fall time*. It is common to use *rise time* to refer to both rise and fall times. This nomenclature will be used in the rest of the book, and any deviations will be pointed out. Figure 5.7 illustrates the definition of rise, fall, and delay times.

To obtain the total delay in a memory or logic circuit where the RC constants of interconnections may be significant, the on-resistance of the driver transistor R_{tr}, the resistance and capacitance of the wire R_{int}, C_{int}, and the input capacitance of the transistors that form the load C_L must be taken into account. A simple circuit model that includes all the significant parameters is illustrated in Fig. 5.8. As shown earlier, under step voltage excitation, the times (T) required for the output voltage of distributed and lumped RC networks to rise from 0 percent to 90 percent of their final values are $1.0RC$ and $2.3RC$, respectively. Accordingly a good approximation for the 90 percent delay of the circuit in Fig. 5.8 is obtained by combining the resistive and capacitive terms and weighting them by 1.0 when they are distributed and by

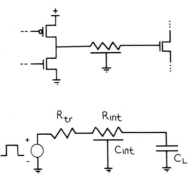

FIGURE 5.8   Interconnection delay model

2.3 when they are lumped,

$$
\begin{aligned}
T_{90\%} &= 1.0R_{int}C_{int} + 2.3(R_{tr}C_{int} + R_{tr}C_L + R_{int}C_L) \\
&\approx (2.3R_{tr} + R_{int})C_{int} \qquad \text{for} \quad C_L \ll C_{int}.
\end{aligned} \tag{5.10}
$$

This approximation is in agreement with Sakurai's [5.29] expression, which is reported to have less than 4 percent error over the entire range of parameters. A similar equation can be formed for the 50 percent delay

$$
\begin{aligned}
T_{50\%} &= 0.4R_{int}C_{int} + 0.7(R_{tr}C_{int} + R_{tr}C_L + R_{int}C_L) \\
&\approx (0.7R_{tr} + 0.4R_{int})C_{int} \qquad \text{for} \quad C_L \ll C_{int}.
\end{aligned} \tag{5.11}
$$

Let us investigate the components of this delay equation: transistor resistance, interconnection resistance, and interconnection capacitance.

Transistor Resistance

A first-order approximation for the on-resistance of an MOS transistor is

$$
R_{tr} \approx \frac{L/W}{\mu C_{gox}(V_{DD} - V_T)}, \tag{5.12}
$$

which remains approximately constant with ideal scaling. To obtain a more accurate value, the resistance can be averaged to obtain values for 50 and 90 percent delay calculations. In order to calculate an average transistor resistance, the delay equation described in Chapters 2 and 4 can be extended to an integral form

$$T = \frac{C_L \Delta V}{I}$$

$$= \int_{V_1}^{V_2} \frac{C_L dV}{I} \tag{5.13}$$

$$= R_{tr} C_L,$$

which yields the following definition for R_{tr}:

$$R_{tr} = \int_{V_1}^{V_2} \frac{dV}{I}. \tag{5.14}$$

Here V_1 and V_2 are the end points of the voltage range the transistor resistance is being averaged over, and I is a function of V as defined by the I_D-V_{DS} characteristics of a transistor with $V_{GS} = V_{DD}$. For example, to calculate the equivalent pull-down resistance of an nMOS transistor for a 50 percent delay calculation, the limits of integration are selected as $V_1 = V_{DD}$ and $V_2 = V_{DD}/2$. Transistor resistance as defined by Eq. 5.14 includes the multiplicative constants, such as 0.7 for 50 percent and 2.2 for 90 percent delays. Accordingly the delay is expressed as $R_{tr}C_L$ without a multiplicative constant. Of course, R_{tr} for 50 percent and 90 percent delays will be different because the limits of the integral will be different.

As shown in Fig. 5.9, the relative rise times of input and output waveforms can have significant effects on transistor resistance. Different input and output rise times are usually a result of different capacitive loadings at the input and output nodes. A transistor exhibits a smaller effective resistance if the input rise time is much smaller than the output rise time because the transistor is fully turned on during the entire period ($V_{GS} = V_{DD}$). Here the shape of the input is not important; it can be assumed to be a step. On the other hand, if the input waveform is slower than the output waveform, the transistor is only partially turned on while it is charging the output node. The effective resistance is larger than the value obtained from Eq. 5.14 because the assumption that $V_{GS} = V_{DS}$ is not true. Here the shape of the input is important; however, the main issue is the accuracy of the total delay. Consequently the most significant parameter is the time it takes for the input to reach the switching point of the gate. The slope of the input becomes secondary because it determines the rise time of the faster output waveform, which is a small portion of the overall delay. The complicated case occurs when the input and output rise times are comparable; then the shape of the input is significant, and it is difficult to calculate the delay accurately [5.30]. In most CAD tools, many circuit simulations under different scenarios are performed, and empirical curve-fitting methods are used to obtain accurate resistance values for a given technology. These refinements are important when the equations are used to calculate the absolute delays in order to determine the critical paths of

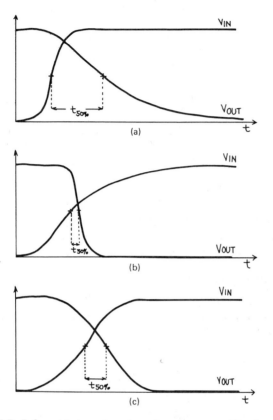

FIGURE 5.9   Various input and output combinations: (a) fast input, slow output, (b) slow input, fast output, (c) comparable input and output rise times

specific chips. In this chapter, only a comparative delay analysis is performed, and representative transistor resistances of 1 kΩ and 10 kΩ are used in the calculations.

Interconnection Resistance

Interconnection resistance is

$$R_{int} = \rho \frac{l_{int}}{W_{int} H_{int}}, \tag{5.15}$$

where ρ is the resistivity, and W_{int}, H_{int}, and l_{int} are the width, thickness, and length of the interconnection. As shown in Table 5.2, wire resistance increases as $S^2 S_C$ when ideal scaling is applied.

Interconnection Capacitance

The capacitance of the center one of three adjacent lines above a ground plane is expressed as

$$\frac{C_{int}}{\epsilon_{ox}} = 1.15 \left(\frac{W_{int}}{t_{ox}}\right) + 2.80 \left(\frac{H_{int}}{t_{ox}}\right)^{0.222}$$

$$+ \left[0.06 \left(\frac{W_{int}}{t_{ox}}\right) + 1.66 \left(\frac{H_{int}}{t_{ox}}\right) - 0.14 \left(\frac{H_{int}}{t_{ox}}\right)^{0.222}\right] \left(\frac{t_{ox}}{W_{sp}}\right)^{1.34}$$

$$(5.16)$$

with an error of less than 10 percent over a wide range of H/t_{ox}, W_{int}/t_{ox}, and W_{sp}/t_{ox} ratios [5.10]. Here C_{int} is the capacitance per unit length, and t_{ox} and ϵ_{ox} are the thickness and dielectric constant, respectively, of the insulator. Various analytical models for two-dimensional interconnection capacitance calculations can be found in references [5.10]–[5.16].

As an example of a global wire, a 1 cm aluminum line in a current commercial microprocessor may be 2 μm wide and 0.5 μm thick and may have a 1.0 μm thick oxide beneath it. Its resistance is 300 Ω, and its capacitance is 2.0 pF. As a result it has a distributed RC constant of 0.6 nsec, which produces a 90 percent delay of 0.6 nsec and a 50 percent delay of 0.24 nsec. These are smaller than but still in the same order of magnitude as a typical gate delay of 1 to 2 nsec. In larger submicron CMOS chips, the RC delays of wires can be even more significant. For example, a 4 cm long aluminum interconnection,[3] which is 1 μm wide and 0.3 μm thick, and has a 0.5 μm of interlevel insulator, produces a resistance of 4000 Ω and a capacitance of 8.0 pF. Its distributed RC constant is 32 nsec, which is much larger than the subnanosecond gate delays of submicron technology. Obviously measures must be taken to avoid such large distributed RC delays. Interconnection resistance becomes even more important when a long wire drives multiple large capacitance loads at a far corner of the chip. Then the overall delay just due to wire resistance times the wire and load capacitance $T_{50\%} = R_{int}(0.4C_{int} + 0.7C_L)$ can be many nanoseconds, even at 2 μm design rules. This delay cannot be reduced by making the driver stronger. Overall delay can be somewhat reduced by increasing wire width, but this stops being beneficial when the RC constant of the wires becomes dominant. Then as the wire is made wider, its resistance is reduced, but its capacitance also increases by the same amount, keeping the total delay constant. Buffers or repeaters that divide the line into smaller subsections may be required. This will be described in the following sections.

Table 5.3 lists the sheet resistances of aluminum, tungsten silicide, and polysilicon lines as a function of the feature size. Interconnection thickness

---

[3]Interconnections with such lengths inevitably exist on "superchips" that are being developed with 1.4 × 1.8 inch dice [5.31].

TABLE 5.3   Sheet Resistance of Interconnections as a Function of the Feature Size

$$R_{\square int} = \frac{\rho_{int}}{H_{int}}, \qquad H_{int} = \frac{W_{int}}{3}$$

W_{int}	Sheet Resistance $R_{\square int}$ $(\Omega/\square)$		
	Aluminum $\rho = 3\mu\Omega\text{-cm}$	WSi$_2$ $\rho = 130\mu\Omega\text{-cm}$	Polysilicon $\rho = 1000\mu\Omega\text{-cm}$
3.00 μm	0.03	1	10
2.00 μm	0.04	2	15
1.00 μm	0.09	4	30
0.75 μm	0.12	5	40
0.50 μm	0.18	8	60
0.25 μm	0.36	16	120

H_{int} is assumed to be one-third the interconnection width W_{int}. Resistivities of tungsten silicide and polysilicon are adjusted to obtain an effective value. Silicides are usually formed on higher-resistivity regions and do not penetrate down the entire thickness of the line. Polysilicon resistivity depends on the doping profile and can give misleading results when plugged into the simple sheet resistivity formula. Consequently, the values for tungsten silicide and polysilicon in Table 5.3 are meant to demonstrate the basic trends and are not as accurate as the aluminum sheet resistivities. At small feature sizes, even the aluminum lines become highly resistive. Table 5.4 shows the critical interconnection lengths for aluminum, tungsten-silicide, and polysilicon wires, where critical length is defined as the point the interconnection resistance goes above 500 Ω. Critical length is inversely proportional to the square of the design rule. Even for aluminum wires, the critical length can become as small as 3 mm at 0.75 μm design rules.

Based on Eq. 5.10, Fig. 5.10 plots the 90 percent delay for various chip dimensions as a function of minimum feature size and for two driver resistances (1 kΩ and 10 kΩ) that correspond to a W/L of 10 and 1 in an nMOS transistor. It can be seen that delay becomes dominated by $R_{int}C_{int}$ as dimensions are scaled down. With the existing aspect ratios and design rules, the capacitive term is mainly determined by the capacitance between the interconnection and substrate ($\approx \epsilon_{ox}l_{int}\frac{W_{int}}{t_{ox}}$). For short wire lengths, the resistive term is dominated by R_{tr}, and for longer wires, it is dominated by R_{int}. A striking feature of all the curves in Fig. 5.10 is the existence of a knee or a break point beyond which the delay increases rapidly. To ensure operation below the knee,

TABLE 5.4   Critical Interconnection Length as a Function
of the Feature Size

$$R_{int,crit} = \rho_{int} \frac{l_{int,crit}}{W_{int}H_{int}} = 500 \ \Omega, \qquad H_{int} = \frac{W_{int}}{3}$$

W_{int}	Critical wire length $l_{int,crit}$ (mm)		
	Aluminum $\rho = 3\mu\Omega$-cm	WSi$_2$ $\rho = 130\mu\Omega$-cm	Polysilicon $\rho = 1000\mu\Omega$-cm
3.00 μm	50	1.2	0.15
2.00 μm	22	0.5	0.07
1.00 μm	6	0.13	0.02
0.75 μm	3	0.07	0.01
0.50 μm	1.4	0.03	0.004
0.25 μm	0.4	0.01	0.001

the design requirement for total interconnection resistance is

$$R_{int} = \rho \frac{l_{int}}{W_{int}H_{int}} < R_{tr}. \qquad (5.17)$$

Table 5.4 lists critical interconnection lengths for $R_{tr} = 500 \ \Omega$. The critical
wire lengths for bipolar and GaAs ICs are even shorter than in CMOS chips
because bipolar and GaAs transistors have smaller impedances (smaller R_{tr}).

The projected long-distance interconnection propagation delays of future
integrated circuits with "ideally" scaled metal layers are plotted in Fig. 5.11 .
Here the minimum feature size F is assumed to drop by a factor of two every
six years. In addition, the chip size and, accordingly, global interconnection
lengths l_{int} are assumed to double every eight years as

$$\begin{aligned} F &= 2.0 \times 2^{-\frac{t-1983}{6}} \quad \mu m \\ l_{int} &= 0.35 \times 2^{\frac{t-1983}{8}} \quad cm. \end{aligned} \qquad (5.18)$$

It can be seen in Fig. 5.11 that in the early 1970s, all interconnections were
pure capacitive loads, and time delay was determined by the size of the driver
and the length of the interconnection. As chips grew larger and the minimum
feature size shrank, R_{int} gained importance. Currently aluminum lines are beginning to exhibit RC behavior, whereas the resistance of polysilicon lines has
been significant for some time. If the conventional methods for interconnecting transistors are not improved, longer interconnection lengths and smaller
widths and thicknesses will result in unacceptably long propagation delays.

As interconnection resistance gains more importance, the design trade-offs
also change. In the past the channel of the transistor dominated the resistive
component of delay, and the interconnections were virtually perfect conductors

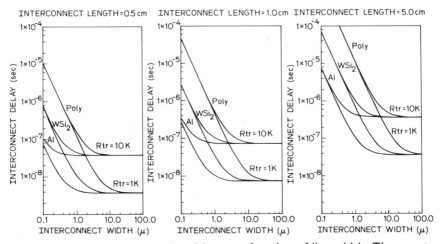

FIGURE 5.10   Interconnection delay as a function of line width. Three materials (Al, WSi$_2$, polysilicon), two driver resistances 1 kΩ (W/L=10) and 10 kΩ (W/L=1), and three lengths (0.5, 1.0, 5.0 cm) are considered. It is assumed that $W_{sp} = W_{int}$, $H_{int} = W_{int} = 3$, $t_{ox} = W_{int}/5$, and ideal scaling is applied; also, $\rho_{Al} = 3 \ \mu\Omega$-cm, $\rho_{WSi_2} = 30 \ \mu\Omega$-cm, and $\rho_{PolySi} = 500 \ \mu\Omega$-cm.

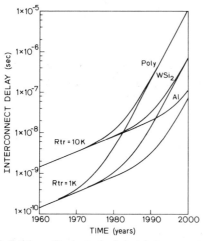

FIGURE 5.11   Projected global interconnection propagation delays in future integrated circuits with "ideal" scaling. Equation 5.18 is used to determine F and l_{int}. It is assumed that $W_{int} = \frac{3}{2}F$, $W_{sp} = \frac{3}{2}F$, $H_{int} = \frac{1}{4}F$, $t_{ox} = \frac{1}{6}F$, and ideal scaling is applied. Here F is the minimum feature size.

with respect to the channel. As a result the parasitic capacitances of all components (gate, diffusion, polysilicon, aluminum) have been significant, and the major challenges were to minimize these capacitances and to drive large capacitive loads. With smaller feature sizes and greater chip dimensions, the parasitic resistances of interconnections are becoming comparable to channel resistance, and the present design challenges are to reduce parasitic resistances as well as capacitances and to drive large RC loads in addition to large capacitive loads.

5.4   Methods for Improving Interconnection RC Delay

Two approaches to shortening distributed RC delays will be described. The first is to reduce interconnection resistance by using only aluminum lines for global communication and by forming a hierarchy of wiring levels with thicker and wider lines in the upper levels. The second is using repeaters that divide the interconnection into smaller subsections [5.4]–[5.5].

5.4.1   Multilayer Interconnections

It is not possible to interconnect hundreds of thousands transistors on VLSI chips with only one level of aluminum without polysilicon or diffusion wires; moreover even aluminum lines will introduce excessive delay when chips get larger and minimum feature size is reduced. Multilayers of interconnections are partial solutions to both problems. First, layers of aluminum in the x- and y-directions interconnected through vias between the levels enable global communication without the need for polysilicon or diffusion wires. Second, based on Eq. 5.17, the cross-sectional dimensions of the upper layers can be optimized to reduce their RC constants. The local interconnections can use the scaled-down first-level metal, and the global interconnections can use the upper levels with wider and thicker lines that yield shorter propagation delays. Figure 5.12 illustrates a possible scheme. Because much of the chip area is occupied by interconnections, multilayers of metal can also reduce chip size and further improve the time delay because the average interconnection length is inversely proportional to the number of levels [5.32]. Wider and thicker wires at the upper levels also provide low-resistance power distribution lines. In many cases, it is desirable to reserve the top layer entirely for power distribution. Currently almost all commercial CMOS processes have at least two levels of aluminum, and the trend is toward even larger numbers of interconnection levels.

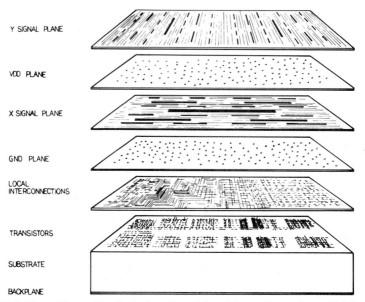

Y SIGNAL PLANE

VDD PLANE

X SIGNAL PLANE

GND PLANE

LOCAL
INTERCONNECTIONS

TRANSISTORS

SUBSTRATE

BACKPLANE

FIGURE 5.12 Hierarchically designed multilevel interconnections. Thicker and wider lines at the upper layers provide low-resistivity interconnections.

5.4.2 Repeaters

When the resistance of the interconnection is comparable to or larger than the on-resistance of the driver, propagation delay increases proportionally to the square of the interconnection length because both capacitance and resistance increase linearly with length. The use of repeaters makes time delay linear with length by dividing the interconnection into smaller subsections.

The principle behind the repeaters can be best illustrated by an example. Let us take an interconnection that is 5 units long and has 5 units of resistance and 5 units of capacitance. Its RC constant is $5 \times 5 = 25$ units. When this wire is divided into five equal sections by buffers, its cumulative RC constant is reduced to $1 + 1 + 1 + 1 + 1 = 5$ units. Of course, the additional delay due to buffers should be taken into account, as will be shown in the following calculations. The repeaters require additional area but do not increase the power dissipation significantly[4] when implemented in CMOS technology.

Based on the model developed in the previous section, the 50 percent propagation delay of an interconnection with k minimum-size inverters as

---

[4]The additional capacitance due to CMOS buffers increases the power dissipation somewhat; however, the cumulative capacitance of the buffers is smaller than the line capacitance.

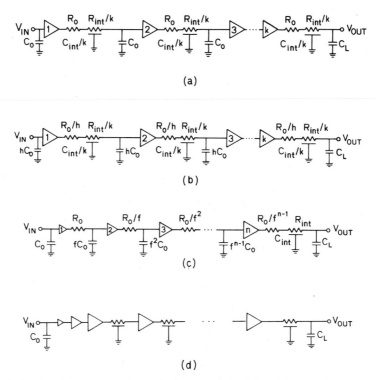

FIGURE 5.13   Interconnection driving methods: (a) minimum-size repeaters, (b) optimal repeaters, (c) cascaded drivers, (d) optimal repeaters with a cascaded first stage

repeaters (Fig. 5.13(a)) can be expressed as

$$T_{50\%} = k\left[0.7R_o\left(\frac{C_{int}}{k} + C_o\right) + \frac{R_{int}}{k}\left(0.4\frac{C_{int}}{k} + 0.7C_o\right)\right] \qquad (5.19)$$

where C_o and R_o are the input capacitance and output resistance of the minimum-size inverter. By setting $dT/dk = 0$,

$$0.4\frac{R_{int}C_{int}}{k^2} = 0.7R_oC_o. \qquad (5.20)$$

To achieve the shortest total delay, therefore, the delay of the segments connected by the repeaters should be equal to that of a repeater. The number of repeaters is

$$k = \sqrt{\frac{0.4R_{int}C_{int}}{0.7R_oC_o}}. \qquad (5.21)$$

Total delay is then expressed as

$$T_{50\%} = 0.7R_oC_{int} + 0.7R_oC_o\sqrt{\frac{0.4R_{int}C_{int}}{0.7R_oC_o}} + 0.4R_{int}C_{int}\sqrt{\frac{0.7R_oC_o}{0.4R_{int}C_{int}}}$$
$$+0.7R_{int}C_o$$

$$= 0.7R_oC_{int} + 1.1\sqrt{R_oC_oR_{int}C_{int}} + 0.7R_{int}C_o.$$

$$(5.22)$$

Let us compare this with the delay in Eq. 5.11. Here R_o and C_o correspond to R_{tr} and C_L. Both equations have $0.7R_oC_{int}$ and $0.7R_{int}C_o$; these terms remain unchanged. The delay without repeaters has separate $0.7R_oC_o$ and $0.4R_{int}C_{int}$ terms. The delay with the repeaters, on the other hand, has twice the geometric mean of these two terms: $1.1\sqrt{R_oC_oR_{int}C_{int}}$. Because the geometric mean is always smaller than arithmetic mean,[5] repeaters improve the delay.

The value of k in Eq. 5.21 must be at least two in order for this method to reduce the overall delay (the first "repeater" is the driver). Consequently, for the minimum-size repeaters to be useful, the RC constant of the wire ($R_{int}C_{int}$) must be at least seven times the delay of a minimum-size buffer (R_oC_o).

Propagation time can be further improved by increasing the size of the repeaters because the current drive capability of a buffer is directly proportional to the W/L ratios of its devices. When the W/L ratios of the transistors are increased by a factor h, the output resistance and input capacitance become R_o/h and hC_o (Fig. 5.13b). The delay expression then takes the form of

$$T_{50\%} = k\left[0.7\frac{R_o}{h}\left(\frac{C_{int}}{k} + hC_o\right) + \frac{R_{int}}{k}\left(0.4\frac{C_{int}}{k} + 0.7hC_o\right)\right]. \quad (5.23)$$

By setting dT/dk and dT/dh to zero, optimal values for k and h are obtained as

$$k = \sqrt{\frac{0.4R_{int}C_{int}}{0.7R_oC_o}}$$

$$(5.24)$$

$$h = \sqrt{\frac{R_oC_{int}}{R_{int}C_o}},$$

and resulting delay expression becomes

$$T_{50\%} = 2.5\sqrt{R_oC_oR_{int}C_{int}}, \quad (5.25)$$

---

[5]Geometric mean is always smaller than arithmetic mean because

$$\sqrt{xy} < (x+y)/2 \quad \Leftrightarrow 4xy < x^2 + 2xy + y^2$$
$$\Leftrightarrow 0 < x^2 - 2xy + y^2$$
$$\Leftrightarrow 0 < (x-y)^2.$$

which is proportional to the geometric mean of the repeater and interconnection delays. Note that the delay in Eq. 5.25 is linearly proportional to wire length. As expected, this is shorter than the delay achieved by minimum-size repeaters in Eq. 5.22 because $R_o C_{int} \gg R_{int} C_o$. For global interconnections, C_{int} is in the order of picofarads and C_o is on the order of femtofarads $(C_{int} \gg C_o)$. The on-resistance of a minimum-size transistor R_o is approximately 10 kΩ, which is usually larger than R_{int} $(R_o > R_{int})$. Therefore $R_o C_{int} \gg R_{int} C_o$. For the repeaters to be useful, the RC constant of the wire $(R_{int} C_{int})$ must still be at least seven times the delay of a minimum-size buffer $(R_o C_o)$.

The effectiveness of the inverters in reducing the distributed RC delays can be illustrated by returning to the numerical example discussed in Section 5.2. First to be considered is the aluminum global line in a commercial microprocessor. The line is 1 cm long, 2 μm wide, and 0.5 μm thick and has a 1.0 μm insulator beneath it. The resistance of this line is 300 Ω, and its capacitance is 2.0 pF. As a result it has a distributed RC constant of 0.6 nsec. Using 2 μm CMOS device parameters $(R_o = 10,000~\Omega,~C_o = 20~\text{fF},~R_o C_o = 0.2~\text{nsec})$, the optimal number of sections k is calculated to be fewer than two, and as a result repeaters cannot reduce delay under these conditions. For repeaters to be useful, buffer delay should be less than one-seventh of the interconnection RC delay. The second example (a 1 μm wide, 0.3 μm thick, and 4 cm long aluminum interconnection with 0.5 μm thick insulator) has a resistance of 4000 Ω and a capacitance of 8.0 pF. Its distributed RC constant is 32 nsec. This produces a 90 percent delay of 32 nsec and a 50 percent delay of 13 nsec, which are much larger than the subnanosecond gate delays of submicron technology. Using 0.7 μm CMOS device parameters $(R_o = 8000~\Omega,~C_o = 10~fF,~R_o C_o = 0.08~nsec)$, k and h are calculated to be 15 and 40, respectively. Total 50 percent delay is 4 $nsec$—one-third the 50 percent delay of the distributed RC line. The improvement would be a factor of five with faster quarter-micron CMOS repeaters.

5.4.3   Cascaded Drivers

Just as repeaters are suitable for RC loads, cascaded drivers described in Chapter 4 are very effective for driving large capacitive loads. Instead of a single minimum-size driver, a chain of drivers can be used that increase in size until the last device is large enough to drive the load as illustrated in Fig. 5.14 [5.33]–[5.36]. This is necessary because if the load is driven by a large transistor, which in turn is driven by a small device, the turn-on time of the large transistor dominates overall delay. Optimal delay is obtained with a sequence of drivers that increase gradually in size. When applied to interconnections (Fig. 5.13(c)), this method optimizes the sum of the delay caused by charging the input capacitances of the cascaded drivers and the

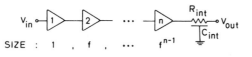

SIZE :   1   ,   f   ,   ... f^{n-1}

DELAY: fR_oC_o, fR_oC_o,   ... , fR_oC_o

FIGURE 5.14   Cascaded drivers

interconnection propagation delay. Total delay is then expressed as

$$T_{50\%} = 0.7(n-1)fR_oC_o + \left(\frac{0.7R_o}{f^{n-1}} + 0.4R_{int}\right)C_{int} + \left(\frac{0.7R_o}{f^{n-1}} + 0.7R_{int}\right)C_L$$

$$(5.26)$$

and, by setting dT/dn and dT/df to zero,

$$f = e$$

$$n = \ln\left(\frac{C_{int} + C_L}{C_o}\right)$$

$$(5.27)$$

$$T_{50\%} = 0.7eR_oC_o\ln\left(\frac{C_{int} + C_L}{C_o}\right) + 0.4R_{int}C_{int} + 0.7R_{int}C_L$$

where e is the base of the natural logarithm. Because total time delay increases slowly with larger f, in practice a size ratio f greater than e (for example, 4–6) is used to save chip area with little increase in propagation time. Note that cascaded drivers minimize the $0.7R_{tr}C_{int}$ and $0.7R_{tr}C_L$ terms in Eq. 5.11, but the $0.4R_{int}C_{int}$ and $0.7R_{int}C_L$ terms remain unchanged. As a result this method is useful when R_{int} is small and and R_{tr} is dominant; however, it is not adequate when R_{int} is comparable to or larger than R_{tr}. As illustrated in Fig. 5.13(d), the first stage of the optimal-size repeaters must be a cascaded driver to lower the input capacitance of the structure and to optimize the total propagation delay when it is driven by a minimum-size transistor.

Figures 5.15–5.18 compare the driving schemes. In Figs. 5.15–5.17, propagation delay is plotted for four driving mechanisms as a function of interconnection length and for three widths (0.25, 0.5, 1.0 μm). It is assumed that $W_{sp} = W_{int}$, $H_{int} = W_{int}/3$, $t_{ox} = W_{int}/5$, $t_{gox} = W_{int}/75$, and the gate length is $2W_{int}/3$. As a result, $R_o = 10$ kΩ, $C_o = 1.17 \times 10^{-15}\frac{F}{\mu m} \times W_{int}$, $C_{int} = 3.0 \times 10^{-12}\frac{F}{cm} \times l_{int}$, and $R_{int} = 3.0 \times \rho l_{int}/W_{int}^2$.

It can be seen that when repeaters are used, delay increases linearly with l_{int}. Cascaded drivers are preferred when l_{int} is short because interconnection resistance is low for small l_{int}; however, for longer l_{int}, delay increases rapidly because cascaded drivers do not improve the $R_{int}C_{int}$ term. Under some conditions, repeaters are not effective unless their sizes are optimized.

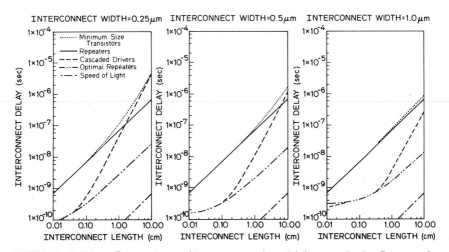

FIGURE 5.15   Comparison of interconnection driving methods. Propagation delay is plotted for four driving mechanisms as a function of interconnection length and for three widths (0.25, 0.5, 1.0 μm). The interconnection material is aluminum.

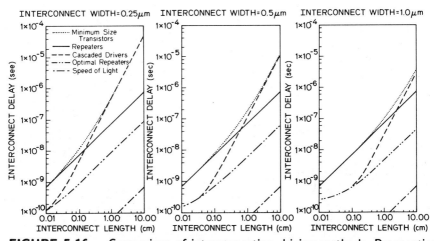

FIGURE 5.16   Comparison of interconnection driving methods. Propagation delay is plotted for four driving mechanisms as a function of interconnection length and for three widths (0.25, 0.5, 1.0 μm). Interconnection material is tungsten silicide.

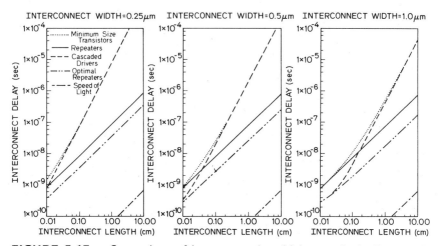

FIGURE 5.17 Comparison of interconnection driving methods. Propagation delay is plotted for four driving mechanisms as a function of interconnection length and for three widths (0.25, 0.5, 1.0 μm). Interconnection material is polysilicon.

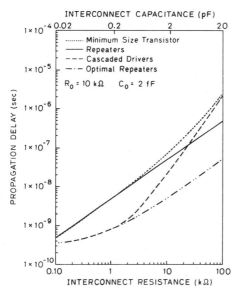

FIGURE 5.18 Comparison of interconnection driving methods as a function of R_{int} and C_{int}

Optimal size repeaters with a cascaded first stage obtain the shortest delay under all conditions, at times reducing the overall delay by more than an order of magnitude. The ultimate lower limit is the propagation delay of a lossless transmission line shown at the lower right-hand corner of the plots. Here, transmission speed is

$$v = \frac{c_0}{\sqrt{\epsilon_r}}, \tag{5.28}$$

where c_0 is the speed of light in vacuum and ϵ_r is the dielectric constant of the medium in which the line is buried. Transmission lines will be analyzed in the next chapter, and an upper limit for interconnection resistance will be established to ensure lossless transmission line behavior.

5.5   Additional Considerations

Following are brief descriptions of second-order effects that may influence the practicality of the methods described in the previous sections.

Area and Yield Penalty

All of the above methods introduce area penalties to improve performance, and the models reported here do not include the possible increase in the average interconnection length and resulting complications. In addition to interconnection delay, the designer must consider the total area required to implement the circuit and the yield degradation caused by process complexity introduced by some of the methods. For example, the repeater circuits will require some additional area. It is also challenging to obtain high yields with a process that has many levels of interconnections with wider and thicker lines at the upper layers. When the number of interconnection levels is increased, some period of time is required for such a process to mature and reach the yield levels of earlier and simpler technologies.

Source/Drain and Contact Resistances

As described in Chapter 2, source/drain and contact resistances of MOS transistors become major concerns as minimum feature size and junction depths are scaled down. These problems can be partially resolved by silicidation of the source/drain areas and by new contact materials and methods [5.37]–[5.38]. These parasitic resistances can also be included in the delay model described in this chapter by combining them with the on-resistance of the transistor.

Step Coverage Resistance

Interconnections are thinner at the steps because of step coverage problems described in Chapter 2. This reduction of the cross-sectional areas at the steps causes the interconnection resistance to be larger than the value obtained in Eq. 5.15. The amount of increase depends on the specific metal deposition system. The delay model can be adjusted to include this effect by experimentally determining a factor by which R_{int} can be multiplied to account for the reduced cross-sectional area at the steps.

5.6   Clock Skew

In the model developed in this chapter, a single transistor drives one interconnection line; however, a single transistor must often drive many signal lines. A dramatic example is the clock driver and clock lines (Fig. 5.19). As chip size increases, the network of clock lines constitutes a large RC load that may produce excessive clock skew and set an upper limit on clock frequency. Clock frequency determines the overall performance of the chip (cycle time, access time, instructions per second) because, despite its fast switching speed, a transistor must normally wait until the next clock cycle before it can change its state. In MSI or LSI circuits, maximum clock frequency is determined by the switching speed of the gates. In advanced VLSI circuits, clock skew caused by interconnection delay, if not controlled, may constitute a significant portion of the total cycle time. Ideally system level clock skew should be less than 5–10 percent of the total cycle time. Radical changes in design methods (such as self-timed systems) have been proposed as possible solutions to this problem [5.39]; however, it would be difficult to address the design complexity of VLSI circuits without a clock signal that serves as a convenient sequence and time reference [5.40]. To enhance system performance, an intermediate approach may be a highly partitioned IC with a collection of clock signals

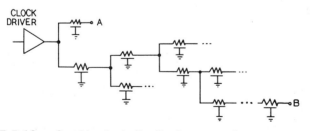

FIGURE 5.19   On-chip clock distribution network

(fastest clock for the smallest partition, slowest clock for overall chip communication). Chapter 8 focuses on high-speed clock distribution and analyzes the clock skew and effects of distributed RC delays among other factors.

5.7   Power Distribution

On-chip power distribution becomes more challenging as integration density increases and devices get faster. When minimum feature size is scaled down, resistance of the lines goes up. In addition, the total current that must be supported by these lines rises due to increased circuit speeds and increased device counts. This results in two problems: higher IR voltage drops along the power lines and increased electromigration rate due to large current densities. Since the supply voltage is scaled down, the magnitude of the voltage drops that can be tolerated are also smaller, which makes the situation worse. Methods that improve the electromigration immunity of the materials were described in Chapter 2.

The IR voltage drops along the power lines degrade the circuit noise margins. In Fig. 5.20, the ground level at point 2 is higher than the level at point 1 by an amount IR. This difference will be reflected at the output of the inverter I2; the low value of V_X will be IR above the ground level at point 1. If I2 feeds a precharged circuit located near point 1 and V_X receives a large glitch due to this IR drop, the precharged node can be inadvertently discharged. This also reduces the noise margin of a static circuit such as inverter 1. The glitches also can cause switching problems even in nonprecharged circuits because I1

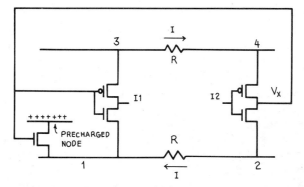

FIGURE 5.20   Effects of IR voltage drops along the power lines. The ground level at point 2 is higher than the level at point 1 by IR. This can cause precharged circuits to fail and reduces the noise margins of static circuits.

TABLE 5.5   Scaling of IR Voltage Drops at Power Lines

Parameter	Ideal Scaling	Improved Scaling
Total chip current	$S^2 S_C^2$	$S^2 S_C^2$
Conductor thickness (H_{int})	$1/S$	S
Sheet resistance ($R_{int\square}$)	S	$1/S$
Number of power planes	1	S
Number of power connections	1	$S S_C^2$
Effective resistance	S	$1/S^3 S_C^2$
IR voltage drop	$S^3 S_C^2$	$1/S$
Signal-to-noise ratio	$1/S^4 S_C^2$	1

and I2 can be part of a clock generator, and a glitch in the clock signal can cause latches to switch inadvertently or lose their data. Similarly the positive power supply level at point 4 is below the level at point 3 by IR, and the high value of V_X is $(V_{DD} - IR)$.

As will be shown in Chapter 7, the total current consumption of the chip grows by a factor between $S S_C^2$ to $S^2 S_C^2$. In addition, the sheet resistance of the power lines rises by S if the aluminum thickness is reduced. This causes the IR voltage drops to increase by $S^3 S_C^2$ if no structural changes are introduced to control them. Because the power supply and signal levels are scaled down, signal-to-noise ratio is degraded by an additional factor of $1/S$ (Table 5.5).

Such large increases in the IR drops are unacceptable. To curb the growth of the IR voltage drops, the topmost and thickest metal level can be used solely for power distribution. As illustrated in Fig. 5.21, interweaved comb structures laid out in this low-resistivity layer can supply a power distribution network with minimal IR drops [5.42]–[5.43].

Table 5.5 also summarizes an improved power distribution scheme. This is more acceptable than the straightforward "ideal scaling," which keeps the basic chip structure unchanged and introduces no major changes to reduce the IR drops. With improved scaling, the aluminum thickness is increased, and its sheet resistance goes down. The effective resistance of the power planes can be further reduced by adding more power planes and by increasing the number of power connections between the die and the package. By combining all these techniques (thicker aluminum layers, dedicated power planes with comblike structures, increased number of power planes, and more power leads between the dice and the package), a satisfactory power distribution tree can be constructed. In addition to IR voltage drops, LdI/dt inductive drops must also be minimized. Chapter 7 addresses inductive problems in power distribution.

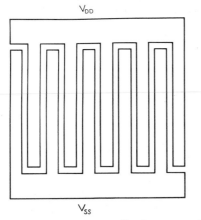

FIGURE 5.21   Power distribution network with low resistance

5.8   Optimal Chip Size

By using the model for interconnection propagation delay, optimal chip size for a given multichip package can be determined. The on-chip wire delay is usually kept much smaller than the chip-to-chip delay. As a result the best performance will be achieved when the interconnection delay on the chip is roughly equal to 10 percent of the delay of the chip-to-chip interconnections. If the chips are made larger, the system will become slower. If the chips are made smaller, the complexity of the package will increase with no improvement, and possible degradation, in speed. Assuming driving schemes of optimized repeaters for on-chip lines and cascaded drivers for package wires, chip and package level 50 percent delays can be plugged into the following equation to obtain

$$T_{chip} = 0.1 T_{package}$$

$$2.5\sqrt{R_o R_{int} C_o C_{int}} = 0.07 e R_o C_o \ln\left(\frac{C_{INT} + C_L}{C_o}\right) + 0.04 R_{INT} C_{INT}$$

$$+ 0.07 R_{INT} C_L. \tag{5.29}$$

Because the package level R_{INT} is usually negligible, Eq. 5.29 can be simplified to yield

$$\sqrt{R_{int}C_{int}} = 0.08\sqrt{R_oC_o}\ln\left(\frac{C_{INT}+C_L}{C_o}\right)$$

$$l_{int} = 0.08\sqrt{\frac{R_oC_o}{R_{int}C_{int}}}\ln\left(\frac{C_{INT}l_{INT}+C_L}{C_o}\right),$$

(5.30)

where R_{int}, C_{int}, and C_{INT} are the resistance and capacitances per unit length, and the lower- and upper-case subscripts refer to the chip and package levels, respectively. Substituting $l_{int} = \sqrt{A}/2$, where A is the area of the die or package [5.41], the optimal chip size is obtained as a function of package size and characteristic interconnection parameters at the chip and package levels,

$$\sqrt{A_{chip}} = 0.16\sqrt{\frac{R_oC_o}{R_{int}C_{int}}}\ln\left(\frac{C_{INT}\sqrt{A_{package}}/2+C_L}{C_o}\right). \qquad (5.31)$$

Chapter 9 describes a much more detailed packaging model that is based on the same basic principles as the above derivation.

5.9   Summary

The interconnection delay is becoming a major factor in determining the performance of VLSI circuits because the RC time constant of wires increases rapidly as chips get larger and device dimensions are reduced. This behavior affects bipolar circuits and GaAs-based ICs even more than it does MOS circuits. The performance advantage of bipolar transistors and GaAs devices over MOS transistors is their low output impedance, which is significant if the load is capacitive; however, it is less important if the transistor drives an RC load with a resistance larger than the output resistance of the transistor.

Multilevels of interconnections improve propagation delay significantly and are essential for high-performance VLSI. With multilevels, all the global interconnections can be formed from low-resistivity aluminum lines, and the cross-sectional dimensions of the upper layers can be adjusted to reduce the distributed RC delays. Multilevels also occupy less chip area, which reduces interconnection length and again improves propagation delay.

Distributed RC delay of interconnections is proportional to the square of the wire length, but *repeaters* reduce this to a linear dependence. *Cascaded drivers* optimize the driving of a capacitive load, but the $R_{int}C_{int}$ term remains unchanged. *Optimal-size repeaters* yield the shortest delay under all conditions but they require additional area. The ultimate lower limit for the time delay is set by the propagation speed of a signal in a lossless transmission line, and this limit is approached as parasitic resistances of wires and transistors are eliminated.

In addition to increased signal propagation delays, wiring resistance also gives rise to larger *clock skews* and higher IR *voltage drops along the power lines*. A combination of special layout techniques, multilevel interconnections with thicker upper levels, and increased number of power connections can resolve most of these problems.

6

TRANSMISSION LINES

In Chapter 4, interconnections were treated as lumped *capacitive* loads. In Chapter 5, wires on scaled-down ICs were shown to have significant *resistance* and were analyzed as distributed RC lines. If the interconnections are sufficiently long or the circuits are sufficiently fast such that rise times of the waveforms are comparable to the time of flight across the line (determined by the speed of light), the *inductance* also becomes important, and the wires then must be modeled as transmission lines.

This is the case at the package level for all IC technologies. To obtain accurate chip-to-chip delay and cross-talk noise estimates, bonding wires, package pins, and board interconnections should be modeled as transmission lines. At the board level, significant reflections can be generated from the end of the line and from capacitive and inductive discontinuities due to connectors, package pins, vias, and corners in board wiring. The power distribution network, consisting of connectors, wires, pins, lead frames, and decoupling capacitors, must also be modeled as a collection of transmission lines, including all inductive, capacitive, and resistive effects.

As chip size increases and signal wavelengths approach wire lengths, transmission line properties of on-chip interconnections also gain importance. Currently CMOS chips are not sufficiently big or fast to warrant transmission line analysis of on-chip wires. Gallium arsenide ICs with 100 psec rise times, however, are sufficiently fast to require such analysis.

In general, if the time of flight is much shorter than the signal rise times, then the wire can be modeled as a capacitive load, and transmission line effects are not important. The lossy nature of IC interconnections and the wires on thin film hybrids distinguishes them from more traditional transmission lines [6.1]–[6.2]. If the wire resistance is greater than the characteristic impedance of the line, the resistive effect dominates the inductive phenemona, and transmission line behavior is not significant, then the wire can be modeled as a distributed-RC line.

The silicon substrate also introduces unique problems by behaving like a conductor for capacitive effects and like an insulator for inductive effects because of the peculiar behavior of magnetic and electric fields in semiconductors known as the *slow-wave effect* [6.3]–[6.7]. These phenomena are strongly dependent on substrate resistivity and operating frequency and make it desirable to place a metal return path in close proximity to high-performance interconnections [6.1], [6.8]–[6.10]. Numerous studies of microstrips for microwave and logic applications have been reported in the literature [6.11]–[6.19].

In this chapter, transmission line theory is presented as applied to package and chip interconnections. As a function of signal rise times and wire lengths, guidelines are developed for when the transmission line analysis should be used and when the interconnection resistance overwhelms the inductive effects. Effects of capacitive and inductive discontinuities are investigated. Various termination methods are reviewed, and circuits are described that prevent reflections while maintaining the low dc power-dissipation property of CMOS. Point-to-point transmission lines, as well as large fan-out bus structures, are analyzed.

6.1   Transmission Line Structures

Printed circuit board interconnections, package pins, lead frames, bonding wires, TAB frames, and solder bumps all have large cross-sections. The large cross-sections yield sufficiently low resistances; therefore these structures can be treated as lossless transmission lines. The required detail of modeling depends on the problem at hand. If the detailed electrical characteristics of a high-speed chip carrier are being calculated, pins, lead frame, and bonding structure should be treated as transmission lines, and computer-aided two-dimensional analysis should be applied. For an approximate chip-to-chip delay calculation, board wiring, which is the dominant component, can be modeled as a uniform transmission line, and the package pins, lead frame, and bonding structure can be included as lumped capacitive and inductive loads (mainly capacitive).

A power distribution network has different requirements than a signal distribution network. In a power distribution network, the voltage level is

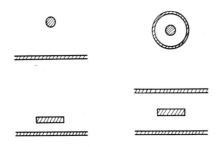

FIGURE 6.1 Cross-sections of typical transmission line structures present in digital systems. Clockwise from top left: wire above ground, coaxial line, triplate strip line, and microstrip.

constant, and large currents may be switching. As a result, a high capacitance is desirable because it helps to keep the voltage level stable. In addition, minimizing the inductance is important because of LdI/dt switching noise. Here, two- and three-dimensional analysis may be required to obtain accurate inductance values. Low-noise power distribution will be covered in more detail in Chapter 7.

Figure 6.1 illustrates the cross-sections of transmission line structures that can be used to approximate digital components. For example, wire above ground can model a bonding wire; microstrip can be used to calculate the characteristics of an IC interconnection; and triplate strip line accurately describes a PC board wire sandwiched between two reference planes. Coaxial cables are widely used to connect boards and in local area networks (LANs). Transmission line properties of these structures have been studied in detail [6.20]–[6.26], and are summarized in the Appendix at the end of this chapter starting on page 274.

A problem with most of the published results is that they are derived for older hybrid microwave technologies where the thickness of the wire is much smaller than its width and the insulator thickness. Modern IC and PC board wires, however, can have thickness-to-width ratios as high as 0.5 or even 1. Then the fringing fields become significant, and older formulas cannot be used. Either new formulas must be employed that take into account the wire thickness (see Appendix on page 274), or computer programs must be used that facilitate numerical analysis to calculate two- and three-dimensional capacitances and inductances.

A uniform transmission line is shown in Fig. 6.2. Regardless of their structures (microstrip, coaxial, etc.), such lines are schematically represented as two parallel lines: signal and ground. Sometimes they are also pictured like a coaxial cable irrespective of their actual physical structure (see Fig. 6.9). The primary electrical parameters of a transmission line are resistance along the line $\mathcal{R}$, inductance along the line $\mathcal{L}$, conductance shunting the line $\mathcal{G}$, and ca-

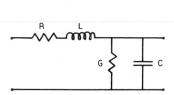

FIGURE 6.2   Electrical model of a uniform transmission line. The primary parameters are series resistance $\mathcal{R}$; series inductance $\mathcal{L}$; shunt conductance $\mathcal{G}$; and shunt capacitance C; all per unit length.

pacitance shunting the line C, all specified per unit length [6.21]–[6.26]. Series components are the cumulative effect of both signal and ground. For example, in calculating the resistance of a microstrip, the resistance of the signal line as well as that of the ground plane must be taken into account. Resistance is very small in package components and becomes significant only in thin film interconnections. Inductance and capacitance are defined by the geometric shape of the transmission line; they are independent from the actual size and are determined by the ratio of the cross-sectional dimensions. An upscaled model of a transmission line has the same characteristic impedance and per unit length inductance and capacitance as the original one.[1] Shunt conductance $\mathcal{G}$ is almost always negligible because low loss dielectrics are used.

If the line is short or the rise time of the signal is long, transmission line effects are not significant. Then inductance of the line is negligible, and the line can be approximated as a lumped capacitor. The delay is determined by the impedance of the driver and the capacitance of the line and can be decreased by reducing the source resistance of the driver. All the points on the wire are equipotential; they rise and fall simultaneously.

The inductance becomes important if the line is long and as a result has a large inductance or if the rise times get faster and $\mathcal{L}dI/dt$ grows. Then the transmission line effects surface, and both the distributed inductance and capacitance must be taken into account. Because of the inductance, the current into the line cannot be increased indefinitely by reducing the source resistance of the driver; inductors resist changes in the current by generating a reverse electromotive force. This limits the amount of current pumped into

---

[1]This is sometimes taken advantage of to build upscaled models of transmission line structures and measure their reflection and cross-talk parameters. It is usually easier to build and measure big structures than integrated microelectronic components.

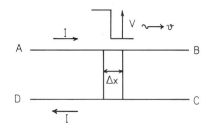

FIGURE 6.3   A voltage and current step propagating along a transmission line

the line and introduces a fundamental limit to how fast a voltage/current waveform can travel down the line, because a limited current can charge up only a certain length of a capacitive line at a given time period. The line is not an equipotential; it accommodates a traveling wave. A more quantitative derivation of these transmission line concepts will be presented in the next section.

6.2   Lossless Transmission Lines

As described in the previous section, most transmission lines of interest have negligible resistance, except IC interconnections and thin-film wires on some packages. Lossless transmission lines are analyzed in this section, and lossy lines are described in Section 6.6.

To calculate the TEM-mode[2] properties of a lossless transmission line, consider a voltage and current step propagating along a line as shown in Fig. 6.3 [6.27]. This transmission line can be a coaxial cable extending between two boards, a microstrip line on a PC board, or an IC interconnection on a high-speed GaAs IC. No change in current or voltage takes place except at the point the propagating step is at a given moment. At every point to the right of the step, the voltage between the signal and ground lines and current into the lines is zero. At every point to the left of the step, the voltage between

---

[2]The simplest and most "ideal" mode an electromagnetic wave can propagate along a transmission line or a wave guide is the case when the electric and magnetic fields have no components in the direction of propagation. Then the electric and magnetic field vectors are perpendicular to the velocity vector and travel in transversal planes. This is referred to as *transverse electromagnetic* (TEM) mode. To achieve perfect TEM mode requires lossless conductors and dielectrics and a homogenous propagation medium. TEM-mode equations cannot account for nonideal effects, such as the frequency dependence of propagation speed, which gives rise to dispersion—the degradation of waveform fronts when the Fourier components of a step travel at slightly different speeds. Most nonideal effects can be included using quasi-TEM-mode calculations.

the two lines is $V = V_{AD}$, and the current in the direction of AB is I and the current in the direction of DC is $-I$.

The magnetic field flux Φ is defined as

$$\Phi = \int \mathbf{B} \cdot d\mathbf{S} \tag{6.1}$$

where $\mathbf{B}$ is the magnetic field vector, $d\mathbf{S}$ is the incremental area vector, "$\cdot$" is the vector dot product, and the integral is taken over the surface (closed or open) for which Φ is defined. Inductance of a device is defined as the proportionality constant between the magnetic flux it holds and the electric current it carries:

$$L = \frac{\Phi}{I}. \tag{6.2}$$

Similarly, in a transmission line

$$\mathcal{L} = \frac{\phi}{I} \tag{6.3}$$

where $\mathcal{L}$ is inductance and ϕ is the magnetic field flux, both per unit length of the line. In time Δt, the voltage/current step advances by Δx such that $v = \Delta x/\Delta t$. Here v is the propagation velocity of the electromagnetic disturbance. The change in the magnetic flux is

$$\begin{aligned} \Delta \Phi &= \Delta(IL) \\ &= I\mathcal{L}\Delta x. \end{aligned} \tag{6.4}$$

Faraday's law ($EMF = -d\Phi/dt$) around the loop ABCD yields the back electromotive force, which is also equal to the voltage step applied to the line:

$$\begin{aligned} V_{AD} &= \frac{\Delta \Phi}{\Delta t} \\ &= I\mathcal{L}\frac{\Delta x}{\Delta t} \\ &= I\mathcal{L}v. \end{aligned} \tag{6.5}$$

Given the magnitude of the voltage step, this equation expresses how the speed and magnitude of the current step traveling down the line are limited by line inductance.

Transmission line phenomenon is an electromagnetic effect; it is a result of the interaction between inductive (magnetic field) and capacitive (electric field) effects. Analogous to inductance, capacitance of a device is defined as a proportionality constant between the electric charge it holds and the potential difference across it,

$$C = \frac{Q}{V}, \tag{6.6}$$

which can be used to calculate the magnitude of the current wave traveling down the line,

$$I = \frac{\Delta Q}{\Delta t}$$

$$= \frac{\Delta(CV)}{\Delta t} \tag{6.7}$$

$$= V_{AD} C v,$$

where C is the capacitance per unit length of the interconnection. Given the current going into the line, this equation expresses how the magnitude and speed of the voltage wave traveling down the line are limited by the line capacitance. Substituting Eq. 6.5 in Eq. 6.7, the propagation speed is obtained as

$$v = \frac{1}{\sqrt{\mathcal{L}C}}. \tag{6.8}$$

Dividing Eq. 6.5 by Eq. 6.7, the characteristic impedance Z_0 is obtained as

$$Z_0 = \frac{V}{I} = \sqrt{\frac{\mathcal{L}}{C}} \tag{6.9}$$

which relates the voltage step traveling down the transmission line to the current traveling down the line by an Ohm's law type of relationship. A uniform transmission line can be completely defined by its characteristic impedance Z_0 and velocity of propagation v.

From Maxwell's laws, it can also be shown that the propagation velocity along a homogenous lossless uniform transmission line can be derived from the material properties of the medium

$$v = \frac{1}{\sqrt{\epsilon\mu}} = \frac{c_0}{\sqrt{\epsilon_r \mu_r}}, \tag{6.10}$$

where ϵ and μ are the dielectric constant and magnetic permeability of the propagation medium; ϵ_r and μ_r are its relative permittivity and permeability; and c_0 is the speed of light in vacuum, equal to 3.0×10^{10} cm/sec (approximately 1 foot/nsec). For nonmagnetic materials μ_r is approximately 1. The propagation speeds for various insulators used in ICs and electronic packages are listed in Table 6.1.

Using Eqs. 6.8, 6.9, and 6.10, the characteristic impedance and inductance of a homogenous lossless line can be obtained as a function of its capacitance and material constants

$$Z_0 = \frac{\sqrt{\epsilon\mu}}{C} = \frac{\sqrt{\epsilon_r}}{c_0 C}$$

$$\mathcal{L} = \frac{\epsilon\mu}{C} = \frac{\epsilon_r}{c_0^2 C}. \tag{6.11}$$

TABLE 6.1   Propagation Speeds in Various Dielectrics

$$v = \frac{c_0}{\sqrt{\epsilon_r}} = \frac{30 \; cm/nsec}{\sqrt{\epsilon_r}}$$

Dielectric	Relative Dielectric Constant (ϵ_r)	Propagation Speed (v) (cm/nsec)
Polyimide	2.5–3.5	16–19
Silicon dioxide	3.9	15
Epoxy glass (PC board)	5.0	13
Alumina (ceramic)	9.5	10

Accordingly C is sufficient for describing lossless lines in the TEM mode.[3] Both the line impedance and inductance are inversely proportional to capacitance. It is important to note that these relationships do not apply to lossy lines, multiconductor structures, and inhomogenous dielectrics. In many calculations the relationship in Eq. 6.11 is applied to lossy lines and inhomogenous structures as a reasonable approximation. It must be verified that the error introduced by such approximations is within the acceptable range.

The following set of equations is useful in transmission line calculations. Propagation speed is given by

$$
\begin{aligned}
v &= \frac{c_0}{\sqrt{\epsilon_r}} \\
&= \frac{1}{\sqrt{\mathcal{L}\mathcal{C}}}.
\end{aligned}
\tag{6.12}
$$

Time-of-flight delay from one end of the line to the other end is

$$
\begin{aligned}
t_f &= \frac{l}{v} \\
&= l\sqrt{\mathcal{L}\mathcal{C}} \\
&= \sqrt{LC}.
\end{aligned}
\tag{6.13}
$$

Here l is the length of the line, and L and C are the *total* inductance and capacitance of the line.

---

[3] Capacitance is usually relatively easier to determine than the other transmission line parameters because more analytical methods and CAD tools have been developed to calculate capacitances and electric fields than to compute inductances and magnetic fields.

Line impedance can be expressed as

$$Z_0 = \sqrt{\frac{\mathcal{L}}{\mathcal{C}}}$$

$$= \frac{1}{v\mathcal{C}} \tag{6.14}$$

$$= v\mathcal{L}.$$

6.3 Circuit Models for Transmission Lines

Figure 6.4 shows how a transmission line can be modeled at various points along the line.

At the driver end, the transmission line can be modeled as a resistor with a value Z_0. Consequently the initial voltage step at point A can be calculated using the resistive divider formed by the source resistance of the driver R_S and the impedance of the line Z_0

$$V_A = \frac{Z_0}{R_S + Z_0} V_S. \tag{6.15}$$

In a CMOS circuit, the voltage swing V_S is usually equal to the power supply V_{DD}.

At an intermediate point, the transmission line to the left of point P can be modeled as a voltage source with a source resistance Z_0 and driven by a voltage step of $2V_i$ where V_i is the magnitude of the step traveling along the line. The factor of two may seem nonintuitive, but it is necessary to account

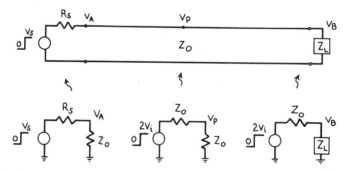

FIGURE 6.4 Circuit models for transmission lines. The equivalent circuits at the source end, at an intermediate point along the line, and at the receiving end are shown.

for the reflections and to satisfy the continuity of current and voltage, as will be shown in Section 6.4. The transmission line to the right of point P is modeled as a resistor Z_0. As expected, the model implies that the step will travel along the line with no attenuation:

$$V_P = \frac{Z_0}{Z_0 + Z_0} 2V_i = V_i. \tag{6.16}$$

This equation is useful in calculating the portion of the incident voltage step transferred across a discontinuity, such as a transmission line with an impedance Z_1 feeding another line with an impedance Z_2. Then the voltage step that is transferred to the second line is given by

$$V_t = \frac{Z_2}{Z_1 + Z_2} 2V_i, \tag{6.17}$$

where V_i is the incident step in line 1 and V_t is the transmitted wave in line 2. The reflected wave is obtained as

$$\begin{aligned} V_r &= V_t - V_i \\ &= \frac{Z_2 - Z_1}{Z_1 + Z_2} V_i \end{aligned} \tag{6.18}$$

(see Section 6.4).

The load end can be modeled as a voltage source with a step of $2V_i$ and a source resistance Z_0 feeding the load impedance Z_L. As a result the voltage step observed at the output is

$$V_B = \frac{Z_L}{Z_0 + Z_L} 2V_i, \tag{6.19}$$

and the reflected wave is

$$V_r = \frac{Z_L - Z_0}{Z_0 + Z_L} V_i. \tag{6.20}$$

If the receiving end is an open circuit ($Z_L = \infty$), then the incident voltage step doubles at the output ($V_B = 2V_i, V_r = V_i$). If the load impedance is the same as line impedance ($Z_L = Z_0$), then $V_B = V_i$ and $V_r = 0$. If the load impedance is less than Z_0, then the polarity of the reflection is opposite of the incident wave ($V_B < V_i, V_r < 0$).

6.4 Discontinuities in Transmission Lines

In practice, transmission lines are not perfectly uniform. There are two main kinds of discontinuities: changes in line impedance and individual inductive and capacitive discontinuities along a uniform line. Both kinds of discontinuities are generated when two transmission lines on a PC board are connected. The impedances of the x and y wiring planes are not exactly equal, and the connection via forms a small inductive and capacitive discontinuity. Inductive and capacitive discontinuities are also formed by the connector cables that attach to PC boards and by the pins of IC packages, capacitors, and terminating resistors. In addition, the wires on the PC board are not perfect straight lines; they take corners and turns that generate discontinuities. These discontinuities must be controlled in order to keep the resulting reflections to a minimum.

6.4.1   Discontinuities in Line Impedance

When two boards are connected via an edge connector, the impedance of the wire on one board, the impedance of the connector, and the impedance of the wire on the other board can all have slightly different values. Similarly, in a PC board or a multilayer ceramic substrate, wires at different levels do not have exactly the same impedance. Such mismatches of line impedance can cause reflections from the junction point.

Arrival of an incident voltage and current step (V_i, I_i) at a junction between transmission lines of different impedances, as illustrated in Fig. 6.5, results in a reflected wave (V_r, I_r) and a transmitted wave (V_t, I_t). The incident wave satisfies the following condition,

$$V_i = I_i Z_1, \tag{6.21}$$

and conservation of charge yields

$$I_i = I_r + I_t. \tag{6.22}$$

The continuity of voltage across the junction requires that

$$V_i + V_r = V_t. \tag{6.23}$$

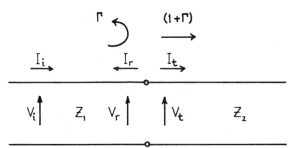

FIGURE 6.5   Discontinuity in a transmission line

The reflected and transmitted waves must satisfy

$$V_r = I_r Z_1 \qquad V_t = I_t Z_2. \tag{6.24}$$

As a result Eqs. 6.21, 6.22, and 6.24 give

$$\frac{V_i}{Z_1} = \frac{V_r}{Z_1} + \frac{V_t}{Z_2} \tag{6.25}$$

and combining Eqs. 6.23 and 6.25, V_r and V_t are obtained as

$$\begin{aligned}
V_r &= V_i \frac{Z_2 - Z_1}{Z_2 + Z_1} \\[2mm]
V_t &= V_i \frac{2Z_2}{Z_2 + Z_1}.
\end{aligned} \tag{6.26}$$

It is customary to define a reflection coefficient $\Gamma = V_r/V_i$,

$$\Gamma = \frac{Z_2 - Z_1}{Z_2 + Z_1}, \tag{6.27}$$

and a transmission coefficient $(1 + \Gamma) = V_t/V_i$,

$$(1 + \Gamma) = \frac{2Z_2}{Z_2 + Z_1}. \tag{6.28}$$

6.4.2   Inductive and Capacitive Discontinuities

Package pins, wire bonds, vias between two wiring levels, and line width variations on a single level can be modeled as individual inductive and capacitive discontinuities. Table 6.2 summarizes the capacitances and inductances of various package components, including dual-in-line package (DIP), pin grid array (PGA), and surface mount technology (SMT) [6.31].

Figure 6.6 illustrates reflections from inductive and capacitive discontinuities along a transmission line such as a via between two wiring levels or the capacitance of a package pin on a multisink data bus [6.32]–[6.33]. The transmission line is driven by a signal generator with $R_S = Z_0$, and the reflected waveforms are observed at the source end. The line is either infinitely long or is terminated with $R_L = Z_0$ at the far end.[4] In the first set of waveforms, the input is assumed to be a step function. The second set assumes an input waveform with a finite rise time.

The inductor acts as an open circuit immediately after the waveform reaches the discontinuity, and the step doubles. Later the inductance transforms into a short circuit as current builds up through it and the voltage disturbance decays with a time constant of $L/2Z_0$.

---

[4]The drive and termination assumptions are chosen such that there will be no reflections from the source or the far end; all reflections are due to the discontinuity.

TABLE 6.2   Capacitances and Inductances of Various Package Components

Component	Capacitance (pF)	Inductance (nH)
68 pin plastic DIP pin[†]	4	35
68 pin ceramic DIP pin[††]	7	20
68 pin SMT chip carrier lead[†]	2	7
68 pin PGA pin[††]	2	7
256 pin PGA pin[††]	5	15
Wire bond	1	1
Solder bump	0.5	0.1

[†] No ground plane; capacitance is dominated by wire-to-wire component.
[††] With ground plane; capacitance and inductance are determined by the
distance between the lead frame and the ground plane, and the
lead length.

The capacitor acts as a short circuit immediately after the waveform reaches the discontinuity and shorts down the line. Later the capacitor transforms into an open circuit as a potential difference builds across it, and the voltage disturbance decays with a time constant of $Z_0 C/2$.

In the second set of waveforms, the input is assumed to have a rise time larger than the time constants of the discontinuities. Then the disturbances

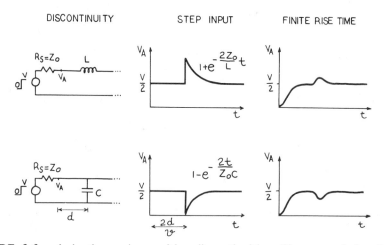

FIGURE 6.6   Inductive and capacitive discontinuities. The transmission line is driven by a signal generator with $R_S = Z_0$, and the reflected waveforms are observed at the source end. The line is either infinitely long or is terminated with $R_L = Z_0$ at the far end. Two types of reflected waveforms are plotted for each discontinuity: one assuming a step input (left) and another assuming a finite-rise-time input (right).

do not get a chance to reach their maximum possible amplitudes because the inductor builds up a current and the capacitor charges up while the incident waveform rises. In CMOS circuits the rise times are usually sufficiently slow to screen out the effects of most inductive discontinuities, and capacitive discontinuities usually increase the rise time rather than generate reflections. In faster bipolar and GaAs systems with short rise times, however, the reflections can be significant. The amplitude of the reflections is

$$V_r = \frac{L_D V_i}{2Z_0 t_r}$$

$$V_r = -\frac{C_D Z_0 V_i}{2t_r}$$

(6.29)

where

$$
\begin{aligned}
L_D &= \text{inductance of the discontinuity} \\
C_D &= \text{capacitance of the discontinuity} \\
V_r &= \text{peak voltage of the reflection} \\
V_i &= \text{incident voltage magnitude} \\
t_r &= \text{rise time of the incident wave.}
\end{aligned}
$$

The areas under the reflections are $V_i L_D/2Z_0$ and $V_i C_D Z_0/2$ respectively [6.33]. Capacitive discontinuities will be further analyzed in Section 6.9, which describes buses.

6.5 When to Use Transmission Line Analysis

Transmission line behavior becomes significant when the rise time t_r of a signal is less than or comparable to the transmission line time-of-flight delay t_f. Here the rise time is defined as the time required for the signal to move from 10 percent to 90 percent of its final value, and time of flight is expressed as

$$t_f = \frac{l}{v}.$$

(6.30)

where l is the line length, and v is the propagation speed. As a rule of thumb, transmission line phenomona become significant when

$$t_r < 2.5 t_f,$$

(6.31)

and the line acts as a lumped capacitor[5] when

$$t_r > 5 t_f.$$

(6.32)

---

[5]Transmission line analysis always gives the correct answer irrespective of the rise time; however, the same answer can be obtained with the same accuracy using lumped approximation when $t_r > 5t_f$. The lumped element calculations are much simpler than transmission line analysis.

In between is the gray area where either transmission line analysis or lumped approximations can be used, depending on the application and required accuracy:

$$2.5t_f < t_r < 5t_f. \tag{6.33}$$

A signal with a 0.5 nsec rise time and SiO_2 as the insulator ($\epsilon_r = 4$) yields $l = 3$ cm as the interconnection length when $t_r = 2.5t_f$. Table 6.3 lists critical line lengths for a number of rise times. To put these line lengths in perspective, typical rise times for various IC technologies and dimensions of chips and package components are also listed. At the board level (50 cm), transmission line properties are significant for signals with rise times shorter than 8 nsec. For a multichip module (10 cm), rise times have to be faster than 1.5 nsec. Single-chip carriers (3 cm) require transmission line analysis when rise times are 0.5 nsec or better, and on-chip interconnections (1 cm) can be treated as lumped elements until rise times reach 150 psec. In present CMOS systems, transmission line properties are important only at the module and board levels; bipolar circuits require transmission line analysis at the chip carrier level and beyond; GaAs technology may require transmission line calculations even for on-chip interconnections.

The rise time of a signal is determined by two factors:

1. The rate at which the driver is turned on. The rise time of the signal at the transmission line cannot be much faster than the rise time of the signal that feeds the driver. The exact relation depends on the driver circuit characteristics.

2. The ratio of the driver source resistance to line impedance. This determines the size of the initial step generated on the line and the number of round-trips required for the output to reach its final value. If the driver resistance R_S is much larger than the line impedance Z_0, the initial step is small, and many trips are required. Then the rise time is determined by the driver resistance and the total capacitance at the output node, including the line capacitance and parasitics. If R_S is comparable or smaller than Z_0, the initial step is large, and the rise time is determined by the signal that feeds the driver, driver circuit characteristics, and parasitic pin capacitances.

Let us show how the ratio of signal rise time to time-of-flight delay is closely related to the ratio of the source resistance of the driver R_S to the characteristic impedance of the line Z_0. Assume that the driver is fed by a fast signal and the rise time is limited by the transmission line and source resistance of the driver. When the transmission line in Fig. 6.7 is modeled as

TABLE 6.3 Critical Transmission Line Lengths, Signal Rise Times, and Chip and Package Dimensions

Critical Transmission Line Lengths for Various Rise Times

$$v = \frac{c_0}{\sqrt{\epsilon_r}} = 15 \text{ cm/nsec}$$

Rise Time t_r (psec)	Critical Line Length $l_{crit} = vt_r/2.5$ (cm)
50	0.3
100	0.6
250	1.5
500	3.0
750	4.5
1000	6.0

Typical Rise Times for IC Technologies

Technology	On-Chip Rise Times	Off-Chip Rise Times
CMOS	0.5–2 nsec	2–4 nsec
Bipolar	50–200 psec	200–400 psec
GaAs	20–100 psec	100–250 psec

Typical Chip and Package Dimensions

Component	Dimension
Die	1 cm
Chip carrier (single-chip)	3 cm
Module (multichip)	10 cm
PC Board	50 cm

a lumped capacitor, the rise time t_r of the resulting RC network is expressed as

$$t_r = 2.2R_SC = 2.2R_SCl. \qquad (6.34)$$

The time-of-flight delay is

$$t_f = \frac{l}{v}. \qquad (6.35)$$

In order for transmission line properties to be negligible, the round-trip delay across the line must be much less than the rise time $(2t_f \ll t_r)$. This implies

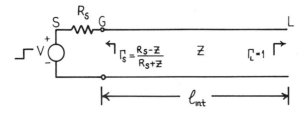

FIGURE 6.7   Unterminated transmission line. The line is driven by a source resistance R_S, and the end of the line remains an open circuit.

that

$$\frac{l}{v} \ll R_S C l$$

(6.36)

$$\frac{1}{vC} \ll R_S.$$

From Eqs. 6.10 and 6.11, Z_0 can be solved as a function of line capacitance and propagation speed,

$$Z_0 = \frac{1}{vC},$$

(6.37)

which can be substituted in Eq. 6.36 to yield

$$Z_0 \ll R_S.$$

(6.38)

Accordingly for the transmission line behavior to be negligible, source resistance of the driver must be much greater than the characteristic impedance of the line. If the source resistance is comparable or less than the line impedance, reflections and other transmission line phenomena become important.

Figure 6.8 compares several schemes with different R_S/Z_0. The waveforms are calculated using the equations derived in Sections 6.3 and 6.4. In Fig. 6.8(a), $R_S = 10Z_0$, and the waveform resembles an RC exponential rise and the transmission line properties are hidden in the details of the staircase-like waveform (capacitive behavior). In Fig. 6.8(c), $R_S = 0.1Z_0$ and the transmission line properties are evident in the ringing of the waveform (inductive behavior). As can be seen in Fig. 6.8(b), an intermediate point with $R_S = Z_0$ is a convenient driving condition for a lossless line not terminated at its receiving end. This is called "termination at the driving end" and has additional power-saving benefits for CMOS applications, as will be described in Section 6.8.

Actually the input waveform is never a perfect step, and consequently the finite rise time of the input waveform has a smoothing effect on the output (Fig. 6.9). When the source resistance R_S is much larger than the characteristic impedance of the line Z_0, the output is an exponential wave. When R_S

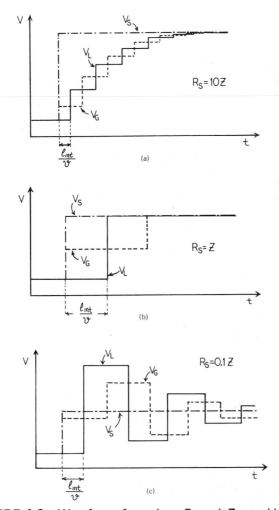

FIGURE 6.8 Waveforms for various R_S and Z_0 combinations in the circuit in Fig. 6.7. As can be seen by the l_{int}/v markers, the time axes of the plots are not drawn at the same scale for clarity. The top plot $R_s = 10Z_0$ has the longest delay. The receiver initially switches at identical times for $R_s = Z_0$ and $R_s = 0.1Z_0$, but the bottom waveform rings and causes the receiver to switch on and off a few times before the waveform settles. This ringing can be disastrous in a clock or asynchronous line.

is smaller than Z_0, the output exhibits overshoot and ringing. The plots in Fig. 6.9 correspond to Fig. 6.8(a) and (c) but with a finite rise time input waveform instead of perfect step excitation. As can be seen in these figures, driving an unterminated line with a low-impedance buffer can increase the

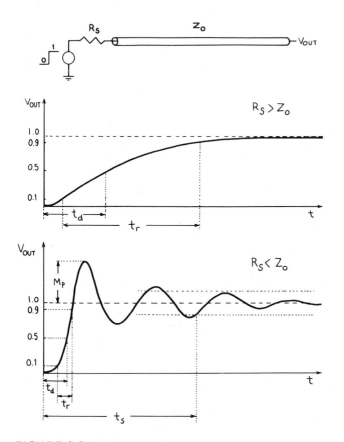

FIGURE 6.9 Waveforms for various R_S and Z_0 combinations with a finite input rise time. The top plot shows the case when $R_S > Z_0$; the output is an exponential waveform. In the bottom plot $R_S < Z_0$; the waveform rings. The delay time t_d, rise time t_r, and settling time t_s are also shown.

settling time, defined as the time it takes for the output to settle in an envelope around the final value of the signal. Ringing adds delay in synchronous signal lines and can cause logic failures in clock and asynchronous signal lines if the machine state is altered inadvertently. It is especially problematic in clock and asynchronous signal lines because the glitches can be observed as a transition and can cause the circuit to transit to a wrong state. Unlike the additional delay in synchronous lines due to reflections, the clock and asynchronous signal line reflection problems cannot be resolved by slowing down the system clock frequency. For such lines, if the source impedance is small, the line can be terminated by a resistor or diode at the far end to eliminate the ringing.

6.6   Lossy Transmission Lines

The board and module wires are sufficiently wide and thick to be treated as lossless transmission lines. On-chip interconnections and some thin film package wires, however, have significant resistance, and they should be treated as lossy transmission lines. When the line resistance becomes larger than the characteristic impedance, resistance dominates the electrical behavior, and the inductive effects disappear. Then the wire should be modeled as a distributed RC line. There are three major factors that cause attenuation in transmission lines: *resistive loss, skin effect loss,* and *dielectric loss.*

Consider a uniform line with per unit length series resistance, series inductance, shunt capacitance, and shunt conductance of $R, \mathcal{L}, C, \mathcal{G}$ (Fig. 6.2). The ratio of the amplitude of a waveform at a distance l from the beginning of the line to the amplitude of the input waveform is given by

$$\frac{V(x=l)}{V(x=0)} = e^{-\alpha l}, \tag{6.39}$$

where α is the attenuation constant made up of two components: conductor loss due to line resistance and skin effect (α_C), and dielectric loss due to shunt conductance (α_D), all per unit length. The attenuation constant is expressed as

$$\begin{aligned}
\alpha &= \alpha_C + \alpha_D \\[2mm]
&= \frac{R}{2}\sqrt{\frac{C}{\mathcal{L}}} + \frac{\mathcal{G}}{2}\sqrt{\frac{\mathcal{L}}{C}} \\[2mm]
&= \frac{R}{2Z_0} + \frac{\mathcal{G}Z_0}{2}.
\end{aligned} \tag{6.40}$$

These losses will be analyzed in the following subsections.

6.6.1   Conductor Losses—dc Resistance

A uniform transmission line with per unit length inductance, capacitance, and resistance of $\mathcal{L}, C, R$ can be characterized by the following set of equations:

$$\begin{aligned}
\frac{\delta V}{\delta x} &= RI + \mathcal{L}\frac{\delta I}{\delta t} \\[2mm]
\frac{\delta I}{\delta x} &= C\frac{\delta V}{\delta x}.
\end{aligned} \tag{6.41}$$

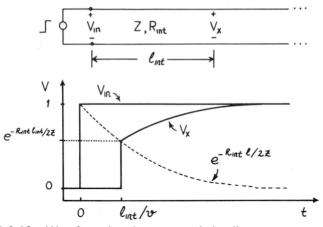

FIGURE 6.10   Waveforms in a lossy transmission line

The response of this line to a unit step input at a distance x from the beginning of the line is [6.1]:

$$V(x,t) = \left\{ e^{-Rx/2Z_0} + \frac{Rx}{2Z_0} \int_{t=x\sqrt{LC}}^{t} \left[\frac{e^{-Rt/2L}}{\sqrt{t^2 - x\sqrt{LC}}} \frac{I_1 R}{2L} \sqrt{t^2 - x\sqrt{LC}} \right] dt \right\}$$

$$\times u\left(t - x\sqrt{LC}\right).$$

(6.42)

Here the first term in the curly brackets is an attenuated step function, and the size of the step is exponentially dependent on the resistance and characteristic impedance of the line. The rest of the expression in the curly brackets represents an RC-like behavior with a slow rise time. The term $u\left(t - x\sqrt{LC}\right)$ is a unit step function that is 0 for $t < x\sqrt{LC}$ and 1 for $t > x\sqrt{LC}$.

As illustrated by Eq. 6.42, the response of a uniform lossy line to a unit step input has two components. The first is a step function that travels down the line with transmission line speeds, and its magnitude is attenuated exponentially along the length of the line; the second is a slow-rising waveform that exhibits an RC-like behavior. Figure 6.10 shows a typical solution. In Fig. 6.11, the potential response to a unit step is captured at successive points along a lossy transmission line. This figure illustrates how the height of the initial step is attenuated as the wave travels along the line. Far down the line, the response resembles that of a distributed RC line.

If $Rl \gg 2Z_0$, the fast-rising portion of the waveform is negligible, and the transmission line behaves like an RC line. The delays and rise times are then determined by the distributed RC constants rather than by the time-of-flight

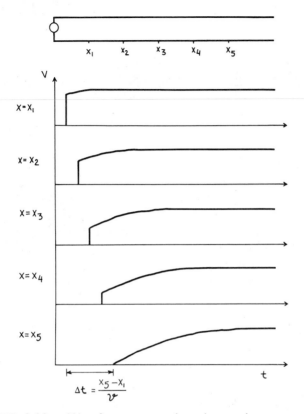

FIGURE 6.11   Waveform propagating along a lossy transmission line. The input to the transmission line is a unit voltage step, and the response is captured at successive points along the line. The size of the step is attenuated as the waveform travels down the line, and at a point far enough from the source, the response is like that of a distributed RC line.

delays. To achieve high-speed signals, the ratio of the fast-rising portion of the waveform to the input,

$$\frac{V(x=l)}{V(x=0)} = e^{-Rl/2Z_0}, \tag{6.43}$$

must be close to 1. For example, if the total resistance of the line R is equal to $Z_0/2$, then at the end of the line $\frac{V(x=l)}{V(x=0)} = 0.78$. As a result a good portion of the initial step reaches the end of the line, and the transmission line behavior is dominant when

$$R < \frac{Z_0}{2}. \tag{6.44}$$

If the line resistance is $5Z_0$, then only 8 percent of the original voltage step

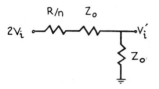

FIGURE 6.12   Subsection of a lossy transmission line

reaches the end of the line. As a result, if

$$R > 5Z_0, \tag{6.45}$$

then the line can be more appropriately modeled as a distributed RC line.

The attenuation along a resistive transmission line can be derived using the transmission line model described in Section 6.3. A lossy transmission line and the circuit model of one of its subsections is shown in Fig. 6.12. Here R is the total line resistance between the source ($x = 0$) and the point of observation ($x = l$), and n is the number of subsections. The voltage transfer across a subsection is expressed as

$$V_i' = \left(\frac{2Z_0}{2Z_0 + R/n} \right) V_i. \tag{6.46}$$

Accordingly the voltage at the point of observation is

$$V(x = l) = \left(\frac{2Z_0}{2Z_0 + R/n} \right)^n V(x = 0). \tag{6.47}$$

Using one of the elementary results of calculus,

$$\lim_{n \to \infty} \left(1 + \frac{x}{n} \right)^n = e^x, \tag{6.48}$$

where e is the base of the natural logarithm, the value of $V(x = l)$ can be calculated as n goes to infinity

$$
\begin{aligned}
\frac{V(x = l)}{V(x = 0)} &= \lim_{n \to \infty} \left(\frac{2Z_0}{2Z_0 + R/n} \right)^n \\
&= \lim_{n \to \infty} \frac{1}{\left(1 + \frac{R/2Z_0}{n} \right)^n} \\
&= e^{-R/2Z_0}.
\end{aligned}
\tag{6.49}
$$

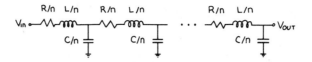

FIGURE 6.13   Approximate circuit for a lossy transmission line to be used in a circuit simulator

Using a similar approach, a lossy transmission line can be simulated with a circuit simulator such as SPICE. Figure 6.13 illustrates the approximate transmission line circuit with n subsections. Here,

$$Z_0 = \sqrt{\frac{L}{C}}$$

$$t_f = \sqrt{LC} \tag{6.50}$$

and L, C, R are the total inductance, capacitance, and resistance of the line. Time-of-flight delay can be calculated from the line length and dielectric constant of the insulator:

$$t_f = \frac{l\sqrt{\epsilon_r}}{c_0}. \tag{6.51}$$

Given the line impedance and time-of-flight delay, the total line inductance and capacitance can be derived as follows:

$$L = t_f Z_0$$

$$C = \frac{t_f}{Z_0}. \tag{6.52}$$

6.6.2   Conductor Losses—Skin Effect

A steady current is distributed uniformly throughout the cross-section of the conductor through which it flows; however, time-varying currents concentrate near the surfaces of conductors. This is known as the *skin effect* [6.24]–[6.26]. An incident electromagnetic field generates currents on the surface of a conductor in the direction of the electric component of the field. If the conductor is perfect, the current is confined to an infinitesimally thin layer at the surface, and the electric field set up by this current cancels the incident field, resulting in a zero field in the conductor. In reality, all conductors have a finite resistance, which causes the field and the current to penetrate in the conductor, and this gives rise to resistive losses. The higher the conductivity and frequency are, the thinner the penetration is. Because of this effect, at high frequencies, a hollow tube is as good a conductor as a solid rod of the same diameter. Therefore the effective resistance of an interconnection for high-frequency signals does not decrease by making it thicker than a critical value.

As a consequence of this electromagnetic induction phenomenon, the magnitude of the current density drops exponentially with distance away from the surface. The distance at which current density becomes a fraction $1/e$ of its value at the surface is called *skin depth,* denoted by δ_S. The skin depth is expressed as

$$\delta_S = \sqrt{\frac{\rho}{\pi \mu f}}, \tag{6.53}$$

where f is frequency and μ and ρ are the permeability and resistivity of the material.[6] For nonmagnetic materials, μ is approximately equal to $\mu_0 = 1.257 \times 10^{-8} = 4\pi \times 10^{-9}$ H/cm. Making the conductor thicker than approximately $2\delta_S$ will not reduce the effective resistance of the interconnection. Skin depth is plotted in Fig. 6.14 as a function of frequency for aluminum at room temperature. As can be seen, skin depth at 1 GHz is 2.8 μm.

To calculate the conductive loss (α_C) of a transmission line, $\mathcal{R}$ for low-frequency components of the signal is determined by a simple resistance calculation $\mathcal{R} = \rho / W_{int} H_{int}$, where W_{int} and H_{int} are the conductor width and thickness. For high-frequency components, the effective wire thickness is δ_S because of skin effect. As a result high-frequency Fourier components of step or pulse signals are attenuated more, and the steepness of the wave fronts degrades.

Figure 6.15 shows a microstrip line. The low-frequency loss factor per unit length α_R is given by the dc resistance of the line

$$
\begin{aligned}
\alpha_R \quad &= \quad \frac{\mathcal{R}}{2Z_0} \\[2mm]
&= \quad \frac{\rho}{2W_{int} H_{int} Z_0}.
\end{aligned}
\tag{6.54}
$$

Because of the skin effect, the high-frequency components of the current are confined to the shaded areas of Fig. 6.15, and the high-frequency attenuation

---

[6]Skin depth being thinner for better conductors may seem to be contradictory since a smaller cross-sectional area will give rise to a larger resistance. Because the skin depth is proportional to the square root of the resistivity, however, the product of the conductivity and the effective cross-sectional area is still larger for better conductors. Based on Eq. 6.53, if the wire is thick enough or the frequency is high enough such that the skin effect is important, when the resistivity of the wire is reduced by a factor of k, the overall resistance improves by $\sqrt{k}$.

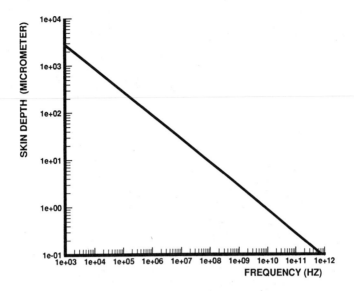

FIGURE 6.14   Skin depth as a function of frequency for aluminum at room temperature. Making the interconnection thicker than twice the skin depth will not reduce the effective resistance of the line for the corresponding frequency.

factor per unit length α_S is given by

$$\alpha_S = \frac{2\mathcal{R}_{SKIN}}{2Z_0}$$

$$= \frac{\rho/W_{int}\delta_S}{Z_0} \qquad (6.55)$$

$$= \frac{\sqrt{\pi\mu_0 f\rho}}{W_{int}Z_0}.$$

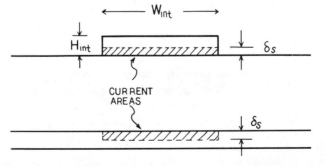

FIGURE 6.15   Microstrip line current areas with skin effect. The high-frequency component of the current can travel only at the shaded areas determined by the skin depth.

Note that the resistance in the high-frequency attenuation factor is multiplied by two because the signal and ground plane are assumed to contribute equally to the series resistance. The calculation of low-frequency attenuation constant in Eq. 6.54 neglects the contribution of the ground plane because the ground plane is much wider than the signal line and the current can spread sideways at low frequencies.

The conductor loss α_C is determined by the larger one of α_R and α_S.

6.6.3   Dielectric Losses

Loss tangent $\tan \delta_D$ is a common measure of the dielectric losses and is defined as

$$\tan \delta_D = \frac{\mathcal{G}}{\omega C} = \frac{\sigma_D}{\omega \epsilon_r}, \tag{6.56}$$

where σ_D is the conductivity of the dielectric, and ω is the angular frequency ($\omega = 2\pi f$). The dielectric loss per unit length is then expressed as[7]

$$
\begin{aligned}
\alpha_D &= \frac{\mathcal{G} Z_0}{2} \\[2mm]
&= \frac{2\pi f C \tan \delta_D \sqrt{\frac{\mathcal{L}}{C}}}{2} \\[2mm]
&= \pi f \tan \delta_D \sqrt{\mathcal{L}C} \\[2mm]
&= \frac{\pi \sqrt{\epsilon_r} f \tan \delta_D}{c_0}.
\end{aligned}
\tag{6.57}
$$

In low-loss substrates, such as alumina, $\tan \delta_D$ is 10^{-3} or lower and can be neglected with respect to resistive losses. In silicon or gallium arsenide substrates, however, dielectric loss can be significant.

6.6.4   Superconducting Transmission Lines

Superconducting materials discussed in Chapter 3 provide conductors with zero dc resistance and near-perfect transmission lines. Superconducting transmission lines have very low attenuation and dispersion, which makes a greater bandwidth possible.

---

[7]The final step in Eq. 6.57 assumes that $v = 1/\sqrt{\mathcal{L}C} = c_0/\sqrt{\epsilon_r}$, which is true only for homogenous transmission lines. For inhomogenous lines, the expression before the final one is more appropriate.

6.7   Transmission Line Termination Conditions

In the previous sections, transmission lines were assumed to be infinitely long. The end of the line introduces a discontinuity that can generate reflections. The reflection coefficient Γ in Eq. 6.27 also applies to end of the line. A terminating resistor can be placed at either the driving or receiving end to eliminate the reflections. In this section, schemes with various combinations of source and load impedances are analyzed, and the next section describes bipolar and CMOS circuit implementations of these termination methods.

6.7.1   No Termination: $R_S = 0$, $R_L = \infty$

Figure 6.16 shows a hypothetical case of zero source resistance with an open circuit at the end of the line. The plots at the top are snapshots of the potential

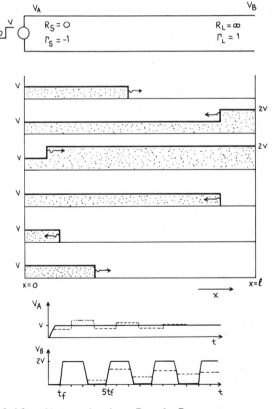

FIGURE 6.16   No termination: $R_S = 0$, $R_L = \infty$

distribution along the line taken at a time sequence. Earlier snapshots are at the top, and time progresses as one moves to lower plots. At the bottom, potential at the driving and receiving ends is plotted as a function of time.

The reflection coefficents at the source and load are $\Gamma_S = (0 - Z_0)/(0 + Z_0) = -1$ and $\Gamma_L = (\infty - Z_0)/(\infty + Z_0) = 1$. When the signal source turns on, a full potential swing starts traveling down the line, and when it reaches the end, it doubles because $\Gamma_L = 1$. The doubled waveform travels back to the source end and gets inverted as it is reflected because $\Gamma_S = -1$. This negative step again doubles at the receiving end and discharges the line completely as it travels back to the source. The step is inverted at the source end, and the entire sequence repeats itself with a period of $4t_f$. No heat is dissipated because there are no resistive elements in the circuit. The energy oscillates.

If there were any resistance at the source or the load end, the absolute value of the reflection coefficients would be smaller than one. The energy will be dissipated slowly at this resistive element. The reflections would slowly decay, and the line potential would stabilize at the source potential. Broken lines at the bottom of Fig. 6.16 show how the waveforms at the driving and receiving ends would look with $R_S = 0.1Z_0$.

6.7.2   Termination at the Receiving End: $R_S = 0$, $R_L = Z_0$

This is also referred to as *parallel termination* because the termination resistor at the receiving end is in parallel with the transmission line. Here, $\Gamma_L = 0$,

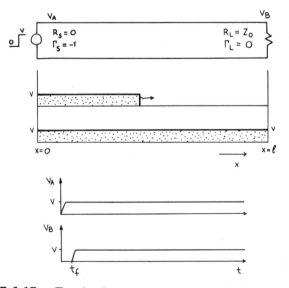

FIGURE 6.17   Termination at the receiving end: $R_S = 0$, $R_L = Z_0$

and no reflections are generated from the load (Fig. 6.17). The line behaves as if it were infinitely long with no discontinuities. This is the simplest and most common termination method (especially in bipolar circuits) but requires low-impedance drivers and dissipates dc power.

In practical applications, R_L is selected to be somewhat larger than Z_0 to minimize the power dissipation. Still it must be kept close to Z_0 to prevent excessive undershoots and overshoots that can switch the receiver back to the wrong state and cause additional settling delay in synchronous signals, or cause the machine to transit to a wrong state in clock and asynchronous signals.

6.7.3   Termination at the Source End: $R_S = Z_0$, $R_L = \infty$

Both the driver impedance requirements and power consumption of parallel termination can be improved by placing the resistor at the driving end. This is also referred to as *series termination* because the termination resistor is in series with the transmission line (Fig. 6.18).

In this scheme, the on-resistance of the driver is equal to the characteristic impedance of the line. As a result the driver forms a series potential divider with the transmission line, and when the driver turns on, a voltage step with

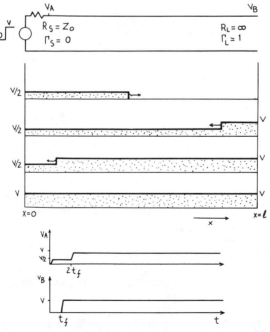

FIGURE 6.18   Termination at the source end: $R_S = Z_0$, $R_L = \infty$

an amplitude of $V/2$ begins to travel toward the receiver, as illustrated in Fig. 6.18. At the load end, the reflection coefficient is $\Gamma_L = (\infty - Z_0)/(\infty + Z_0) = 1$. As a result the signal doubles to a value of V when it reaches the load and starts traveling back toward the source. When the doubled waveform arrives at the source end, no reflections are generated because the source reflection coefficient is $\Gamma_S = (Z_0 - Z_0)/(Z_0 + Z_0) = 0$. The driver current turns off, and there is no dc power dissipation. All the current is used to charge and discharge the wire and other parasitic capacitances. This circuit is the natural choice for CMOS.

Because this scheme depends on the reflected waveform's reaching back to the source, $2t_f$ must be less than the cycle time of the buffer. The reflected waveform must have time to return to the driver before it is switched to a new state. As a result this method is not desirable for paths where the chip-to-chip delay takes most of the cycle. That is when the signal is launched from a latch in a chip, takes the entire cycle to reach another chip, and is captured there by a latch. Fortunately there are very few such paths. Most paths combine chip crossings with some logic. In addition, delay in a source-end terminated bus is determined by the receiver farthest from the driver because the receivers in between must wait for the reflected signal to reach them before their inputs "see" the doubling effect. As a result the receiver closest to the driver will have the longest delay.

In practical applications, the source resistance as well as line impedance can significantly deviate from their nominal values because of semiconductor device process variations. If R_S is larger than Z_0 due to variations in R_S or Z_0, multiple trips along the line may be required to switch the receiver. To prevent such excessively long delays, the nominal source resistance is set to a value *smaller* than the line impedance so that under any process variation, no more than one trip is required to switch the far end. If the source end is too underterminated, however, the receiver can switch back due to ringing. To prevent this in a rising waveform, the first undershoot under worst conditions must be sufficiently small to keep the receiver on. Similarly in a falling waveform, the first overshoot must also be sufficiently small to keep the receiver off (Fig. 6.19).

The first worst-case condition occurs when R_S attains its maximum and Z_0 is at its minimum. Then multiple trips may be required to switch the load. As shown in Fig. 6.20, in order to ensure that the receiver switches at the first trip, R_S/Z_0 must be smaller than five-thirds. That way the first step at the load will be

$$V_{B1} = 3/8 + 3/8 = 6/8 = 0.75. \tag{6.58}$$

This first step is sufficiently large to switch a CMOS receiver. Then the output potential follows this geometric series:

$$V_{B2} = 3/8 + 3/8 + 3/32 + 3/32 = 15/16 = 0.94$$
$$V_{B3} = 3/8 + 3/8 + 3/32 + 3/32 + 3/128 + 3/128 = 63/64 = 0.98.$$

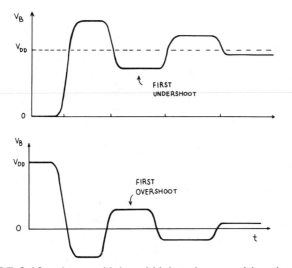

FIRST
UNDERSHOOT

FIRST
OVERSHOOT

FIGURE 6.19   Low-to-high and high-to-low transitions in a circuit with under-terminated source-end termination $R_s < Z_0$. The voltage waveforms shown are at the receiving end.

Here V_{B1} is the first value of the step at the receiving end, and V_{B2} and V_{B3} are the subsequent values.

The second worst-case condition occurs when R_S attains its minimum and Z_0 is at its maximum. Then, ringing at the receiving end occurs. To ensure proper operation, the circuit parameters should be selected such that, with process variations, the minimum value of R_S/Z_0 should be larger than one-third. In this way, the first step will switch the receiver, and the primary undershoot or overshoot will not be large enough to switch the receiver back. The waveforms at the source (V_A) and receiving end (V_B) are shown in

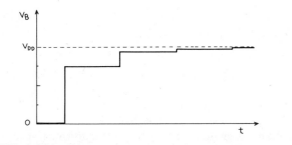

FIGURE 6.20   Worst-case source-end overtermination $R_S = 5Z_0 = 3$

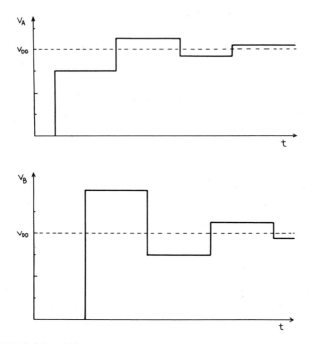

FIGURE 6.21 Worst-case source-end undertermination $R_S = Z_0/3$. V_A is the potential at the source end, and V_B is the potential at the receiving end.

Fig. 6.21. The source end goes through a geometric series with terms

$$
\begin{aligned}
V_{A1} &= 3/4 = 0.75 \\
V_{A2} &= 3/4 + 3/4 - 3/8 = 9/8 = 1.12 \\
V_{A3} &= 3/4 + 3/4 - 3/8 - 3/8 + 3/16 = 15/16 = 0.94 \\
V_{A4} &= 3/4 + 3/4 - 3/8 - 3/8 + 3/16 + 3/16 - 3/32 = 33/32 = 1.03.
\end{aligned}
$$

The receiving end goes through a geometric series with terms

$$
\begin{aligned}
V_{B1} &= 3/4 + 3/4 = 6/4 = 1.5 \\
V_{B2} &= 3/4 + 3/4 - 3/8 - 3/8 = 6/8 = 0.75 \\
V_{B3} &= 3/4 + 3/4 - 3/8 - 3/8 + 3/16 + 3/16 = 18/16 = 1.12 \\
V_{B4} &= 3/4 + 3/4 - 3/8 - 3/8 + 3/16 + 3/16 - 3/32 - 3/32 = 30/32 = 0.94.
\end{aligned}
$$

As a result the initial step at the source is 0.75, and the minimum value at the receiving end is also 0.75; both are acceptable.

Of course, if the driver is switched again before the output settles down, the initial condition at the line can be different from V_{DD} or 0 V. This must be taken into account to ensure proper operation, but it is usually not a problem because in addition to chip-to-chip transmission delay, there is receiver delay and latch setup time requirements. These provide enough additional time for the line to settle down before the clock period ends and the off-chip drivers are switched again.

In summary, the window of operation for a series terminated network is

$$1/3 < R_S/Z_0 < 5/3. \tag{6.59}$$

For optimal performance, it is best to keep the circuit as close as possible to the $R_S/Z_0 = 1/3$ point, without going below it.

6.7.4 Capacitive Load

The capacitive load C_L in Fig. 6.22 represents input capacitance of a receiver, including the pin, module, and bonding wire capacitances in a chip-to-chip net. For the effects of this load capacitance to be negligible, the rise time of the disturbance caused by C_L must be much smaller than the transmission line delay. The current a transmission line can supply is limited by its characteristic impedance. As a result, looking from the receiver end, the line behaves like

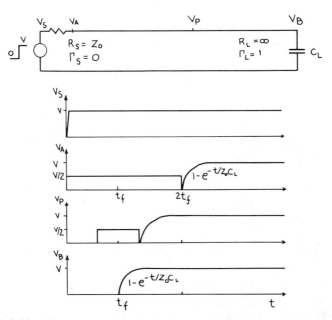

FIGURE 6.22 Capacitively terminated transmission line

a resistor with a value Z_0. The time constant at the load end is $\tau = Z_0 C_L$. For the disturbance to be smaller than the propagation delay, the following condition must be met:

$$Z_0 C_L \ll t_f. \tag{6.60}$$

Substituting

$$Z_0 = \frac{\sqrt{\epsilon\mu}}{C}$$

$$v = \frac{1}{\sqrt{\epsilon\mu}}$$

$$t_f = \frac{l}{v}$$

and

$$C = Cl$$

in Eq. 6.60, the following condition is obtained:

$$\frac{\sqrt{\epsilon\mu}}{C} C_L \ll \frac{C/C}{1/\sqrt{\epsilon\mu}} \tag{6.61}$$

$$C_L \ll C.$$

As long as the input capacitance of the receiver circuit is much smaller than the total interconnection capacitance, the rise time of the wave front at the receiving end will be a small portion of the total transmission delay, and the effect of C_L can be neglected.

If the capacitive load is significant (for example, a signal pin with 5 pF capacitance) the time delay from point A to B as measured at the 50 percent point is [6.35]

$$T_{50\%} = t_f + \frac{Z_0 C_L}{2}. \tag{6.62}$$

6.7.5   RC Termination at the Receiving End: $R_S = 0$, $R_L = Z_0$, $C_L \gg t_f/Z_0$

The dc power dissipation of parallel termination can be eliminated by placing a capacitor in series with the resistor at the receiving end, as shown in Fig. 6.23.

Because $R_S = 0$, a full potential step travels down the line, and when it reaches the end, it first sees a load impedance of $R_L = Z_0$ because at time $0+$ all capacitors act like short circuits. A perfect termination is obtained, and no reflections are generated. What happens next depends on the value of the capacitor. If the capacitor is very small, it will rapidly transform from short circuit mode to open circuit mode, and the voltage step will double. The resulting waveforms will look like the case with no termination ($R_S = 0$, $R_L = \infty$) shown in Fig. 6.16. If the capacitor is sufficiently large, when a

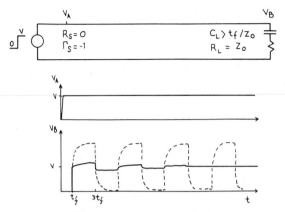

FIGURE 6.23   RC termination at the receiving end. $R_S = 0$, $R_L = Z_0$, $C_L \gg t_f/Z_0$. Dashed lines show the case when the capacitance is too small. The solid line shows an appropriate value such as $C_L = 25t_f/Z_0$.

positive step of V_i arrives at the capacitor, potential difference between the plates will remain zero, and both plates will jump to a potential of V_i. Then the top plate of the capacitor will start slowly being charged up to $2V_i$, and the bottom plate will start being discharged to 0, both with a time constant of $Z_0 C_L$. If this time constant is much larger than the time of flight along the line, the negative reflections from the source end will cancel the charge buildup at the top plate, and doubled waveforms will never appear. As a result this termination technique can provide all the advantages of parallel termination without the dc power dissipation. The main disadvantages are that it requires two terminating components rather than one, and the source resistance must be much lower than Z_0—hard to achieve in CMOS.

To satisfy the $Z_0 C_L > t_f$ condition, a good rule for C_L is[8]

$$C_L > \frac{25t_f}{Z_0}. \tag{6.63}$$

---

[8]The termination impedance as a function of frequency can be expressed as

$$Z_L = R_L + \frac{1}{j\omega C_L}$$

When the condition defined by Eq. 6.63 is satisfied, the impedance due to the load capacitance at the frequency $\omega = \pi/t_f$ is less than 1 Ω. As a result the termination condition is satisfied because $Z_L \approx R_L = Z_0$. This assumes that the rise time of the signal is within the same order of magnitude as $1/t_f$ or shorter. This is appropriate because if the rise time is long, transmission line effects do not show up, the line acts as a capacitor, and reflections are nonexistent.

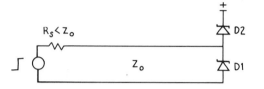

FIGURE 6.24   Schottky diode termination

In addition, any parasitic pin or receiver capacitance to ground should be small.[9] The capacitance C_L must be large, but if it is too large, the power dissipation will increase. In the extreme limit, an infinitely large capacitance will act like a short circuit, and this termination scheme will be equivalent to parallel termination ($R_S = 0$, $R_L = Z_0$) shown in Fig. 6.17.

6.7.6   Schottky Diode Termination

In circuits with low-impedance drivers, such as high-speed bipolar, Schottky diodes can be used to terminate the line. Figure 6.24 illustrates a typical diode termination network. When the output switches to a low value, the lower diode D1 clamps the output potential to a diode drop below ground. When the output switches to a high value, the upper diode D2 clamps the output to a diode drop above the plus power supply. The Schottky diode drop $V_{Schottky}$ is typically 0.3–0.4 V. As a result the output potential is clamped at both ends, and undershoots and overshoots are minimized. This method is especially suitable for environments where line impedances are not well defined, such as breadboards and backplanes. Diodes will clamp the output irrespective of the line impedance.

6.7.7   Termination of Resistive Lines

The lossy nature of resistive transmission lines can be used to avoid reflections [6.1]. As described previously, a voltage step gets attenuated while propagating down a lossy transmission line. If the line end is an open circuit, the waveform will double when it reaches the line end. This doubling effect can compensate for the resistive loss. Because the step height is attenuated expo-

---

[9]It can be shown that the parallel parasitic capacitance C_P should be kept small by observing the total impedance at the end of the line

$$Z_L = \cfrac{1}{\frac{1}{\frac{1}{j\omega C_P}} + \frac{1}{R_L + \frac{1}{j\omega C_L}}} = \cfrac{1}{j\omega C_P + \frac{j\omega C_L}{1 + j\omega C_L R_L}}$$

If C_P is large, it may shunt the termination at high frequencies and cause reflections. The maximum allowable parasitic capacitance depends on the signal rise times, as illustrated by the above equation.

nentially with distance, the reflection from the line end can restore the pulse amplitude and yet not cause excessive ringing and distortion. This is because the step is already attenuated when it reaches the receiving end, and the reflected step is further attenuated as it travels back to the driving end. By the time it gets back to the driver, it will have lost most of its amplitude. As a result when the line length, line resistance, and Z_0 satisfy certain conditions, the reflection from the line end can restore the pulse amplitude and still avoid ringing and distortion at the receiving end. If the total line resistance falls in the following range,

$$\frac{2Z_0}{3} < \mathcal{R}l < 2Z_0, \tag{6.64}$$

the pulse at the receiving end is very similar to the transmitted pulse [6.1]. To satisfy this condition, the line length should be in the range

$$\frac{2Z_0}{3\mathcal{R}} < l < \frac{2Z_0}{\mathcal{R}}. \tag{6.65}$$

In practical applications, minimum interconnection length is not a consideration because the input waveform rise times are usually longer than the time of flight across the minimum line length. The maximum line length limit must be satisfied to obtain full amplitude at the receiving end.

6.8 Driver and Termination Circuits for Transmission Lines

This section describes bipolar and CMOS implementations of termination circuits that can be placed at either the driving or receiving end to avoid undesirable reflections.

6.8.1 Termination at the Receiving End

Figure 6.25(a) illustrates a typical bipolar implementation of the parallel termination technique [6.27]. The line is driven by an open-collector pull-down transistor $Q1$, and a pull-up resistor R_L at the receiving end satisfies the termination condition. The capacitance C_L is the parasitic capacitance of the package pins. When $Q1$ is off, the interconnection is charged to V_{CC} through R_L, and at the steady state no dc current flows. Because $Q1$ is designed to sink large currents, the potential at its collector falls rapidly when it is turned on. As a result, when $Q1$ is switched on, a voltage and current wave of ΔV and $\Delta V/Z_0$ begins to travel down the transmission line with a velocity of $v = c_0/\sqrt{\epsilon_r}$. The impedance of $Q1$ is much smaller than Z_0; consequently the voltage step $\Delta V = Z_0 V_{CC}/(Z_0 + R_S)$ is approximately equal to V_{CC}. Because R_L is equal to Z_0 of the line, when the signal reaches the receiving end, R_L

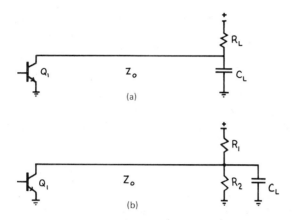

FIGURE 6.25   Parallel terminated bipolar transmission line drivers: (a) with rail-to-rail voltage swings, (b) with reduced voltage swings

supplies the correct amount of current $(I = \Delta V/R_L = \Delta V/Z_0)$ to satisfy the boundary conditions, and no reflections occur. In other words, the magnitude of the current wave traveling down the line is equal to the current that the potential step draws from R_L when it reaches the end. After this, the resistor conducts a steady-state current of $\Delta V/Z_0 \approx V_{CC}/Z_0$ until $Q1$ switches off. The resistor and its leads must be kept short to minimize the parasitic inductances and capacitances that may cause reflections.

When $Q1$ is turned off, the current sink at the driving end disappears, and this time a negative voltage/current step of $-\Delta V$ and $-\Delta V/Z_0$ begins to travel down the line. Because R_L is equal to Z_0, the step reaching the receiving end exactly cancels the voltage across R_L and the current being supplied by it. Again there are no reflections. The transistors $Q1$ and R_L should be ratioed to obtain a valid low level when they are both on; in other words, the effective resistance of $Q1$ must be much smaller than Z_0. In addition, due to the dc path at the receiving end, the driver supplies a current of $V_{CC}/(Z_0+R_S)$ 50 percent of the time, and as a result average power dissipation is $V_{CC}^2/2(Z_0 + R_S)$.

If the transistor $Q1$ cannot sink a current V_{CC}/Z_0, the termination circuit can be modified as shown in Fig. 6.25(b). The resistors must be selected such that in parallel they satisfy the termination condition

$$Z_0 = \frac{R_1 R_2}{R_1 + R_2}. \tag{6.66}$$

The Thevenin equivalent of the power supply as seen at the receiving end is

$$V_{THEV} = V_{CC}\frac{R_2}{R_1 + R_2}. \tag{6.67}$$

Consequently $Q1$ has to sink only a current of V_{THEV}/Z_0 rather than V_{CC}/Z_0.

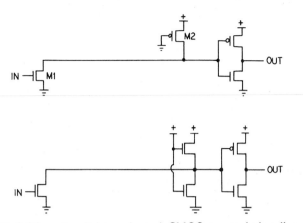

FIGURE 6.26   Parallel terminated CMOS transmission line drivers: (a) with rail-to-rail voltage swings, (b) with reduced voltage swings

This increases the power consumption because the current dissipation is $V_{CC}/(R_1 + R_2)$ when $Q1$ is off and V_{CC}/R_1 when $Q1$ is on.

Figure 6.26 shows the CMOS versions of these termination methods. In Figure 6.26(a), the line is driven by a pull-down transistor $M1$, and a pull-up transistor $M2$ (or resistor) at the receiving end satisfies the termination condition. A resistor would be preferred for better match. The characteristic impedance of an interconnection ranges between 10 and 200 Ω, and achieving a high current drive capability with MOS transistors is difficult. The circuit in Fig. 6.26(b) can be used to reduce the voltage swing and current drive requirements. These circuits are not used in practical CMOS applications because of current drive and matching difficulties.

6.8.2 Termination at the Source End

As described before, both the required on-resistance of the driver and the average power consumption can be improved by terminating the line at the driving end (series termination). In this scheme, the on-resistance of the driver is equal to the characteristic impedance of the line, which reduces the area required by the driver transistor because the aim is for an $R_S = Z_0$ ($\approx 50\ \Omega$) source impedance rather than $R_S \ll Z_0$. In addition, as shown in Figs. 6.18 and 6.27, the interaction between incident and reflected waveforms eliminates continuous power dissipation; power is consumed during logic transitions only. This circuit is the natural choice for CMOS.[10]

---

[10]The improvement factor t_f/PW is analogous to the duty factor in CMOS; as the time of flight t_f approaches pulse width PW, the efficiency of source-end termination is

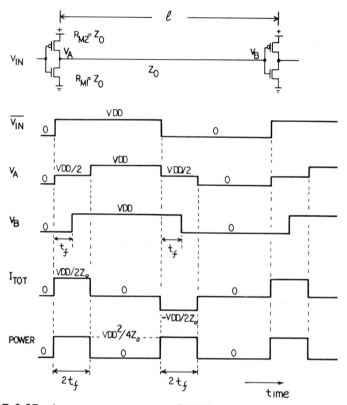

FIGURE 6.27 Low-power source-end CMOS transmission line termination

Because the effective on-resistance of the driver is equal to the impedance of the line, the driver forms a series potential divider with the transmission line, and when the driver turns on, a voltage step with an amplitude of $V_{DD}/2$ begins to travel toward the receiver, as illustrated in Fig. 6.18. The receiver behaves like a capacitive termination, and the voltage step doubles with a time constant of $\tau = Z_0 C_L$ where C_L is the input capacitance of the receiver. With a 50 Ω line and 0.1 pF input capacitance (typical for a big on-chip buffer), the rise time is 11 psec. Even with a 5 pF load (typical for package pins), the rise time is 550 psec. As a result, for most practical purposes, the line behaves as though there were an open circuit at the receiving end and the doubling of the voltage step were instantaneous. If the rise time due to load capacitance is significant, a delay of $Z_0 C_L/2$ should be added to the time-of-flight delay. This

degraded (similar to the reduced advantages of CMOS when the rise times of signals become comparable to the clock period and dynamic current flows most of the time). In almost all cases, however, the time of flight is much shorter than the cycle time (pulse width).

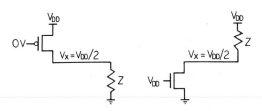

FIGURE 6.28 Circuit model for source-end termination: pulling up (left) and pulling down (right)

doubled waveform then travels back and turns off the driver when it arrives at the source end because the potential drop between the source and drain of the driving transistor becomes zero, and as a result there is no dc power dissipation. Figure 6.18 illustrates the traveling waveform at four consecutive times, and Fig. 6.27 plots the potential as a function of time at several points along the transmission line.

The termination condition needed at the source end can be calculated by using the model in Fig. 6.28. The current into the transmission line is expressed as

$$I = \frac{V_{DD}/2}{Z_0}. \tag{6.68}$$

Because the transistor is in the linear region, it can supply a current of

$$I = \mu C_{ox} \frac{W}{L_{eff}} \left[(V_{DD} - V_T) \frac{V_{DD}}{2} - \frac{V_{DD}^2}{8} \right]. \tag{6.69}$$

Transistor sizes are obtained by setting the currents in Eqs. 6.68 and 6.69 equal; then,

$$\frac{W}{L_{eff}} = \frac{1}{\mu C_{ox} Z_0 \left(\frac{3}{4} V_{DD} - V_T \right)}. \tag{6.70}$$

Here, μ is carrier mobility, C_{ox} is gate capacitance per unit area, V_T is threshold, and W and L_{eff} are the width and effective length, respectively, of the transistor. If the transistor operates in the velocity saturation mode, Eq. 6.69 must be modified accordingly.

One disadvantage of terminating at the driver end using the effective resistance of the transistor is the sensitivity of the circuit to the on-resistance of the driver. When the transmission line is terminated at the receiving end, such as the circuits shown in Figs. 6.25 and 6.26, the on-resistance of the driver is much smaller than the characteristic impedance of the line. As a result if the source resistance varies by 50 percent or even 100 percent, the circuit still functions properly because when an R_S that is much less than Z_0 is doubled, it is still much less than Z_0. The goal in far-end termination is to make sure that the worst-case R_S is much less than Z_0. The terminating

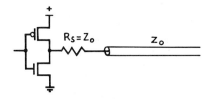

FIGURE 6.29   Source-end termination with a resistor. The value of the resistors is equal to the impedance of the line, and the transistor resistance is much smaller than Z_0. This circuit takes more area but has better termination properties across the process variations and for a wide operation temperature range. The resistor can be fabricated on the chip or can be a separate discrete component.

resistor R_L must be closely matched to Z_0, but compared with transistors, it is relatively easier to build precise resistors that stay constant over a large temperature range. In source-end termination, on the other hand, it is crucial that the source resistance satisfy $1/3 < R_S/Z_0 < 5/3$. Controlling the current drive of the transistor across semiconductor device process variations and for a wide operation temperature range is difficult. Usually diffused resistors can be controlled better than transistors; as a result, for very-high-speed applications, the driver may be more reliably built as a resistor equal to Z_0 (or slightly smaller) in series with a transistor with an on-resistance much smaller than the resistor (Fig. 6.29). This, of course, takes more area, and the driver buffer delay increases because more stages are required in the cascaded buffer structure. Another alternative is to use a discrete resistor in series with the driving pin, which increases the component count and, consequently, board cost (but not much more than parallel termination with discrete resistors).

Series termination is further complicated by the fact that the initial waveform is only half the power supply, and the operation of the circuit depends on the reflected waveform's reaching back to the source. The round-trip across the wire $2t_f$ must be less than the cycle time allocated for chip-to-chip delay; the reflected waveform must have time to return to the driver before it changes

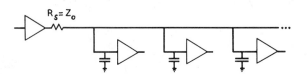

FIGURE 6.30   Source-end termination as applied to a data bus. The driver has to be either at one end of the bus or exactly in the middle. In addition, the receiver closest to the driver will switch the last.

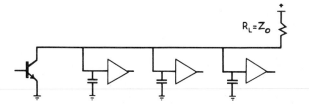

FIGURE 6.31   Data bus with a single driver and multiple sinks. Parallel termination is applied at the far end of the bus.

state. In addition, delay in a source-end terminated bus is determined by the receiver farthest from the driver because the receivers in between must wait for the reflected signal to reach them before their inputs "see" the doubling effect (Fig. 6.30). As a result, the receiver closest to the driver will have the longest delay. If this receiver is very close to the driver, its delay will be twice as long as the far-end receiver delay of the circuit in Fig. 6.31, where the bus is parallel terminated. This may be undesirable for certain data bus applications with multiple loads, and care must be taken when designing source-end terminated nets with fan outs greater than one. Therefore, terminating at the source-end has some disadvantages, but in certain applications it can save die area, improve power dissipation, and eliminate the need for termination resistors. It is perfect for driving point-to-point nets, is suitable for "not-the-fastest" applications such as CMOS, and serves very well for clock distribution using a *symmetric* tree where all the nets are point to point and all the loads are equidistant from the driver, as described in Chapter 8.

6.9   Transmission Lines with Fan-out: Buses

Some high-end computers have only point-to-point nets between chips. If any signal has to go to more than one chip, multiple off-chip drivers and pins are used, one for each sink. This is very costly because it increases the number of output buffers/pins and requires an expensive board to satisfy the wiring demand. It also increases power dissipation, especially in bipolar systems. In most applications, one signal pin feeds multiple chips. This section describes issues specific to transmission lines with a fan-out greater than one.

When parallel termination is used in a data bus with multiple loads, various termination configurations are possible depending on the physical placement of the wires and system performance requirements. If many long nets are present, a terminating resistor for each sink may be required, which would increase the power dissipation. This topology is not very common because it

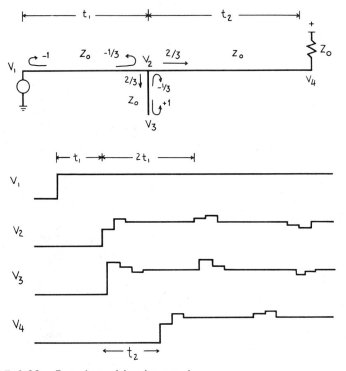

FIGURE 6.32 Data bus with a long stub

is not an efficient way of routing wires. Usually buses are formed by one long line with multiple short loads branching from it. Then a termination resistor may be placed at both ends of the bus. The driver transistor should be able to sink twice the current when two termination resistors are used. If the driver is at one end of the bus or one of the two halves of the bus is much shorter than the other half, only one termination resistor is required. In this case, the resistor must be attached to the load that is electrically farthest away from the source. This is common and is shown in Fig. 6.31.

When series termination is used, no resistors are required at the sinks, but, as mentioned earlier, the receiver closest to the driver will have the longest delay, among other complications. The following derivations apply to both parallel and series terminated buses.

6.9.1 Long Stubs

If the stubs that tie the loads to the bus are long and are unterminated, they generate reflections with a pulse width equal to the round-trip delay of the stub [6.28]. Figure 6.32 illustrates the effects of a long stub on a transmission line that is driven by a low-impedance driver and terminated by a resistor (parallel termination). Looking into the branching point from the source end, the effective impedance is $Z_0/2$ because the line and stub are driven in parallel. As a result the reflection coefficient is $\Gamma = (Z_2 - Z_1)/(Z_2 + Z_1) = (Z_0/2 - Z_0)/(Z_0/2 + Z_0) = -1/3$, and the transmission coefficient is $(1 + \Gamma) = 2/3$. The end of the stub is an open circuit, and the reflection coefficient is $\Gamma = (\infty - Z_0)/(\infty + Z_0) = +1$. When the driving wave arrives at the branching point, it divides into three parts: one going into the stub, one going down the line, and one reflecting back toward the driver. The portion that goes into the stub reflects back and forth a few times and finally stabilizes, as can be observed from the waveforms labeled V_2 and V_3. End of the line is terminated, and as a result V_4 is a delayed version of V_2. The pulse that was reflected toward the driver continues being reflected back and forth with a period of $2t_1$ and eventually dies away.

The above calculations assumed that the rise time of the signal was much shorter than the transmission delay through the stub. If the rise time of the signal is greater than the round-trip delay of the stub, the discontinuity will not generate a noise pulse, but it will degrade the rise time instead.

6.9.2 Short Stubs

If the connecting stubs are kept short, the fan-out loads on the bus can be modeled as capacitors. The details of the calculation also depend on the distance between the stubs.

Closely Spaced Short Stubs

When the spacings between the loads are small, their effect is to increase the total line capacitance. The line inductance stays the same, and the characteristic impedance is reduced because $Z_0 = \sqrt{\mathcal{L}/\mathcal{C}}$. In addition, the propagation speed also slows down[11] because $v = 1/\sqrt{\mathcal{L}\mathcal{C}}$. The line can be treated as a distributed net if the spacings between the loads are short such that the round-trip time-of-flight delay across a segment is less than the rise time of the disturbance caused by the discontinuity $(2t_f < t_r)$. Here the rise time is dominated by the load capacitance and is defined as $t_r = 2.2Z_0C_D$ where C_D is the capacitance of a single discontinuity. Under these conditions, the line capacitance $C' = C + C_L$ increases, the effective impedance of the loaded line Z_L is reduced, and the propagation delay T_L increases. Accordingly, the

---

[11]Actually adding capacitive stubs to a transmission line is a popular way to construct delay lines in microwave circuits.

following impedance and delay expressions are obtained [6.35]–[6.36]:

$$Z_L = \frac{Z_0}{\sqrt{1 + C_L/C}}$$

$$T_L = T_0\sqrt{1 + C_L/C}$$

(6.71)

where

$$
\begin{aligned}
Z_L &= \text{loaded line impedance} \\
Z_0 &= \text{unloaded line impedance} \\
C_L &= \text{added load capacitance per unit length} \\
C &= \text{unloaded line capacitance per unit length} \\
T_L &= \text{loaded line propagation delay} \\
T_0 &= \text{unloaded line propagation delay.}
\end{aligned}
$$

The delay adder due to capacitive loading can be expressed as

$$T_\Delta = T_0\left(\frac{Z_0}{Z_L} - 1\right).$$

(6.72)

Widely Spaced Short Stubs

When the loads are widely spaced, a simple distributed net approximation cannot be used; the loads and reflections must be treated individually. If the rise time of the signal is less than the round-trip delay between two discontinuities $(2t_f > t_r)$, the following expressions can be used for delay and reflection calculations [6.35]–[6.36]. Here the rise time is not necessarily dominated by the load capacitance.

$$V_r = -\frac{C_D Z_0 V_i}{2t_r}$$

$$T_{W50} = t_r$$

$$T_{W0} = t_r + 1.5 Z_0 C_D$$

(6.73)

$$T_\Delta = \frac{C_D Z_0}{2},$$

where

$$
\begin{aligned}
C_D &= \text{capacitance of the discontinuity} \\
V_r &= \text{peak voltage of the reflection} \\
V_i &= \text{incident voltage magnitude} \\
T_{W50} &= \text{width of the reflection at the 50\% points} \\
T_{W0} &= \text{width of the reflection at the baseline} \\
T_\Delta &= \text{delay adder due to the discontinuity.}
\end{aligned}
$$

This is same as the reflection equation for a capacitive discontinuity described in Section 6.4.2.

The reflections from the capacitive discontinuities can increase the delay significantly by either slowing down the propagation (closely spaced short stubs) or by requiring multiple reflections from the end of the line before the circuits switch (widely spaced or long stubs). In addition, if multiple reflections are superimposed—for example, when many stubs are *equally* spaced—false switching can be generated at the line end or at the intermediate nodes. Therefore reflections must be limited to avoid excessive delays and false switching.

6.10   Summary

As the operating frequency increases, the transmission line behavior of package and on-chip interconnections becomes important. In this chapter, basic lossless and lossy transmission line theory are reviewed and quantitative expressions are developed that describe the regions where IC and board interconnections must be treated as transmission lines, distributed RC lines, and lumped capacitive loads. The drive and termination of transmission lines are discussed, and CMOS circuits are presented that eliminate dc power dissipation and do not require low-impedance drivers.

Transmission line behavior is visible when the time-of-flight delay along the line is longer than the rise time of the signals

$$2.5t_f > t_r,$$

which also implies that Z_0 is not much smaller than the source resistance R_S.

If the source resistance R_S is much greater than Z_0 or the input rise time is much larger than the time-of-flight delay, the line can be modeled as a lumped capacitance.

Integrated-circuit transmission lines are lossy. If the total resistance of the line is greater than or equal to its characteristic impedance, the wire behaves like a distributed RC line. To achieve high speeds, the fast-rising portion of the waveform $e^{-Rl/2Z_0}$ must be close to 1, which requires that

$$R < \frac{Z_0}{2}.$$

Reflections can be generated from the discontinuities along the transmission lines or from the end of the line. Inductive and capacitive discontinuities must be kept to a minimum, and transmission lines must be appropriately terminated to avoid reflections. Bipolar circuits traditionally use a low-impedance open collector device to drive the line and terminate it at the receiving end with a resistor R_L equal to Z_0 (also referred to as *parallel termination*). This is the simplest and cleanest method, but it dissipates dc power. CMOS circuits can mimic this technique by employing an open drain driver. Instead, however, CMOS circuits are usually terminated at the source end by taking advantage of the intrinsic resistance of the driver or by placing a termination resistor at the output of the buffer (also referred to as *series termination*). This eliminates the undesirable reflections without dissipating any dc power but also requires more careful design, especially for bus structures, because the operation of the circuit depends on the first reflection from the line end. The condition that must be satisfied by the driver in order to guarantee first-incidence switching is

$$\frac{1}{3} < \frac{R_S}{Z_0} < \frac{5}{3}.$$

Other termination methods that use Schottky diodes or RC loads at the receiving end or take advantage of the lossy nature of thin film lines are also possible.

6.11 Appendix: Useful Impedance Formulas

In the following formulas Z_0 is the characteristic impedance, C and $\mathcal{L}$ are the capacitance and inductance per unit length, and ϵ and μ are the electric permittivity and magnetic permeability of the insulator. Typically, $\epsilon = \epsilon_r \epsilon_0 = \epsilon_r \times 8.854 \times 10^{-14}$ F/cm, and $\mu = \mu_0 = 1.257 \times 10^{-8}$ H/cm.

Coaxial cables (Fig. 6.33) are used to connect local area networks, com-

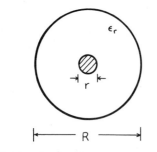

FIGURE 6.33 Coaxial cable

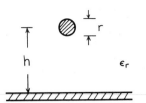

FIGURE 6.34 Wire above a plane

puter subcomponents, and sometimes PC boards:

$$Z_0 = \frac{1}{2\pi}\sqrt{\frac{\mu}{\epsilon}}\ln\left(\frac{R}{r}\right)$$

$$C = \frac{2\pi\epsilon}{\ln\left(\dfrac{R}{r}\right)} \qquad\qquad (6.74)$$

$$\mathcal{L} = \frac{\mu}{2\pi}\ln\left(\frac{R}{r}\right).$$

Wire above a plane (Fig. 6.34) is useful in bonding wire, package pin, and board wiring calculations. For $r \ll h$ the line parameters are obtained as

$$Z_0 = \frac{1}{2\pi}\sqrt{\frac{\mu}{\epsilon}}\ln\left(\frac{4h}{r}\right)$$

$$C = \frac{2\pi\epsilon}{\ln\left(\dfrac{4h}{r}\right)} \qquad\qquad (6.75)$$

$$\mathcal{L} = \frac{\mu}{2\pi}\ln\left(\frac{4h}{r}\right).$$

Parallel plates in Fig. 6.35 ($h \ll W$) can be used as a zeroth-order check for more complex impedance, capacitance, and inductance formulas for strip

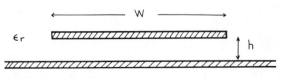

FIGURE 6.35 Parallel plates

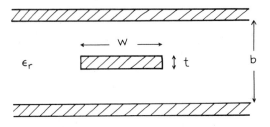

FIGURE 6.36   Triplate strip line

lines and microstrips.

$$Z_0 = \sqrt{\frac{\mu}{\epsilon}}\left(\frac{h}{W}\right)$$

$$C = \frac{\epsilon}{\left(\frac{h}{W}\right)} \tag{6.76}$$

$$\mathcal{L} = \mu\left(\frac{h}{W}\right).$$

Triplate strip line in Fig. 6.36 describes accurately PC board wires sandwiched between two reference planes:

$$Z_0 = \frac{1}{4}\sqrt{\frac{\mu}{\epsilon}}\ln\left(\frac{1+W/b}{W/b+t/b}\right)$$

$$C = \sqrt{\epsilon_r}\frac{1}{c_0 Z_0} \tag{6.77}$$

$$\mathcal{L} = \sqrt{\epsilon_r}\frac{Z_0}{c_0}.$$

This formula is accurate to 1–2 percent for $W/b \geq 1$ and to 5–6 percent for $W/b \geq 0.75$, provided that $t/b \leq 0.2$ [6.20].

For a microstrip on a PC board (Fig. 6.37) characteristic impedance including the effect of line thickness is expressed as

$$Z_0 = \frac{1}{2\pi}\sqrt{\frac{\mu}{\epsilon_{eff}}}\ln\left(\frac{4h}{d}\right) \tag{6.78}$$

where,

$$\begin{aligned} d &= 0.536W + 0.67t \\ \epsilon_{eff} &= (0.475\epsilon_r + 0.67)\epsilon_0. \end{aligned} \tag{6.79}$$

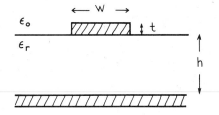

FIGURE 6.37   Microstrip line on a PC board

For $W/h < 1.25$ and $0.1 < t/W < 0.8$, the accuracy is 5 percent for $2.5 < \epsilon_r < 6$. Notice the similarity between the equations for microstrip and wire above a plane.

Microstrip on a semi-insulating substrate, such as gallium arsenide, is shown in Fig. 6.38(a). If the wire thickness is negligible and $W < h$, the line

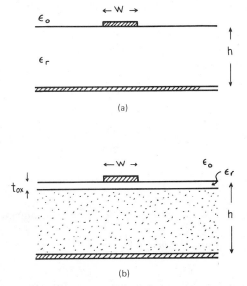

FIGURE 6.38   Microstrip line: (a) semi-insulating substrate, (b) semiconductor substrate

parameters are obtained as [6.8], [6.15]

$$Z_0 = \sqrt{\frac{\mathcal{L}}{C}}$$

$$C = \frac{2\pi\epsilon_{eff}\epsilon_0}{\ln\left(\dfrac{8h}{W} + \dfrac{W}{4h}\right)}$$

$$\mathcal{L} = \frac{\mu}{2\pi}\ln\left(\frac{8h}{W} + \frac{W}{4h}\right)$$
(6.80)

$$\epsilon_{eff} = \frac{\epsilon_r + 1}{2} + \frac{\epsilon_r - 1}{2}\left(1 + \frac{10h}{W}\right)^{-1/2}$$

Microstrip on a semiconductor substrate, such as silicon, is shown in Fig. 6.38(b). If the line thickness is negligible and $W < h$, the line parameters are obtained as [6.8]

$$Z_0 = \sqrt{\frac{\mathcal{L}}{C}}$$

$$C = \frac{2\pi\epsilon_{eff}\epsilon_0}{\ln\left(\dfrac{8t_{ox}}{W} + \dfrac{W}{4t_{ox}}\right)}, \qquad W \leq t_{ox}$$

$$C = \epsilon_r\epsilon_0\left[\frac{W}{t_{ox}} + 2.42 - 0.44\frac{t_{ox}}{W} + \left(1 - \frac{t_{ox}}{W}\right)^6\right], \quad W \geq t_{ox}$$

$$\mathcal{L} = \frac{\mu}{2\pi}\ln\left(\frac{8h}{W} + \frac{W}{4h}\right), \qquad h = t_{ox} + t_{Si}$$

$$\epsilon_{eff} = \frac{\epsilon_r + 1}{2} + \frac{\epsilon_r - 1}{2}\left(1 + \frac{10h}{W}\right)^{-1/2}.$$
(6.81)

A few general observations can be obtained from these formulas:

- If the propogation medium is homogenous,

$$\mathcal{L}C = \epsilon\mu$$

and

$$v = \frac{1}{\sqrt{\mathcal{L}C}} = \frac{c_0}{\sqrt{\epsilon_r}}.$$

- If the propagation medium is not homogenous (for example, microstrip on a SiO_2/Si substrate where Si acts as a conductor for capacitive effects and as an insulator for inductive effects), C and $\mathcal{L}$ are determined by different geometries and

$$\mathcal{L}C \neq \epsilon\mu$$

  usually capacitance is larger than it would have been without the dielectric and

$$v = \frac{1}{\sqrt{\mathcal{L}C}} < \frac{c_0}{\sqrt{\epsilon_r}}.$$

  This is the case in Eq. 6.81.

- The characteristic impedance of a TEM-mode transmission line can be expressed as

$$Z_0 = \sqrt{\frac{\mathcal{L}}{C}}$$

$$Z_0 = v\mathcal{L}$$

$$Z_0 = \frac{1}{vC}.$$

  If the line is a microstrip (or a similar structure that has a dielectric sandwiched between the line and ground plane with air above the line), then its impedance when the dielectric is replaced by air is given by

$$Z_{01} = \sqrt{\frac{\mathcal{L}}{C_1}}$$

$$Z_{01} = c_0\mathcal{L}$$

$$Z_{01} = \frac{1}{c_0 C_1}$$

  $\mathcal{L}$ is the same in both cases, and C and C_1 are the line capacitance with and without the dielectric. Combining the above equations, one obtains

$$Z_0 = \frac{1}{c_0\sqrt{CC_1}}$$

$$v = c_0\sqrt{\frac{C_1}{C}}.$$

  As a result, the properties of a microstrip or a similar structure can be obtained only by calculating the line capacitance with and without the dielectric [6.21].

  Some useful constants are:

$$\sqrt{\frac{\mu_0}{\epsilon_0}} = 377 \ \Omega$$

$$\frac{1}{2}\sqrt{\frac{\mu_0}{\epsilon_0}} = 188 \ \Omega$$

$$\frac{1}{\pi}\sqrt{\frac{\mu_0}{\epsilon_0}} = 120 \ \Omega$$

$$\frac{1}{4}\sqrt{\frac{\mu_0}{\epsilon_0}} = 94 \ \Omega$$

$$\frac{1}{2\pi}\sqrt{\frac{\mu_0}{\epsilon_0}} = 60 \ \Omega$$

$$\frac{1}{8}\sqrt{\frac{\mu_0}{\epsilon_0}} = 47 \ \Omega$$

$$\frac{1}{4\pi}\sqrt{\frac{\mu_0}{\epsilon_0}} = 30 \ \Omega.$$

7

CROSS TALK AND POWER DISTRIBUTION NOISE

In high-speed systems, the distribution of signals and power and the containment of noise are major design issues. Chapters 4, 5, and 6 discussed signal distribution by analyzing on-chip and chip-to-chip propagation delays in capacitive, distributed-RC, and transmission line domains. Chapter 5 also covered power distribution noise due to IR voltage drops along the lines, and Chapter 6 analyzed noise due to reflections from discontinuities along the line and at the end of the line. This chapter covers more noise-related issues, such as cross talk between the signal wires and noise associated with power distribution (inductive voltage fluctuations along the power lines).

Noise generated by off-chip drivers and on-chip circuitry is a major discipline in package and IC design for high-speed systems. Noise is closely related to interconnections. It increases as a result of shorter rise times, larger total chip currents, greater die dimensions, and smaller spacings between circuit components on the chips and boards. This has two consequences. First, as custom and semicustom CMOS chips get faster, issues that were significant only in high-end designs gain importance in all digital systems, both high and low end. Second, certain noise forms, such as cross talk, that previously were significant mostly at the board level become prevalent also at the chip level. With the advent of large-area superchips and multichip package substrates with dense thin film wires, the differences between ICs and their packages are diminishing. Chips are becoming like boards, and boards are becoming like chips.

As die dimensions and clock frequency increase, signal wavelengths become comparable to wire lengths, and this makes interconnections better "antennas." Capacitive, inductive, and resistive couplings between neighboring circuits grow because of smaller spacings [7.1]–[7.10]. Capacitive couplings increase because in order to keep the line resistance small, wire thickness is not reduced as much as line width. The IR drops across the power distribution lines become more significant as current densities are increased and minimum feature size is reduced [7.11]. With larger amounts of currents switching, the inductive noise associated with power lines also increases [7.12]–[7.18]. This is especially important in high-speed bipolar ECL chips and precharged CMOS circuits that require large current surges at the beginning of clock periods. These current transients can generate large potential drops due to the inductance of the power distribution network ($EMF = -d\Phi/dt = -LdI/dt$). This is referred to as power-supply-level fluctuations or as simultaneous switching, delta-I, ΔI, dI/dt, and LdI/dt noise.

Power-supply-level fluctuations presently receive more attention in high-speed, large I/O count bipolar circuits for mainframe computers because of the simultaneous switching of large numbers of fast output buffers in these systems. Until recently, a few decoupling capacitors placed on the board were sufficient to provide a good power distribution in a CMOS-based design. As CMOS becomes faster, delta-I noise gains importance. When CMOS circuits reach the speed and I/O count of bipolar ECL systems, such as an all-CMOS super or mainframe computer at liquid nitrogen temperature, the power-supply-level fluctuations will be just as important in CMOS systems.

In addition to growing noise generation, signal levels and supply voltage are lowered in accordance with the MOS scaling rules. The combined result of increased noise and reduced supply level can be a large degradation in the signal-to-noise ratio of future high-speed CMOS circuits if they are designed with conventional techniques and structures.

In this chapter, cross talk between signal lines and power-supply-level fluctuations caused by simultaneous switching noise are modeled, and the effects of reduced minimum feature sizes and larger chip dimensions on these noise components are analyzed. The models developed can be used for both chip- and package-level calculations. The effectiveness of ground planes, multiple power connections, and off-chip and on-chip decoupling capacitors in reducing noise generation is also evaluated.

7.1   Reflections, Cross Talk, and Simultaneous Switching Noise

Reflections are caused by the discontinuities at the transmission lines (for example, capacitive loading due to the package pins tied to a data bus with multiple sinks on a PC board). Reflections from inductive and capacitive discontinuities and from the end of the line were analyzed in Chapter 6.

Cross talk is a result of mutual capacitances and inductances between neighboring lines. The closer the lines, the higher they are from the ground plane, and the longer a distance they neighbor each other, the larger the amount of coupling is. Cross talk is also proportional to the rate of change of the voltage waveforms.

Simultaneous switching noise is caused by the inductances inherent in the power distribution lines. One of the basic laws of physics (Faraday's law) states that any change in the magnetic flux will be confronted by an opposition, namely a self-induced emf, determined by the rate of change in the total magnetic flux, $EMF = d\Phi/dt = LdI/dt$. Accordingly, when many circuits switch at the same time, the current supplied by the power lines can change rapidly, and the inductive voltage drop along the line can cause the power-supply level to go down. The resulting voltage glitch is proportional to the switching speed, the number of drivers that switch simultaneously, and the effective inductance of the power lines.

All three kinds of noise (reflections, cross talk, and simultaneous switching) can cause increased delays or inadvertent logic transitions. They should be minimized through careful design [7.12]. It is important to clarify the difference between the *delay* and *inadvertent logic transition* faults. Delays of synchronous signals may grow due to reflections (nodes may take a longer time to settle), cross talk (coupling between the lines may increase the effective line capacitance and the propagation delay), or simultaneous switching (a reduction in the supply level diminishes the current drive of a circuit and increases its delay). Accordingly they should be included in the worst-case delay calculations. The clock period should be selected such that the sampling period of synchronous signals takes into account all circuit- and noise-related delays. The important distinction between delay and logic faults is that all the above delay faults can be resolved by increasing the cycle time. In contrast, logic faults cannot be eliminated by increasing the cycle time. The major sources of logic faults are asynchronous signals and clocks, such as chip clocks, RAS and CAS signals to memory, and asynchronous I/O signals. If a sufficiently large voltage pulse is injected into one of these lines (for example, due to simultaneous switching of output buffers or coupling between two lines), the state of the receiving circuit will change. For example, Fig. 7.1 shows how a glitch at the power line can be perceived as a glitch at the clock line even if the clock line is perfectly steady. If the power line that feeds the receiving

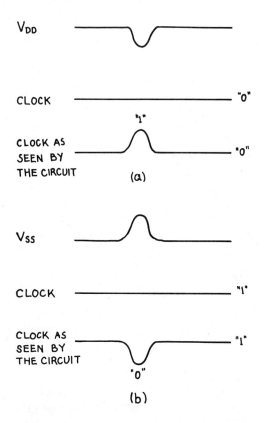

FIGURE 7.1   Perceived clock glitch due to noise at the power lines: (a) negative glitch at V_{DD} perceived as a glitch at the clock line, (b) positive glitch at V_{SS} perceived as a negative glitch at the clock line

latch has a glitch, all the inputs will be perceived as if they shifted because all potential levels are relative to the power-supply level. Consequently there are two kinds of noise at the asynchronous and clock lines that can cause logic faults:

1. Noise at the driver and signal wires. Glitches on the clock and signal lines due to cross talk along the line or power-supply-level fluctuations at the chip output buffer that drives the line.

2. Noise at the receiver. Power-supply-level fluctuation at the receiving chip that is perceived as a glitch at the clock and asynchronous signals because power and ground lines establish the reference levels for all signals.

These problems are a function of the switching speed of the circuits and cannot be resolved by simply reducing the frequency of the system clock.

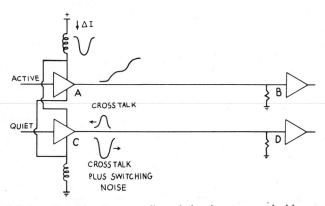

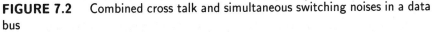

FIGURE 7.2   Combined cross talk and simultaneous switching noises in a data bus

Another difference between the delay and logic failures is that delay faults show up when the ICs are from the slow end of the distribution, and the inadvertent logic transition faults show up when the circuits are from the fast end of the distribution. Delay faults are delay adders; as a result they show up when circuits are slow. Logic faults are due to glitches, and glitches are usually generated more strongly when circuits switch faster than normal.

Precharged circuits form another class susceptible to logic faults. As described in Chapters 2 and 4, in a precharged bus or domino logic, if the data have a glitch during the evaluation period, the precharged node may be discharged inadvertently. Such a noise is more likely to show up if the circuit is from the fast end of the distribution. The problem cannot be resolved by slowing the clock frequency. Unlike static circuits, once a precharged node is discharged, it will not go back to its precharged state until the next clock cycle. Similar cases exist in SRAM and DRAM circuits because they make extensive use of dynamic circuit techniques.

The following example illustrates how reflections, cross talk, and simultaneous switching noise can all add up in a bipolar system. Consider a 32-bit data bus at a state '00···010···00,' and the 31 bits that are '0' switch to '1,' and the line that is a '1' stays high (Fig. 7.2). The plus power supply at the driver chip will dip below V_{DD} because 31 drivers simultaneously start pulling current from the power supply to charge their outputs high. This dip also appears at the power terminal of the quiet driver that is connected to the same power line as the rest of the data bus. Consequently, the output of the quiet driver, which is supposed to stay high, goes somewhat below the high logic level (simultaneous switching noise). In addition, the rising waveforms at the switching lines couple to the quiet line, and a positive pulse is generated that travels toward the quiet driver (cross talk). Because parallel termination at the receiving end is employed in bipolar circuits ($R_S \ll Z_0$, $R_L = Z_0$), the driver has a source resistance much smaller than the characteristic impedance of the line, and the reflection coefficient at the source end is approximately -1. Then the noise pulse is inverted as it reflects from the quiet driver (reflection). As a result the dips due to switching and coupling noises have the same polarity; they add together and start traveling toward the receiver [7.12]. For synchronous lines this pulse will increase the delay and should be included in the worst-case delay calculations. For clocks and asynchronous lines, it may cause inadvertent switching and should be kept small. Designers must make sure that the worst-case sum of these perturbations does not increase the cycle time beyond the design goals for synchronous signals, and it is well within the noise margin of the receiver circuits for asynchronous signals.

The above example is for a bipolar system. In CMOS, source impedance of drivers is large, and the noise pulse is absorbed by the driver rather than being inverted and reflected. The worst-case scenario for asynchronous CMOS lines is the same as the bipolar example described above because that is when the largest glitches are introduced due to power-supply-level collapse. The worst-case delay scenario for synchronous CMOS lines, however, is when the data bus is in a state '1010...1010' and all the 1s switch to 0s, and all the 0s switch to 1s. Then the simultaneous switching increases the delay because the plus power supply V_{DD} dips down when half the drivers switch from 0 to 1, and the negative power supply V_{SS} bumps up when half the drivers switch from 1 to 0. In addition, when the alternating lines on the PC board are switching in opposite directions, the effective capacitance of the lines is at their maximum.[1] This is because one plate of the mutual capacitance goes from V_{SS} to V_{DD} and the other plate goes from V_{DD} to V_{SS}. The effective voltage swing across the coupling capacitor is $2(V_{DD} - V_{SS})$, twice the voltage swing on individual lines.

7.2   Optimal Line Impedance

The impedance of signal wires is very important because it determines the propagation delay, noise level, and power dissipation. In this section, properties of high- and low-impedance transmission lines are compared, and optimal values are investigated for bipolar and CMOS systems.

The smaller the line impedance, the larger its capacitance ($Z_0 = 1/v\mathcal{C}$).

---

[1] This is similar to the Miller effect in analog amplifiers, which is observed when the input of the amplifier is tied to one plate of a capacitor and the amplifier output feeds the other plate of the same capacitor. Then the effective capacitance is $(1 + \beta)C$ where β is the voltage gain of the amplifier. This can increase the effective value of the capacitance tremendously because the amplifiers usually have a large gain. The "gain" in PC board lines is the coupling coefficient, which is less than 1 (usually 0.10 to 0.30). As a result, the increase in the effective capacitance is significant but not as much as in amplifiers.

Accordingly a stronger driver is required to achieve fast propagation delays when the line impedance is small. If the driver resistance is not much smaller than Z_0, driver and line will act as a voltage divider, and the initial potential step on the line will be proportional to $Z_0/(Z_0 + R_S)$. With $R_S > Z_0$, many round-trips and reflections from the end of the line will be required for the output to reach its final value. As described in Chapter 6, source-end termination partially resolves this problem, but even with source-end termination, lines with smaller Z_0 require stronger drivers. In CMOS, stronger drivers have more buffer delay because more stages of cascaded inverters are required to achieve a very small output impedance. Consequently, large Z_0 is desirable because it results in an easy-to-drive line with a small capacitance.

When the line impedance is increased too much, however, circuit speed may degrade. Delays due to capacitive discontinuities along or at the end of the line are proportional to Z_0 because large impedance lines (which do not require large currents) cannot supply as high a current as small impedance lines. The delay adders due to discrete capacitive loads or short stubs are proportional to $Z_0 C_L$. As a result, small Z_0 helps to reduce delays due to capacitive loads attached to the transmission line.

Switching noise goes down with increasing Z_0 because a high-impedance line has a small capacitance and requires smaller current surges ($\Delta I \propto \Delta V/Z_0$). Cross talk, however, increases with Z_0 because a high-impedance line has a small capacitance to the reference plane. When the line and the reference plane are not strongly coupled, the coupling between the lines gains more importance and cross talk increases. In addition, self- and mutual inductances of the lines increase with Z_0 because larger Z_0 implies lines high above the ground plane with a large current loop area and inductance ($Z_0 \propto \mathcal{L}$). Reflected noise also increases with Z_0 because the smaller the line capacitance, the stronger the influence of capacitive discontinuities.

Power dissipation is smaller with large Z_0. In CMOS technology, where series termination is applied and the switching power is dominant, small line capacitance requires less current and dissipates less power ($P \propto fCV^2$). In bipolar technology, where the nets are terminated with resistors that conduct steady-state current, large impedance lines require large termination resistors and consume less power ($P \propto V^2/Z_0$).

In summary, large impedance lines

- have small capacitances
- are easier to drive (and as a result they require smaller buffers with shorter buffer delays)
- generate less switching noise
- dissipate less power in both CMOS and bipolar technologies

On the other hand, small impedance lines

- have large capacitances

- can supply a large current to capacitive loads (and as a result they have less added delay due to capacitive loading of fan-out nodes)

- are strongly coupled to the reference plane (and as a result they generate less cross talk with neighbors)

- receive smaller and narrower reflections from capacitive discontinuities (because a line with a large capacitance is more stable)

Davidson calculated that the optimal transmission line impedance for bipolar systems is 60–80 Ω, and when it is reduced below 40 Ω, many characteristics degrade very rapidly [7.12].

For a CMOS system, the optimal line impedance is somewhat higher (80–100 Ω) because CMOS drivers have higher resistances than bipolar drivers, and high-resistance drivers are more sensitive to line capacitance and perform better when the line impedance is high. The power dissipation is also kept low when the line impedance is high. In addition, CMOS receivers have better noise margins, and no reflections are generated from the source end because it is matched with the transmission line. The fact that high-impedance lines cannot supply a large current to capacitive loads is not as important in CMOS because it does not pay off to make line Z_0 smaller than R_S of the driver. In other words, series-terminated CMOS off-chip drivers with high source impedance may require multiple trips along the line to switch the receiver, and this eliminates the advantages of the fast time constant of a low-impedance line. In CMOS, the delay is determined by the combination of the line and driver impedance. In CMOS systems, therefore, the line impedance is maximized to reduce the delay in synchronous lines. The main disadvantage of high-impedance lines is that they increase coupled noise, and to compensate for it, asynchronous and clock lines must be carefully isolated to avoid inadvertent switching. Noise at the synchronous signal lines acts only as a delay adder, and in most CMOS systems the delay improvement due to smaller wire capacitance outweighs the additional delay due to increased cross talk.

Another intrinsic difference between bipolar and CMOS must be made clear to explain the assumptions behind the above comparisons. It is easy to make low-resistance drivers with bipolar transistors because they have high current drive capability and excellent gain. It is also possible to make very low-resistance drivers with CMOS, but it is not desirable for two reasons. The first reason is the area considerations. A low-impedance CMOS driver with many stages of large transistors requires a large silicon area. The second reason is the low gain of CMOS devices. Many stages of cascaded drivers are required to achieve a final stage with an output resistance of a few ohms. The delay from the output of the buffer to the receiver is small, but the overall delay is large because of the delay of the cascaded driver chain in the output buffer. The overall delay is not optimized. This is not a problem in bipolar because a driver with a few ohms of output resistance can be obtained with only two stages due to the large gain of bipolar devices (see Chap. 2).

The only circuit where the delay of the CMOS cascaded driver chain is inconsequential is the clock buffer. In clock drivers, small skew and a sharp rise time are important; a constant delay has no effect as long as it adds to every copy of the clock signal exactly the same way. Consequently a long clock buffer delay is acceptable in a single-chip system. Even with the above considerations, a short clock buffer delay is still desirable in multichip systems. It is easy to guarantee that the clock buffer delay is the same for all the circuits if there is only one clock buffer in the system but becomes difficult when there are multiple clock buffers. For example, in a multichip system, the *nominal* clock buffer delays on individual chips can be designed to be equal, but they will vary due to semiconductor device process variations. The longer the total clock buffer delay is, the larger the skew due to process variations will be. Consequently, in this case, it is desirable to keep the clock buffer delay short. Phase-locked loops can be used to minimize this skew, as will be shown in Chapter 8.

7.3   Cross Talk

In Chapter 1, a simple model for cross talk based on capacitive charge sharing was described. For long lines or fast-rising waveforms, the simple model is not appropriate; transmission line domain calculations are required that include both capacitive and inductive couplings.

7.3.1   Modeling of Coupled Transmission Lines

In order to analyze the cross talk between package or chip interconnections, wires can be modeled as coupled microstrip lines [7.3], [7.7]. A two-line system can support two modes of propagation with separate impedances and propagation speeds.[2] Any perturbation traveling in a coupled transmission line pair can be expressed as a superposition of these two modes. If the lines are symmetric, the two modes are *even* and *odd*. In the even mode, both lines are positive with respect to ground; they are excited in phase and with equal amplitude. In the odd mode, one of the lines is positive with respect to ground, and the other is negative; they are excited 180 degrees out of phase and again with equal amplitude. Any signal traveling in the coupled transmission line

---

[2]In general, an n line system can support n modes of propagation, each with different characteristic impedance and phase velocity.

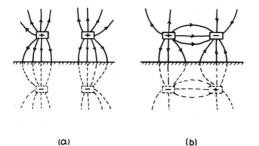

(a)   (b)

FIGURE 7.3 Even- and odd-mode voltage excitations in coupled transmission lines: (a) even mode, (b) odd mode. The lines as well as their images are shown. The ground plane can be replaced by the images.

system can be expressed as a superposition of even and odd modes. Figure 7.3 illustrates the even- and odd-mode electric field patterns. The lines, as well as their images through the ground plane, are shown. A structure consisting of two lines and a ground plane is equivalent to a system of two lines and their images; consequently the ground plane can be replaced by the images of the lines. Sometimes this can be used to simplify the analysis.

The characteristic impedances for the even and odd modes of propagation are Z_{0e} and Z_{0o}. The even-mode impedance can be calculated by virtually tying the two lines together and finding the impedance for the resulting two-conductor system: two lines tied together and a ground plane. The odd mode is harder to conceptualize. Here it is best to replace the ground plane with the images of the lines. In the resulting four-line system, the lines that are in phase are virtually connected together, and again the coupled lines are reduced to a two-conductor system, which can be used to calculate the odd-mode impedance.

The even- and odd-mode parameters are related to line capacitance and inductance as shown below:

$$Z_{0e} = \sqrt{\frac{\mathcal{L}_e}{\mathcal{C}_e}}$$

$$Z_{0o} = \sqrt{\frac{\mathcal{L}_o}{\mathcal{C}_o}}$$

$$v_e = \frac{1}{\sqrt{\mathcal{L}_e \mathcal{C}_e}}$$

$$v_o = \frac{1}{\sqrt{\mathcal{L}_o \mathcal{C}_o}}$$

(7.1)

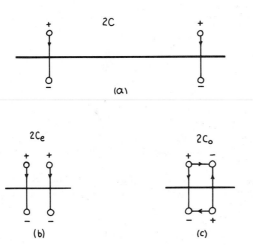

FIGURE 7.4   Intrinsic and even- and odd-mode capacitances of coupled transmission lines: (a) lines are far apart and uncoupled, (b) even-mode capacitance of two coupled lines, (c) odd-mode capacitance of two coupled lines

where
$$\begin{aligned}
Z_{0e} &= \text{even-mode characteristic impedance of a line}\\
Z_{0o} &= \text{odd-mode characteristic impedance of a line}\\
C_e, C_o &= \text{even- and odd-mode capacitances per unit length of a line}\\
\mathcal{L}_e, \mathcal{L}_o &= \text{even- and odd-mode inductances per unit length of a line}\\
v_e, v_o &= \text{even- and odd-mode phase velocities.}
\end{aligned}$$

It can be shown that

$$Z_{0o} < Z_0 < Z_{0e}. \tag{7.2}$$

This can be conceptualized with the aid of Fig. 7.4, which illustrates the capacitance calculations for two uncoupled lines and the even- and odd-mode coupled pairs. Again the ground plane is replaced by the images of the lines to aid the derivation. In Fig. 7.4(a), the wires are infinitely apart and uncoupled. The total capacitance of the system is $2C$ where C is the per unit length capacitance of an isolated line. The even mode in Fig. 7.4(b) has the lines close together. Here the total capacitance is $2C_e$. Because some of the fringing fields will be lost due to the overlap of the area between lines, the total capacitance in Fig. 7.4(b) is less than that in Fig. 7.4(a). Consequently $C > C_e$, which implies that $Z_{0e} > Z_0$ because $Z_0 \propto 1/C$. The odd-mode structure in Fig. 7.4(c) has the largest capacitance because the fields do not overlap as in Fig. 7.4(b), and the opposite polarity of the wires maximizes the efficiency of the capacitive coupling. Therefore $C_o > C$ and $Z_0 > Z_{0o}$. The closer the lines, the stronger the coupling and the larger the difference between Z_{0e}, Z_0, and Z_{0o}. For loosely coupled lines, the following equation applies:

$$Z_0 \approx \sqrt{Z_{0e}Z_{0o}}. \tag{7.3}$$

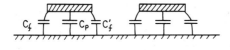

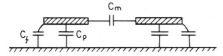

FIGURE 7.5   Coupled microstrip lines. Various components of the even- and odd-mode capacitances are labeled. Even mode is shown at the top and odd mode at the bottom.

This will be proved later.

In coupled microstrip lines, the even- and odd-mode capacitances of the lines can be obtained as shown in Fig. 7.5.

$$
\begin{aligned}
C_i &= C_p + 2C_f \\
C_e &= C_p + C_f + C_f' \\
C_o &= C_p + C_f + C_m
\end{aligned}
\tag{7.4}
$$

where

$\quad C_i \quad = \quad$ capacitance of an isolated line

$\quad C_p \quad = \quad$ parallel plate capacitance between the line and ground plane

$\quad C_f \quad = \quad$ fringing field capacitance between one edge of the line and ground plane

$\quad C_f' \quad = \quad$ reduced fringing field capacitance due to overlap between the fields of neighboring lines

$\quad C_m \quad = \quad$ coupling (mutual) capacitance between the lines,

all per unit length.

Two books that present detailed theories of microstrip lines are [7.5] and [7.6]. An analysis of coupled microstrip lines for IC applications can be found in reference [7.7].

7.3.2   Calculation of the Coupling Constant

The following is a derivation of the coupling constant between two lines as a function of their even- and odd-mode impedances [7.3]. In Fig. 7.6 two transmission lines, an active one A and a quiet one Q, and a ground plane G are schematically represented. The near end of the quiet line is an open circuit, which results in maximum coupling at that end ($\Gamma_{Q1} = 1$).

Assume that a voltage/current step of (V, I) is traveling down the line A.

The coupling between lines A and Q starts when the exciting wave reaches point $A1$. Then the single line transforms into a coupled transmission line pair, and separate even and odd modes are excited. The waves traveling along A and Q can be expressed as a superposition of these modes. In the even mode, both lines A and Q are positive with respect to G, and in the odd mode, A is positive and Q is negative with respect to G. The magnitudes of even- and odd-mode voltages and currents are same in both lines. Consequently the voltages and currents at $A1$ and $Q1$ can be expressed as:

$$\begin{aligned} V_{A1} &= V_e + V_o \\ V_{Q1} &= V_e - V_o \\ I_{A1} &= I_e + I_o \\ I_{Q1} &= I_e - I_o. \end{aligned} \tag{7.5}$$

Assuming that the lines are loosely coupled, the impedance of line A at point $A1$ is not altered significantly by the presence of the line Q. Then the wave travels down line A with negligible change, and the following boundary conditions must be met at points $A1$ and $Q1$. Continuity of voltage across point $A1$ requires that

$$V_{A1} = V_e + V_o = V. \tag{7.6}$$

Because $Q1$ is an open circuit, conservation of charge at point $Q1$ requires that

$$I_{Q1} = I_e - I_o = 0 \Rightarrow I_e = I_o. \tag{7.7}$$

The coupling coefficient between the lines is defined as

$$K_V = \frac{V_{Q1}}{V}. \tag{7.8}$$

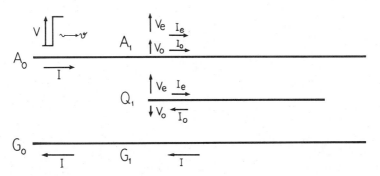

FIGURE 7.6 Calculation of the coupled waveforms between two transmission lines. An active line (A), a quiet line (Q), and ground plane (G) are illustrated. Even- and odd-mode components of the voltage and current at the initial coupling point are also shown.

Substituting the value of V_{Q1} and V from Eqs. 7.5 and 7.6, K_V can be expressed as

$$K_V = \frac{V_e - V_o}{V_e + V_o}. \qquad (7.9)$$

Using the definition of Z_{0e} and Z_{0o}, Eq. 7.9 can be transformed into

$$K_V = \frac{I_e Z_{0e} - I_o Z_{0o}}{I_e Z_{0e} + I_o Z_{0o}}. \qquad (7.10)$$

And using Eq. 7.7, $I_e = I_o$, the coupling coefficient is obtained as

$$K_V = \frac{Z_{0e} - Z_{0o}}{Z_{0e} + Z_{0o}}. \qquad (7.11)$$

The magnitude of the cross talk given by Eq. 7.11 is the maximum that can appear anywhere on the line Q for any values of terminating resistances at its ends. Accordingly this is an appropriate value for worst-case design [7.3]. Even- and odd-mode impedances can be calculated using computer programs or analytical methods and can be used as a measure of cross talk. It is a good practice to keep K_V below 25 percent.

7.3.3 Capacitance and Inductance Matrices

The couplings in an environment with multiple conductors can be determined by capacitance and inductance matrices. The capacitance matrix is defined as

$$\mathbf{Q} = \mathbf{C} \times \mathbf{V}, \qquad (7.12)$$

where Q_i is the charge at node i, V_i is the potential at node i, C_{ii} is the capacitance between node i and the rest of the system when all the other nodes are grounded, and C_{ij} is the coupling capacitance between nodes i and j—all per unit length. For a two-line system, the matrix equation is:

$$\begin{bmatrix} Q_1 \\ Q_2 \end{bmatrix} = \begin{bmatrix} C_{11} & -C_{12} \\ -C_{21} & C_{22} \end{bmatrix} \times \begin{bmatrix} V_1 \\ V_2 \end{bmatrix} \qquad (7.13)$$

where

$$C_{11} = \left. \frac{Q_1}{V_1} \right|_{V_2=0}$$

$$C_{22} = \left. \frac{Q_2}{V_2} \right|_{V_1=0} \qquad (7.14)$$

$$C_{12} = C_{21} = \left. -\frac{Q_1}{V_2} \right|_{V_1=0} = \left. -\frac{Q_2}{V_1} \right|_{V_2=0}.$$

In a symmetric system, C_{11} and C_{22} are same as C defined earlier, and C_{12} and C_{21} are same as C_m.

The inductance matrix is defined as

$$\mathbf{\Phi} = \mathbf{L} \times \mathbf{I}. \tag{7.15}$$

By taking the time derivative of both sides and using $V = d\Phi/dt$, one obtains

$$\mathbf{V} = \mathbf{L} \times \dot{\mathbf{I}}, \tag{7.16}$$

where $\dot{\mathbf{I}}$ is the derivative of $\mathbf{I}$ with respect to time. For a two-line system the inductance matrix equation is:

$$\begin{bmatrix} V_1 \\ V_2 \end{bmatrix} = \begin{bmatrix} L_{11} & L_{12} \\ L_{21} & L_{22} \end{bmatrix} \times \begin{bmatrix} \frac{dI_1}{dt} \\ \frac{dI_2}{dt} \end{bmatrix}. \tag{7.17}$$

In a symmetric system, L_{11} and L_{22} are same as $\mathcal{L}$ defined earlier, and L_{12} and L_{21} are same as $\mathcal{L}_m$. Terms of the capacitance and inductance matrices can be determined by a computer-aided analysis.

In a homogeneous medium, the capacitance and inductance matrices are the inverse of each other,

$$\frac{1}{v^2}\mathbf{U} = \mathbf{L} \times \mathbf{C}, \tag{7.18}$$

where $\mathbf{U}$ is the unity matrix. This is analogous to the one-dimensional equation $v = 1/\sqrt{\mathcal{L}\mathcal{C}}$.

It can be shown that for a symmetric coupled line pair, the even- and odd-mode propagation speeds are given by

$$
\begin{aligned}
v_e &= \frac{1}{\sqrt{(\mathcal{L} + \mathcal{L}_m)(\mathcal{C} - \mathcal{C}_m)}} \\[2mm]
v_o &= \frac{1}{\sqrt{(\mathcal{L} - \mathcal{L}_m)(\mathcal{C} + \mathcal{C}_m)}}
\end{aligned}
\tag{7.19}
$$

where $\mathcal{C}_m$ and $\mathcal{L}_m$ are mutual capacitance and inductance per unit length. In a symmetric system, the inductances and capacitances in Eq. 7.19 can be expressed in terms of the components of the capacitance and inductance matrices as follows:

$$
\begin{aligned}
\mathcal{C} &= C_{11} = C_{22} \\
\mathcal{C}_m &= C_{12} = C_{21} \\
\mathcal{L} &= L_{11} = L_{22} \\
\mathcal{L}_m &= L_{12} = L_{21}.
\end{aligned}
\tag{7.20}
$$

In addition, the even- and odd-mode inductance, capacitance, and impedances can be expressed as

$$
\begin{aligned}
\mathcal{L}_e &= \mathcal{L} + \mathcal{L}_m \\
\mathcal{C}_e &= \mathcal{C} - \mathcal{C}_m \\
\mathcal{L}_o &= \mathcal{L} - \mathcal{L}_m \\
\mathcal{C}_o &= \mathcal{C} + \mathcal{C}_m
\end{aligned}
$$

$$Z_{0e} = \sqrt{\frac{\mathcal{L} + \mathcal{L}_m}{C - C_m}}$$

$$Z_{0o} = \sqrt{\frac{\mathcal{L} - \mathcal{L}_m}{C + C_m}}.$$

(7.21)

Now, the equality in Eq. 7.3, which holds true for loosely coupled lines ($\mathcal{L}_m \ll \mathcal{L}$ and $C_m \ll C$), can be proved:

$$Z_{0e} Z_{0o} = \sqrt{\frac{\mathcal{L}^2 - \mathcal{L}_m^2}{C^2 - C_m^2}} \approx \sqrt{\frac{\mathcal{L}^2}{C^2}} = Z_0^2.$$

(7.22)

Once Z_{0e} and Z_{0o} are solved in terms of $\mathcal{L}$, $\mathcal{L}_m$, C, and C_m, the coupling coefficient K_V in Eq. 7.11 can also be expressed in terms of the same parameters. As a result, all impedance and coupling parameters can be obtained as a function of $\mathcal{L}$, $\mathcal{L}_m$, C, and C_m. They, in turn, are the terms of the capacitance and inductance matrices, which are the usual outputs of most two-dimensional transmission line analysis programs.

7.3.4   Cross Talk in Homogeneous and Inhomogeneous Mediums

If a coupled transmission line pair is embedded in a homogeneous dielectric, both even- and odd-mode waves travel at the same speed:

$$v_e = \frac{1}{\sqrt{\mathcal{L}_e C_e}} = v_o = \frac{1}{\sqrt{\mathcal{L}_o C_o}} = \frac{c_0}{\sqrt{\epsilon_r}}.$$

(7.23)

If the medium is *not* homogeneous, the inductance and capacitance are not true complements of each other, and $v = 1/\sqrt{\mathcal{L}C} \neq c_0/\sqrt{\epsilon_r}$. Then even- and odd-mode waves travel at different speeds, which results in degradation of waveforms and increased noise couplings.

Using Eqs. 7.18 and 7.20, in a homogeneous medium, the electrical and magnetic couplings can be described by the following relation:

$$\frac{C_m}{C} = \frac{\mathcal{L}_m}{\mathcal{L}}.$$

(7.24)

Substituting Eq. 7.24 in Eq. 7.19, it can be shown that in a homogeneous dielectric medium, $v_e = v_o$.

In addition, Z_{0e} and Z_{0o} from Eq. 7.21 can be substituted in Eq. 7.11, and then Eq. 7.24 can be used to obtain

$$K_V = \frac{C_m}{C} = \frac{\mathcal{L}_m}{\mathcal{L}}.$$

(7.25)

Equivalently, the terms of the capacitance and inductance matrices can be used to calculate the coupling coefficient for multiple conductor systems where the number of lines can be more than two. The worst-case capacitive coupling coefficient[3] for node j is given by

$$K_{Cj} = \frac{\sum_{i \neq j} C_{ij}}{C_{jj}}, \qquad (7.26)$$

where C_{ij} are the terms of the capacitance matrix. The inductive coupling coefficient is given by

$$K_{Lj} = \frac{\sum_{i \neq j} L_{ij}}{L_{jj}}, \qquad (7.27)$$

where L_{ij} are the terms of the capacitance matrix. The coupling coefficient in Eq. 7.11 can be expressed as

$$K_V = \frac{K_C + K_L}{2}. \qquad (7.28)$$

In a homogeneous medium, K_L is equal to K_C; as a result,

$$K_V = \frac{K_C + K_L}{2} = K_C = K_L. \qquad (7.29)$$

This gives the same result as Eq. 7.25.

When a pair of coupled lines is in a homogeneous medium, the capacitive and inductive couplings are equal in magnitude and opposite in sign, and they travel at the same speed. As a result there would be no far-end cross talk if the lines were in a homogeneous medium and they were terminated with matched resistors at the source end. In an inhomogeneous medium, however, the proportionality described by Eq. 7.24 does not hold, and the inductive and capacitive couplings are not equal. Consequently noise waveforms are observed at the far end.

An example of an inhomogeneous transmission line is the microstrip where there is a dielectric between the line and the ground plane, and above the line there is air. This forces the electric field lines to concentrate between the line and the plane, but it does not affect the shape of the magnetic field pattern. As a result, the line capacitance and inductance are altered by the existence of the dielectric such that

$$v = \frac{1}{\sqrt{\mathcal{L}\mathcal{C}}} \neq \frac{c_0}{\sqrt{\epsilon_r}}$$

[3] The worst-case capacitive coupling coefficient occurs when all of the other nodes are initially at V_{SS}, and all switch to 1 simultaneously, or equivalently when they are all at V_{DD} and switch to 0 simultaneously.

and

$$\frac{C_m}{C} \neq \frac{L_m}{L}.$$

Another practical example is a transmission line on an SiO_2/Si substrate. The substrate is made of two layers of different dielectrics; it is not homogeneous. Even in PC boards and multilayer ceramic substrates where the line is embedded in a homogeneous dielectric, the medium is not totally homogeneous because of the existence of orthogonal lines and vias. As a result of these conductors, the capacitive coupling between two parallel conductors is reduced, but the mutual inductance remains approximately the same.

7.3.5   Near- and Far-End Noise

Consider two loosely coupled transmission lines[4] as shown in Fig. 7.7. Because the lines are loosely coupled, the waveform traveling on the active line will not be affected significantly, and as it travels down the active line, small capacitive and inductive currents I_C and I_L will be generated in the quiet line [7.1]–[7.2], [7.4]. Both I_C and I_L are proportional to the slope of the driving voltage waveform. The capacitive current divides into two equal parts, one traveling toward each end of the line. As a result capacitively coupled voltage pulses, in the same polarity as the exciting signal, propagate in both directions from the point of coupling. The inductive current travels toward the near end; inductively coupled voltage pulses propagate backward with the same polarity but forward with the opposite polarity. Consequently capacitive and inductive cross talk add in the reverse direction, but they subtract in the forward direction.

At the near end, capacitive and inductive currents add up for a period of time that is equal to the round-trip along the transmission line. This is because I_C and I_L start arriving at the near end at time zero, and the last wave that arrives is generated at the far end at time t_f and takes another t_f to reach the near end. Here t_f is the time-of-flight delay across the line. If the line is sufficiently long such that $2t_f$ is greater than the rise time of the driving signal, the near-end noise pulse has an amplitude that is an attenuated version of the wave that travels on the active line, and the pulse width of the noise waveform is equal to the round-trip delay of the active line,

$$V_{ne}(t) = K_{ne}\left[V_{in}(t) - V_{in}(t - 2t_f)\right], \qquad (7.30)$$

---

[4]Notice that the boundary condition in Fig. 7.7 is different from the condition in Fig. 7.6. In Fig. 7.7, the near end of the quiet line is terminated with Z_0. In Fig. 7.6, the near end is an open circuit.

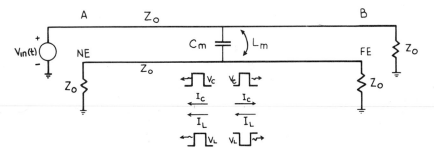

FIGURE 7.7   Forward and backward noise generation in a coupled transmission line pair. The mutual capacitance and inductance are C_m and $\mathcal{L}_m$. The capacitively coupled voltage and current pulses are V_C and I_C, and inductively coupled pulses are V_L and I_L.

where

$$K_{ne} = \frac{v}{4}\left[Z_0 C_m + \frac{\mathcal{L}_m}{Z_0}\right]. \tag{7.31}$$

The amplitude of the near-end noise will be reduced if $2t_f$ is less than the rise time of V_{in} because the waveform will not have sufficient time to reach its potential peak. If the driver resistance were not equal to Z_0, this pulse would be reflected from the source and observed at the far end.

At the far end, the capacitive and inductive currents are in the opposite direction. If the medium is homogeneous, I_C and I_L are equal, and there is no far-end noise.[5] Both the backward and forward coupled noises get generated at the point where the exciting wave is. Backward noise gets separated from the exciting wave because it travels in the opposite direction. The incremental backward pulses continuously arrive at the near end for a time period of $2t_f$. The incremental pulses that form the far-end noise, on the other hand, travel along with the driving wave. The forward pulses get generated and build up as the wave travels down the line, and all arrive at the receiving end together. Consequently the longer the line length, the bigger the forward noise buildup. As described before, line length does not influence the amplitude of the near-end noise but increases its pulse width, which is $2t_f$. Far-end noise, on the other hand, is proportional to the derivative of the driving function and the length of the line,

$$V_{fe}(t) = K_{fe}l\frac{dV_{in}(t-t_f)}{dt}, \tag{7.32}$$

where

$$K_{fe} = \frac{1}{2}\left[Z_0 C_m - \frac{\mathcal{L}_m}{Z_0}\right]. \tag{7.33}$$

---

[5]If the source impedance is not Z_0, near-end noise is reflected and observed at the far end as a reflected noise but not as an intrinsic far-end noise.

If the propagation medium is homogeneous, the capacitive and magnetic couplings have the relationship

$$\frac{C_m}{C} = \frac{\mathcal{L}_m}{\mathcal{L}},$$ (7.34)

which implies that

$$I_C = I_L$$ (7.35)

and

$$
\begin{aligned}
K_{ne} &= \frac{v}{4}\left[Z_0 C_m + \frac{\mathcal{L}_m}{Z_0}\right] \\
&= \frac{1}{4\sqrt{C\mathcal{L}}}\left[\sqrt{\frac{\mathcal{L}}{C}}C_m + \mathcal{L}_m\sqrt{\frac{C}{\mathcal{L}}}\right] \\
&= \frac{1}{4}\left(\frac{C_m}{C} + \frac{\mathcal{L}_m}{\mathcal{L}}\right) \\
&= \frac{C_m}{2C}.
\end{aligned}
$$ (7.36)

The near-end coupling coefficient K_{ne} in Eq. 7.36 is half of the coupling co-efficient K_V in Eqs. 7.11, 7.25, and 7.29 because in calculating K_V, the near end is assumed to be an open circuit, and in calculating K_{ne}, near end has a terminating resistor equal to Z_0.

In a homogeneous medium, the far-end coupling coefficient is obtained as

$$
\begin{aligned}
K_{fe} &= \frac{1}{2}\left[Z_0 C_m - \frac{\mathcal{L}_m}{Z_0}\right] \\
&= 0.
\end{aligned}
$$ (7.37)

The resulting waveforms are shown in Fig. 7.8. Again the far end of the line in a homogenous medium will receive no noise pulses only if the source impedance is Z_0. Otherwise near-end noise will be reflected and observed at the far end as reflected noise. Equation 7.37 shows only the intrinsic far-end coupling coefficient without taking into account any reflections.

The amount of cross talk at the quiet line is strongly influenced by the termination conditions. The worst-case far-end noise is generated when the near end of the quiet line has a low impedance ($R_S \ll Z_0$), and the far end has a high impedance (open circuit or capacitive load). Then the backward cross talk is totally reflected from the near end and adds up with the forward cross talk. In addition, when the noise pulse arrives at the far end, it doubles because of the high-impedance termination. The minimum cross talk is obtained when both near and far ends of the quiet line are terminated in its own characteristic

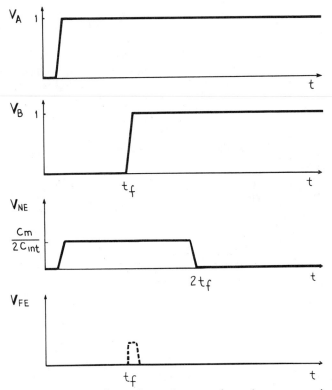

FIGURE 7.8   Near- and far-end coupling waveforms between two homogeneous transmission lines. The pulse shown in broken lines would have been generated at the far end if the medium were inhomogeneous.

impedance; then near- and far-end noise gets absorbed by the terminating resistors, and no reflections are generated.

7.3.6   Coupling Between Inhomogeneous Lines

Figure 7.9 illustrates the coupling between two transmission lines. The lines are microstrips in an inhomogeneous propagation medium; as a result $C_m/C \neq \mathcal{L}_m/\mathcal{L}$. In addition, the lines are fairly well coupled. The lines are assumed to be infinitely long (or, equivalently, terminated by $R_L = Z_0$ at the far end). The waveforms for the active line are at the left, and the ones for the quiet line are at the right-hand side. Two cases are considered. In the first, the active line carries a pulse, and in the second it carries a step [7.3], [7.17]. Both the active and quiet lines are sampled at two points: near the source end and far down the line. The waveforms near the source are similar to those of the homogeneous case. Far down the line, however, the pulses in the active and quiet lines each separate into two pulses. The faster and smaller early pulse is a result of the odd-mode coupling, and the slower and bigger late pulse is a result of the even mode. As described before, in the active line, the even- and odd-mode pulses are both positive. In the quiet line, the even mode is positive, but the odd mode is negative. The step excitation exhibits a similar behavior with a smaller and faster odd wave and a larger and slower even wave. Notice that not only spurious noise pulses are generated at the quiet lines, but also the pulse widths and the steepness of the wave fronts are degraded in the active lines.

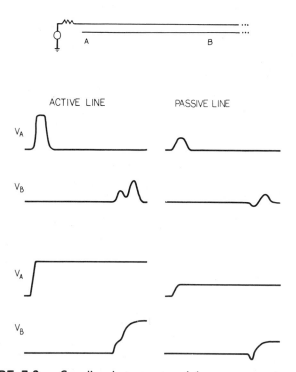

FIGURE 7.9 Coupling between two inhomogeneous transmission lines. On the left-hand side, waveforms on the active line are shown for a pulse and step excitation. V_A is observed near the source end, and V_B is far down the line. On the right-hand side, waveforms at the same locations but for the quiet line are exhibited.

7.3.7 Reducing Cross Talk

The couplings between the lines can be minimized by making sure that no two lines are laid out parallel or next to each other for longer than a maximum length. Noise also can be reduced by keeping the lines apart and by placing a ground plane below the lines such that coupling ratios C_m/C and $\mathcal{L}_m/\mathcal{L}$ are kept small. Ground planes reduce C_m/C because with a ground plane the total capacitance C increases but the mutual capacitance C_m remains approximately the same or is even reduced. Ground planes also reduce $\mathcal{L}_m/\mathcal{L}$ because the current return path goes through the ground plane rather than neighboring lines. This forces most of the magnetic field flux to be between signal lines and the ground plane rather than between the neighboring signal lines.

The coupling can be also reduced by placing grounded lines between signals. This is wasteful of interconnections but in some applications may be preferable to building many reference planes. The couplings should be analyzed in detail using two- and three-dimensional computer-aided inductance and capacitance modeling. Based on these calculations, design rules must be developed that ensure signal integrity and reliable circuit operation.

7.4 Power-Supply-Level Fluctuations

The total capacitive load associated with an IC and its I/O circuitry rises as minimum feature size shrinks, chips grow larger, and their I/O count increases. With improved circuit speed, the average current (I_{av}) needed to charge and discharge these capacitances and the speed at which this current is switched (dI/dt) also increases. As a result, total chip current may change by large amounts within very short time periods. This introduces fluctuations at the power-supply level because of the self-inductance of the power-distribution lines. At present, dI/dt noise caused by simultaneously switching output buffers is more significant than the noise asssociated with on-chip circuitry; however, in the future, both components may become equally important.

Simultaneous switching noise can affect the circuits in three ways [7.16]. First, simultaneous switching noise may increase chip-to-chip delays because when the power-supply level goes down, the output current of *all* active drivers is reduced. Second, it may affect the operation of the receiving chips. The quiet drivers on the sending chip will generate noise pulses due to simultaneous switching of active drivers. As demonstrated by the example in Section 7.1 on page 285, if the sum of this disturbance and cross talk is sufficiently large, it can cause faulty switching in the asynchronous clock lines or introduce additional delay in the synchronous lines. Third, it may affect the gates on the sending chip (the chip with the active drivers). When off-chip drivers turn on and the power-supply level goes down, on-chip gate delays in the sending chip increase due to the reduction of the supply level. In the extreme cases, latches may switch to wrong values due to the collapse of the power-supply level. CMOS latches are more immune to this than bipolar because of their robustness and high noise margins. Disturbance of the internal circuitry is usually avoided by assigning separate power pins for on-chip circuitry and off-chip drivers, as will be described in the next section.

Fluctuations at the power lines can be described by Faraday's law, which states that any change in the magnetic flux will be confronted by an opposition, namely, a self-induced emf, determined by the rate of change in the total magnetic flux Φ,

$$EMF = -\frac{d\Phi}{dt}. \qquad (7.38)$$

The self-inductance L of a circuit is defined as a proportionality constant between Φ and current I,

$$\Phi = LI, \qquad (7.39)$$

which implies that

$$EMF = -L\frac{dI}{dt}. \qquad (7.40)$$

This self-induced emf is the source of power-supply fluctuations on which the rest of this chapter focuses.

7.4.1   Modeling of Power Distribution

An IC and its package are illustrated in Fig. 7.10. At the top is a diagram of the IC die on a chip carrier. Bonding wires, lead frame, and the package pins that form the power connections between the chip and the PC board, as well as the power lines on the board, are highlighted. The electrical model of this structure for a CMOS chip is at the bottom of the figure. It is convenient to separate the circuitry into two groups: on-chip and chip-to-chip. These behave differently. The current driven into the off-chip lines (shown on the right) has only one return path: power and ground pins of the chip carrier. To minimize the effective inductance of the return path and the resulting noise, many power/ground pins must be supplied for off-chip drivers. Any fluctuation at the power lines caused by the switching drivers will also appear at the output of the quiet drivers. On-chip circuitry, on the other hand, can close the loop by small inductance on-chip lines and as a result generates minimal switching noise.

At a given instance, on-chip circuitry of a CMOS chip can be modeled as shown in Fig. 7.11. The transistors are modeled as switches, and the capacitance on the chip is mainly due to thin gate oxides, diffusion layers, and interconnections. Gate oxide of pMOS transistors, $p+$ diffusion layers, and in-

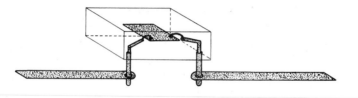

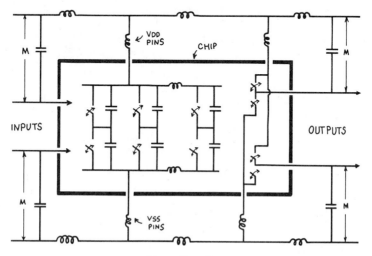

FIGURE 7.10   Power-distribution path in an integrated circuit. The IC die, bonding wire, distribution layer on the chip carrier, and the package pins and power lines on the board are shown. The electrical model of this structure is illustrated at the bottom. Here M represents the mutual inductance between the signal and power/ground planes.

terconnections passing over n-wells have their bottom plates tied to the V_{DD} (power) rail. Gate oxide of nMOS transistors, $n+$ diffusion layers, and interconnections passing over p-wells or p-substrate have their bottom plates tied to the V_{SS} (ground) rail. In the simplified model shown at the bottom, L_{ext} is the inductance of the path that carries the power to the chip; inductance of the on-chip power lines is usually negligible. The capacitor C_2 is the effective switching capacitance, and R_2 is the effective resistance of the transistors that drive these capacitors. (C_2 is the total load being charged, and R_2 is the parallel combination of all the active transistors.) Similarly, C_1 and R_1 correspond to the circuits on the chip that are not switching at that particular clock cycle. It is important to include the nonswitching circuits in the model

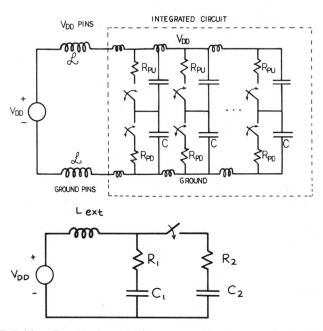

FIGURE 7.11 Electrical model for the on-chip circuitry of a CMOS chip: (top) a detailed model, (bottom) a simplified model

because they reduce the on-chip simultaneous switching noise by sharing their charge with the switching circuits.

At a given cycle, a portion of the circuits will switch. Then the voltage difference between the V_{DD} and V_{SS} lines on the chip will be reduced as the circuits pull current and the power planes on the board try to supply the chip with current through the large-inductance power/ground pins. At this point, the nonswitching circuits on the chip come to the rescue. Since they are tied to the on-chip low-inductance power rails via the on-resistance of the transistors that drive them, nonswitching capacitors can share their charge with the switching capacitors and prevent the on-chip power levels from collapsing. Of course, the charge eventually needs to be supplied by the off-chip power supply but not as immediately as would be required if there were no capacitance on the chip to help out in the process. On-chip capacitance decouples the high-frequency noise. Actually the capacitance and resistance of the chip form a series tank circuit with the inductance of the power/ground pins and can cause oscillations if the resonance frequency is near the clock frequency of the system. This will be described in detail later.

On-chip power-supply-level fluctuations are also small because all the circuits that switch in a cycle are not activated simultaneously. First, a few gates may switch; a fraction of a nanosecond later some other gates may turn on; and so forth. As a result the current fluctuates, attaining its maximum values near the clock edges, but the chip never ceases to draw current from the power supply. This helps to reduce the power-supply-level fluctuations because when a group of circuits turns on simultaneously, the power-supply level is slightly reduced, and previously activated transistors start conducting less current. The current that was being drawn by the already conducting transistors is then channeled to the newly turned on circuits. Consequently the current demand of all circuits is satisfied with little drop at the supply level. The reduction of the current that goes to the already active circuits, of course, degrades their performance, and the circuit delays become harder to predict. In addition to processing, operating temperature, and dc power-supply-level variations, dynamic power-supply-level fluctuations due to simultaneous switching need to be taken into consideration in delay calculations. The worst-case on-chip delta-I noise is generated at the beginning of a clock cycle. At the end of a clock cycle, the nodes are stable, and the current demand of the chip is at its minimum, and when the clock edge rises, many latches are simultaneously activated, and chip current suddenly increases. Later in this chapter, on-chip decoupling capacitors will be investigated as a possible solution to increase the total chip capacitance and provide the electrical charge required for high-speed switching.

This also illustrates additional reasons that the simultaneous switching of output buffers is more problematic than switching of on-chip circuits. As mentioned above, the main problem with off-chip drivers is the lack of a low-inductance return path for the current. In addition, the number of output buffers is relatively small, and there is less of an averaging effect that would reduce the voltage fluctuations. The individual bits of a 32-bit bus are likely to switch simultaneously, and such a wide bus may constitute a sizable portion of the output circuitry of a chip with 100 output buffers. Furthermore, off-chip buffers drive much larger loads and as a result conduct larger currents. The combination of these factors makes off-chip drivers the major source of simultaneous switching noise.

Total Chip Current

As a part of the on-chip switching discussion, it should be noted that even if the total pull-up current were exactly equal to the pull-down current, the chip would still need current from the off-chip power supply because pull-up and pull-down currents do not cancel each other. Figure 7.12 demonstrates this using a simplified example where part of the chip is being pulled up and part of it is being pulled down. Nonswitching circuits are not shown because they do not affect the following derivation. Initially T2 and T3 are on. As a result, C1 and C4 are charged up and have a potential difference of V_{DD} across their

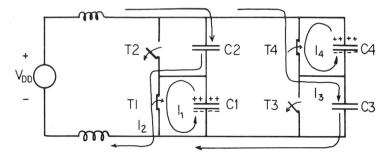

FIGURE 7.12 Pull-up and pull-down currents in a CMOS chip

plates, and C2 and C3 are discharged and have a potential difference of 0 V across them. Later T2 and T3 turn off, and T1 and T4 turn on. The charge built up across C1 and C4 gets dissipated with no effect on the power-supply current as seen at the pins. Capacitors C2 and C3, however, require a net current to charge them up. Although the total pull-up current (I_3+I_4) is equal to the total pull-down current (I_1+I_2), there is a net current at the power pins. Looking from the power pins, the external chip current $I_{ext} = I_2 + I_3$ is half the internal chip current $I_{int} = I_1 + I_2 + I_3 + I_4$. This does not contradict the principle of conservation of energy because exactly half the power is dissipated while a capacitor is being charged up $(\frac{1}{2}CV_{DD}^2)$. The other half is dissipated during discharge. For example, when C_1 is being charged, half the delivered energy gets dissipated in T2, and the other half gets stored in C_1. Later, when T2 turns off and T1 turns on, the energy stored in C_1 gets dissipated in T1 as C_1 is being discharged. As a result the power that is dissipated by $(I_1 + I_4)$ was accounted for in a previous cycle when the external current was charging the capacitors.

Simultaneous Driver Turn-off

One could have problems with drivers that turn off simultaneously just as with drivers that turn on simultaneously. The positive power-supply level (V_{DD}) is reduced, and the ground level (V_{SS}) rises above its normal value when a large group of circuits turns on simultaneously. A similar problem exists when circuits turn off and the current demand suddenly goes away. Then the current built up at the power lines causes the positive power supply to rise and ground level to fall as a reaction, increasing the voltage differential between the V_{DD} and V_{SS} lines. This is an exact dual of the case in which the circuits turn on and accurately describes what happens when a group of open collector bipolar drivers turns off. CMOS circuits do not have a similar problem because when a CMOS driver turns on, it charges or discharges a node and slowly turns itself off as the node reaches its final value. This is also illustrated by the simple model in Fig. 7.11. If the switch were on for a long time and later turned off, none of the currents or voltages would change.

Design Criteria for Simultaneous Switching Noise

The exact solution for the circuit in Fig. 7.11 contains both exponential and sinusoidal functions. Because the inductive effect must be held to a minimum, it can be assumed to be a perturbation in a predominantly RC circuit. For this assumption to be valid, the inductive time constant of the circuit must be much smaller than its capacitive time constant,

$$\frac{L}{R} \ll RC. \tag{7.41}$$

Notice that to increase the switching speed, R needs to be reduced so that the RC time constant improves. This makes it harder to meet the condition of Eq. 7.41.

Equation 7.41 describes the design goal for external power-distribution lines that carry the current to the chip and for internal lines that distribute it on the chip. Presently, on-chip lines meet this criterion because $L_{internal}$ is small. However, the inductance of the board- and package-level power lines and module pins is too high to satisfy Eq. 7.41. One cannot expect to distribute the power on the board simply by using the printed circuit wires and connecting them to the power/ground pins of the chips directly; the wires and pins have too much inductance. The off-chip power distribution network must employ methods that reduce the effects of line and pin inductances. Methods such as decoupling capacitors will be described in the next section.

When the condition in Eq. 7.41 is met, the inductor is considered to be a voltage source with a potential $L(dI/dt)$ across it. To ensure reliable circuit operation, $L(dI/dt)$ should be a small portion of the power-supply voltage,

$$L\frac{dI}{dt} \ll V_{DD}, \tag{7.42}$$

which gives the second design criterion. The equivalence of the conditions in Eqs. 7.41 and 7.42 can be demonstrated by setting $dI = V_{DD}/R$ and $dt = RC$.

7.4.2   Resonance Condition at the Power-Supply Lines

Even if the instantaneous voltage fluctuation is very small, the periodic nature of the digital circuits can cause resonance [7.25]. As shown in Fig. 7.13, the tank circuit formed by the effective module pin inductance L_{mod}, chip

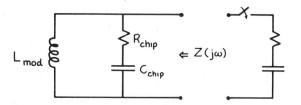

FIGURE 7.13   Tank circuit formed by the pin inductance and chip capacitance and resistance as seen by the switching portion of the chip. If the operating frequency or one of its harmonics is near the resonance frequency and the impedance as seen by the switching circuit is high enough, a large voltage fluctuation can build up at the power lines over many cycles and cause the circuit to fail.

capacitance C_{chip}, and chip resistance R_{chip} has a resonance frequency of

$$f_{chip} = \frac{1}{2\pi\sqrt{L_{mod}C_{chip}}}. \tag{7.43}$$

The effective network as seen by the switching portion of the circuit is shown in Fig. 7.13. The power planes are assumed to be a virtual short for ac analysis because they are usually coupled strongly by large decoupling capacitors. If the operating frequency of the chip or one of its harmonics is near the resonance frequency, and the impedance as seen by the switching circuit at this frequency is sufficiently high, a large voltage fluctuation can build up at the power lines over many cycles and cause the circuit to fail. The impedance seen by the switching circuit is

$$Z(j\omega) \;=\; \frac{1}{\dfrac{1}{j\omega L_{mod}} + \dfrac{1}{R_{chip} + \dfrac{1}{j\omega C_{chip}}}}$$

$$=\; \frac{-\omega^2 R_{chip}C_{chip}L_{mod} + j\omega L_{mod}}{(1 - \omega^2 L_{mod}C_{chip}) + j\omega R_{chip}C_{chip}} \tag{7.44}$$

where $\omega = 2\pi f$. The absolute value of the impedance is

$$|Z(j\omega)| = \sqrt{\frac{(\omega^2 R_{chip}C_{chip}L_{mod})^2 + (\omega L_{mod})^2}{(1 - \omega^2 L_{mod}C_{chip})^2 + (\omega R_{chip}C_{chip})^2}}. \tag{7.45}$$

To obtain the resonant frequency, the real part of the denominator of Eq. 7.44 is set to zero:

$$\omega_{chip} = \frac{1}{\sqrt{L_{mod}C_{chip}}}. \tag{7.46}$$

At this resonance frequency, the impedance attains a maximum of

$$|Z(j\omega_{chip})| = \sqrt{\frac{L_{mod}^2 + L_{mod}C_{chip}R_{chip}^2}{R_{chip}^2 C_{chip}^2}}$$

$$\approx \frac{L_{mod}}{R_{chip}C_{chip}}.$$

(7.47)

The final approximation can be justified by substituting $C_{chip} = 10$ nF, $R_{chip} = 0.1\ \Omega$, and $L_{mod} = 0.5$ nH. However, it should be noted that the neglected term can easily gain importance, and the full expression should be used for the general case.

To prevent oscillations at the power lines, the resonance frequency f_{chip} should be much larger than the system clock frequency, and the resonant impedance $|Z(f_{chip})|$ must be kept small. As expected, reducing the pin inductance increases the resonant frequency and reduces the impedance. The VLSI trend of increasing chip capacitance and reducing chip resistance is the source of the problem. Rising C_{chip} brings the resonance frequency down, and diminishing R_{chip} increases the resonance impedance.

It should be noted that R_{chip} includes not only the resistance of the transistors that feed the plate of the capacitor that is being charged and discharged to V_{DD} and V_{SS} but also the resistance of the path that ties the stable plate of the capacitor to V_{DD} or V_{SS}. The substrate resistance and the resistance of power lines are the major contributors to this resistance. The parasitic resistance of this path gives rise to IR drops and may cause power distribution or latch-up problems; ironically, the same parasitic resistance helps the resonance problem by introducing a damping effect and reducing the resonance impedance. It is better to have a lossy on-chip capacitor because the resonance impedance is reduced due to the series resistance of the capacitor.

7.5   Reducing Power-Supply-Level Fluctuations

A number of methods can be used to minimize dI/dt noise, such as decoupling capacitors, multiple power and ground pins, and tailored driver turn-on characteristics. A large decoupling capacitor placed close to the chip reduces power-supply-level fluctuations because the capacitor charges up during steady state and then assumes the role of power supply during current switching. The energy stored in the capacitor supports the current transients generated during switching and, in effect, spreads the charge restoration from the power supply over the entire cycle time (Fig. 7.14). With multiple power and ground pins, each connection can supply a fraction of the total current. This reduces the effective inductance and can support a larger current spike

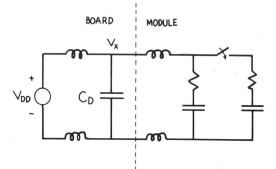

FIGURE 7.14   CMOS chip with a decoupling capacitor. The capacitor screens the inductance of the board wires, but the inductance of package pins and bonding wires is still between the chip and the capacitor.

with little power-supply-level fluctuation. If off-chip drivers are designed such that they do not turn on very sharply, current transients can be reduced.

7.5.1   Decoupling Capacitors

As mentioned above, a decoupling capacitor reduces power-supply-level fluctuations by charging up during steady state and then assuming the role of power supply during current switching (Fig. 7.14). The capacitance should be sufficiently large such that the energy stored in the capacitor can support current transients generated by many simultaneously switching off-chip drivers and any on-chip switching. The leads and the capacitor dimensions must be small to minimize the parasitic inductances. The capacitor must be placed as close to the chip as possible.

A decoupling capacitor next to the chip carrier screens the inductance of the power lines on the board but does not shield the inductances of the pins, distribution layer, and bonding wires. To increase its efficiency, the capacitor can be manufactured on the chip carrier to block the inductances of the pins and the distribution layer on the chip carrier [7.21]. Here the major challenge is to build a large-value capacitor that has a small self-inductance and to locate it very close to the chip in order to minimize the inductance of the connections to the die [7.12], [7.20]–[7.21]. The value of the capacitance must be selected such that it is large enough to minimize the transient noise, but it should also be kept sufficiently small to avoid lowering the resonance frequency near the operating frequency. As mentioned earlier, a lossy capacitor would cut down the resonance peak.

The next logical step is to integrate the decoupling capacitor on the chip. An on-chip capacitor will screen out the bonding-wire/TAB/solder-ball inductances, and the high-frequency current transients will see only the inductance of on-chip power-distribution lines, which are usually very small. Actually the total capacitance of the on-chip power distribution lines and the circuits connected to them may be sufficiently large to store enough energy to support simultaneous switching of on-chip circuits, as described in the previous section. Resonance problems, of course, must be considered when adding decoupling capacitors on the chip.

Capacitors built on the chip and on the chip carrier reduce the ΔI noise generated by on-chip circuitry but are not equally effective in reducing the simultaneous switching noise of output buffers. As off-chip drivers switch, the on-chip capacitor keeps the potential difference across the power lines stable but cannot prevent the on-chip power-supply levels from moving up and down with respect to the board power-supply levels. Additional clamping is required by board-level decoupling capacitors.

7.5.2   System-Level Power Distribution

Usually a hierarchy of decoupling capacitors is utilized to provide a good power-distribution network. In high-end bipolar computers, the power is brought to the board via the frame distribution system, which is an assembly of laminar bus bars. Large amounts of decoupling capacitance are inserted at the interface between the board and power bus, and at this point a feedback path is established to the regulated voltage supply to keep it stable. This large capacitance filters the low-frequency noise. A mainframe board is shown in Fig. 3.21. Smaller capacitors are placed on the board and modules to eliminate the mid- and high-frequency noise. Looking from the chip terminals, a good power distribution network behaves like a resistive load with a very small impedance (less than 0.1 $m\Omega$). This is, of course, the ac behavior of the circuit. The impedance should be flat for a wide frequency range, as shown in Fig. 7.15 [7.12]. The role of decoupling capacitors also can be conceptualized as an agent that reduces the impedance of the power lines. Since the characteristic impedance is $Z_0 = \sqrt{\mathcal{L}/\mathcal{C}}$, adding decoupling capacitors increases the capacitance but does not affect the inductance of the power planes. As a result Z_0 is reduced, and current spikes generate smaller voltage drops because $\Delta V = Z_0 \Delta I$. This helps the pulse response and curbs instantaneous fluctuations. The resonant effects that are related to steady-state conditions that extend over many cycles are discussed later.

In CMOS systems, multilayer PC boards with dedicated power and ground planes are used. To minimize their effective inductance, the power lines are either continuous planes or are laid out in a grid formation as shown in Fig. 7.16. High-frequency 0.1 μF ceramic capacitors are placed at the grid points. The capacitors should be high-K ceramic type with small series inductance and resistance. There should be at least one such capacitor for every other chip. Bulk decoupling to filter out low-frequency current transients is provided by

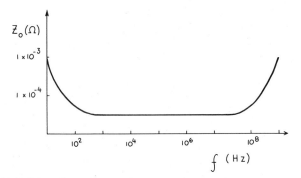

FIGURE 7.15 Impedance of the power distribution network as a function of frequency

large tantalum capacitors placed near the board edge connector, where the power traces on the PC board meet the back plane power-distribution network. One tantalum capacitor of 20–50 μF provides sufficient energy storage for 16 CMOS chips [7.23].

The electrical model of a typical power distribution network in a CMOS

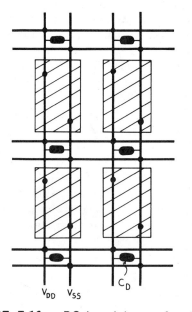

FIGURE 7.16 PC board layout for the power distribution network. V_{DD} (power) and V_{SS} (ground) are in two separate wiring levels.

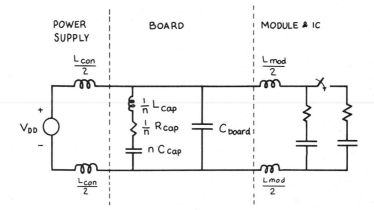

FIGURE 7.17   Electrical model of a typical CMOS power distribution network. The inductance L_{con} is due to the power connectors at the edge of the board. The power and ground planes on the board have an intrinsic capacitance of C_{board}; and each of the decoupling capacitors placed on the board has a capacitance, resistance, and inductance of C_{cap}; R_{cap}; and L_{cap}: The number of decoupling capacitors on the board is n: The chip carrier and the CMOS die are modeled as discussed earlier.

board is shown in Fig. 7.17. On the left-hand side is the power supply and the inductance of the board connectors. PC board is modeled as a capacitor C_{board}, which represents the intrinsic capacitance between the power and ground planes, and a collection of n decoupling capacitors, each with a capacitance of C_{cap}, and a series resistance and inductance of R_{cap} and L_{cap}. The module and chip are modeled the same way as in Fig. 7.11.

The shape of the impedance versus frequency plot in Fig. 7.15 can be explained using the electrical model in Fig. 7.17. The increase in the low-frequency impedance of the power-distribution network is primarily due to the resonance frequency of the board connector L_{con}. The increase in the high-frequency impedance of the power-supply network is primarily due to the resonance frequency of the decoupling capacitor lead inductance L_{cap}. The larger the number of decoupling capacitors, the smaller the effective inductance L_{cap}/n and the larger the resonant frequency. The goal is to keep the low-frequency resonance well below the operating frequency and the high-frequency resonance well above the operating frequency of the system. Not shown in Fig. 7.15 is the resonance frequency of the chip, which can be smaller than the resonance frequency of the decoupling capacitor lead inductance. As described in Section 7.4.2 on page 311, this should also be kept much above the operating frequency.

7.5.3 Decoupling Capacitor Calculations

If decoupling capacitors are used, an upper bound on transient voltage fluctuation can be determined by modeling the power lines behind the capacitor as an infinitely large inductor. Immediately after switching, no current flows through this large inductor, and the circuit can be simplified to a capacitive divider,

$$C_D V_{DD} = (V_{DD} + \Delta V)(C_D + C)$$
$$\Delta V = -\frac{C}{C_D + C} V_{DD}. \tag{7.48}$$

To ensure a small voltage fluctuation ΔV, decoupling capacitor must be much larger than the total capacitor on the chip being charged at a given time.

A typical total chip capacitance for a 32-bit microprocessor is calculated to be 14 nF in Chapter 9. Accordingly a decoupling capacitor of 10 times the chip capacitance $10 \times 14 \ nF = 0.14 \ \mu F$ is required per such chip. The current demand of output buffers must also be taken into account. In a relatively slow CMOS system where the time-of-flight delay across the board is less than the rise time, off-chip wires can be modeled as lumped capacitors charged up by the decoupling capacitor. (If the time-of-flight delay is comparable or greater than the rise time, the board wires must be modeled as transmission lines.) One hundred output buffers each driving a 40 pF load constitutes an additional capacitance of 4 nF. This raises the decoupling capacitance requirement to $10 \times (14nF + 4nF) \approx 0.2 \mu F$. Of course, this is an overestimated calculation because all on-chip circuits and all off-chip drivers never switch on simultaneously. It should be also noted that if the decoupling capacitor is discharged significantly, it may take many cycles to charge it up back to V_{DD}. A conservatively high capacitor value must be selected to prevent the decoupling capacitor from being discharged significantly.

The magnitude and the duration of the current that can be supported by a decoupling capacitor are two other parameters of interest. They can be calculated based on the following basic equation, which describes the time rate change of the potential across a capacitor as a function of the current driven into or supplied by the capacitor:

$$I = C \frac{dV}{dt}. \tag{7.49}$$

As an example, assume that the allowable voltage fluctuation is 0.25 V, decoupling capacitor is 0.1 μF, and the duration that the decoupling capacitor needs to support the current is 5 nsec (a good approximate *upper limit* for the majority of on-chip and chip-to-chip CMOS path delays). The current can then be calculated as

$$I = C \frac{\Delta V}{\Delta t} = 0.1 \mu F \times \frac{0.25 \ V}{5 \ nsec} = 5 \ A.$$

A low impedance buffer driving a 50 Ω transmission line would require a maximum current of

$$I = \frac{V_{DD}}{Z_0} = \frac{5\ V}{50\ \Omega} = 0.1\ A.$$

As a result a 0.1 μF decoupling capacitor can support 50 such buffers for 5 nsec and still have less than 0.25 V of voltage reduction. Actually the power-supply lines are not effectively an open circuit for the entire 5 nsec and, to an extent, charge up the decoupling capacitor while it is supplying current to the off-chip buffers.

7.5.4   On-Chip Decoupling Capacitors

Because the power-distribution lines on a chip are simply an extension of the off-chip power lines, their contribution to the power and ground inductances should also be minimized. Currently the inductance of the lines on the chip is not a problem because of small chip size and low operating frequency; however, as ICs become larger and faster, many of the package-level problems will be transferred to the chip level. In addition, with the advent of very large die sizes and silicon-on-silicon hybrids, the distinctions between the chip and package are becoming vaguer. When the combined effect of increased on-chip latch count, switching speed, and power-line inductance becomes significant, it may be necessary to fabricate decoupling capacitors on the chip. These capacitors should be designed such that they do not occupy excessively large areas or reduce die yields. Currently there is sufficient capacitance on logic chips (due to power lines and devices that are connected to them) to prevent the supply level from collapsing when many latches switch simultaneously at the clock transitions. In DRAM chips, on the other hand, where all bit lines are precharged and many sense amplifiers are activated simultaneously, on-chip-generated noise can be severe. As mentioned in Chapter 4, precharging the bit lines to $V_{DD}/2$ reduces the noise generation. In addition, a "sequential" sense amplifier enabling scheme where the amplifier is turned on gradually rather than sharply can help [7.26]. It is conceivable that future DRAMs will require on-chip decoupling capacitors.

It is important to realize that on-chip decoupling capacitors reduce the ΔI noise generated by on-chip circuitry but cannot help the noise due to simultaneously switching off-chip drivers.[6] Figure 7.18 compares an on-chip and a chip-to-chip path. As can be seen in the figure, on-chip path can receive all its current from the on-chip capacitor. The entire path can be closed on the chip. If the decoupling capacitor is distributed evenly on the die, the ef-

---

[6]This argument also applies to decoupling capacitors on a board and signals that leave a module or a board. Decoupling capacitors on the board cannot help simultaneous switching noise at the active and quiet lines that leave the board. To minimize the noise at these lines, many power and ground connections must be provided at the board connector, as shown in Fig. 7.24.

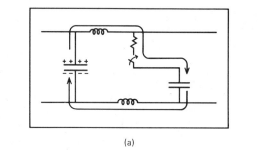

(a)

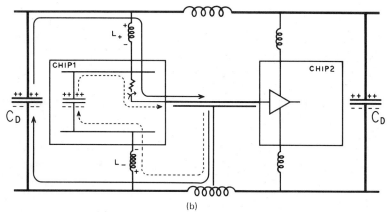

(b)

FIGURE 7.18   Comparison of the effectiveness of on-chip decoupling capacitors reducing the simultaneous switching noise of on-chip and chip-to-chip paths: (a) on-chip path, (b) chip-to-chip path.

fective inductance of on-chip power lines will be very small and the resulting inductive potential drop will be minuscule. In a chip-to-chip path, however, it is not possible to close the loop with only on-chip lines. The current follows the path shown in broken lines when the current is supplied by the on-chip capacitor. Here the current from the positive power supply goes through on-chip wires, but the ground loop must be closed through the high-inductance package pins. This is because off-chip interconnections do not share a common ground with on-chip circuitry. The power/ground pins connect their grounds. As a result a potential drop is developed across L_- and the chip ground is pulled down. The top plate of the on-chip decoupling capacitance will be pulled down by the same amount that will pull down the plus power supply.[7]

---

[7] This is analogous to "bootstrapped" nMOS circuits.

Consequently when an on-chip decoupling capacitor is used to drive off-chip loads, the chip power levels move up and down with respect to the off-chip power supply; a transient ground noise voltage is injected between the chip and system ground, as well as in the outputs of quiet drivers. The stable clock lines will be perceived by the on-chip circuitry as if they are switching. The current loop from the off-chip decoupling capacitor is shown in solid lines. It also goes through a high-inductance package pin L_+. If the off-chip driver were pulling down instead of pulling up, the situation would be similar to the above case. As a result it is best not to use on-chip decoupling capacitors for off-chip drivers. On-chip power distribution network is usually isolated from off-chip drivers, and separate pins are used to feed them. On-chip decoupling capacitor should be connected only to the power tree that feeds on-chip circuitry. Multiple power/ground pins for on-chip power distribution must be provided to keep the chip resonance frequency much above the operating frequency. Transient noise due to off-chip drivers should be minimized by placing many low-inductance decoupling capacitors on the board, providing multiple low-inductance power/ground pins for output buffers, and keeping the chip-to-chip interconnections short to minimize the capacitive loads and resulting current surges.

On-chip decoupling capacitors may not be very effective in curbing the switching noise generated by off-chip drivers but are perfect for on-chip power distribution. This is especially important when the "chip" grows to wafer dimensions or assumes a hybrid nature, as in silicon-on-silicon packaging. A structure with continuous power and ground planes (Fig. 7.19) can be used to minimize the inductance and resistance of power-distribution lines on large chips or thin film substrates.

Solid ground and power planes serve many purposes. The resistive (IR) voltage drops are reduced because the thick and continuous power planes have much less resistance than power-distribution lines using standard interconnection layers. Inductive (LdI/dt) voltage drops are minimized because when the ground and power planes are in close proximity, they act like a capacitor that serves as a distributed decoupling capacitor with an area equal to the size of the substrate or, equivalently, like a transmission line with a very small characteristic impedance. The continuous power/ground planes have a large capacitance and must have a small-inductance connection to the next level of packaging to avoid resonance and excessive quiet line noise in the signal lines that leave the substrate.

Due to continuous power/ground planes, signal lines are supplied with a nearby current return path that improves the transmission line properties. Slow-wave effects are eliminated by not relying on the silicon substrate as the current return path. Cross talk between signal planes and between the lines in the same plane is also reduced.

Continuous power and ground planes are hard to manufacture. If continu-

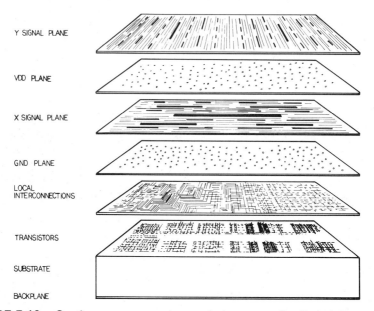

Y SIGNAL PLANE

VDD PLANE

X SIGNAL PLANE

GND PLANE

LOCAL INTERCONNECTIONS

TRANSISTORS

SUBSTRATE

BACKPLANE

FIGURE 7.19 Continuous power and ground planes as a distributed decoupling capacitor

ous planes are used, there will be a large possibility of forming shorts between the planes. To improve the yield, the planes can be constructed as mesh structures with large holes through them rather than as continuous planes.

The effectiveness of on-chip decoupling capacitors can be evaluated by calculating the voltage fluctuation across the capacitor when it must supply a current I for a time period Δt. The capacitance between the power planes is

$$C = \epsilon_{ox} \frac{D_c^2}{H}, \tag{7.50}$$

where D_c is the chip size and H is the spacing between V_{DD} and ground planes. The current I that can be supported by this capacitance for time Δt with a voltage fluctuation of ΔV is expressed as

$$I = \frac{\Delta Q}{\Delta t} = \frac{C \Delta V}{\Delta t} = \frac{\epsilon_{ox} D_c^2 \Delta V}{H \Delta t}. \tag{7.51}$$

This assumes that the inductance of the capacitor is negligible.

As a numerical example, assume that $D_c = 1$ cm and $H = 6$ μm, which results in a capacitance of

$$C = \frac{\epsilon_{ox} D_C^2}{H} = \frac{3.45 \times 10^{-13} \times 1^2}{6 \times 10^{-4}} = 0.6 \; nF.$$

If the allowed voltage fluctuation is $\Delta V = 0.25$ V, for a period of $\Delta t = 1.0$ nsec, this on-chip capacitor can supply a current of

$$I = \frac{\epsilon_{ox} D_c^2 \Delta V}{H \Delta t} = \frac{3.45 \times 10^{-13} \times 1^2 \times 0.25}{6 \times 10^{-4} \times 1.0 \times 10^{-9}} = 0.14 \, A.$$

7.5.5   Minimizing the Inductance

In addition to using decoupling capacitors, the inductance of the package pins and bonding structures (bonding wire/TAB/solder ball) must be minimized by increasing their cross-sectional dimensions, bringing them as close to the ground plane as possible, and keeping them very short. This can be observed in Fig. 7.20, which summarizes the electrical parameters of a wire above a ground plane. The shorter the wire and the closer it is to the ground plane, the smaller is its inductance. When the line is closer to the ground plane, the current loop area and magnetic flux are minimized. Among the existing packages, dual in-line package has the poorest lead inductance (3–50 nH) because of long lead lengths and the lack of a ground plane. Tape bonding results in smaller inductances (1–10 nH) because of shorter and thicker lines and reduced separation from the ground plane. Pin grid arrays that employ multilayer ceramic substrates with power and ground planes also have small inductances due to the power and ground planes. Direct chip attach via TAB or flip-chip technology eliminates one level of packaging and provides connections with very small inductances. Flip-chip technology with solder bumps yields the smallest inductances (0.25–1 nH).

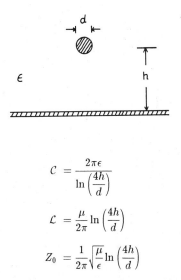

$$C = \frac{2\pi\epsilon}{\ln\left(\frac{4h}{d}\right)}$$

$$\mathcal{L} = \frac{\mu}{2\pi} \ln\left(\frac{4h}{d}\right)$$

$$Z_0 = \frac{1}{2\pi}\sqrt{\frac{\mu}{\epsilon}} \ln\left(\frac{4h}{d}\right)$$

FIGURE 7.20   Electrical parameters of a cylindrical wire above a ground plane

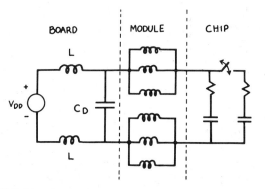

FIGURE 7.21   Electrical model of the chip with multiple power and ground pins

After the inductance per power/ground connection is minimized, multiple power and ground connections may be required to reduce the ΔI noise further. If there are multiple power/ground pins, each connection can supply a fraction of the total current transient, and together they can support a larger current spike. Another way of analyzing multiple power/ground connections is by looking at the collective inductance of the power/ground pins. The effective inductance is reduced when inductances are connected in parallel because $1/L_{eff} = 1/L_1 + 1/L_2 + \cdots + 1/L_n$. Multiple power/ground pins act like parallel inductances, and this reduces the LdI/dt noise (Fig. 7.21). A good rule of thumb for high-speed systems is to provide a pair of V_{DD}/V_{SS} pins for six signal pins (signal pin count/power supply pin count = 3/1). In addition to dedicating multiple power/ground pins, inductances of the connections between the pins and the chip can be minimized by using multiple bonding wires for power pads. Two or three bonding wires are sufficient. If the number of bonding wires is increased beyond three, the mutual inductance between the wires swamps out the benefit of additional wires.

When multiple power/ground connections are used, the power-distribution networks for on-chip circuitry and off-chip drivers can be totally isolated by connecting them to separate pins. In this way, off-chip drivers cannot cause the power-supply level on the chip to collapse, which may increase gate delays or force some latches to lose their states. Since off-chip drivers are more problematic, most of the additional power/ground pins can be reserved for them, and wide off-chip buses can be divided into smaller groups that are powered by separate pins to minimize the simultaneous switching noise. Multiple power/ground pins for off-chip drivers provide a low inductance current return path and minimize the simultaneous switching noise injected into the quiet drivers. In addition, multiple power/ground pins for on-chip circuitry minimize the effective external inductance L_{ext} and help to push the chip resonance frequency much above the operating frequency.

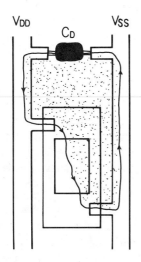

FIGURE 7.22   Ground loop of a DIP

To reduce the effective inductance, the area of the ground loop must be minimized because inductance is proportional to the loop area. It is important that the complete loop is taken into account when different package and pin configurations are compared. Figure 7.22 schematically shows the ground loop of a DIP.

One way to ensure small ground loops is to form two circles around the die as shown in Fig. 7.23. One of the circles is V_{DD} and the other is V_{SS}. Multiple pins form the connections between these circles and the board V_{DD} and V_{SS} lines. These power circles are tied to the die by multiple bonding wires. As a result the current will follow a minimal loop, generating a small inductive voltage drop.

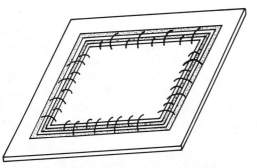

FIGURE 7.23   Power and ground circles around a die

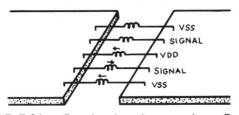

FIGURE 7.24 Board-to-board connections. Power and ground connections must be generously interdispersed within the signal connections to provide a low-inductance current return path and avoid excessive quiet line noise. As shown by the arrows, when the signal propagates, V_{DD} and V_{SS} lines provide current return paths.

Board-to-board connectors are similar to off-chip drivers in the sense that when a signal is driven from one board to another, a low-inductance return path must be provided for the current to close the loop. If a specific path is not provided, the current will find the path of least inductance, which may not be sufficiently good to satisfy the noise requirements. Then when a large number of board-to-board signals switch simultaneously, noise will be injected in the quiet lines. To minimize the noise, board-to-board connectors usually have at least one power or ground wire per two or three signal wires, and the power/ground connections are uniformly interdispersed among the signal wires (Fig. 7.24).

7.5.6 Driver Turn-on Characteristics

The ΔI noise generated by off-chip drivers can be also reduced by controlling the turn-on characteristics of output buffers. The driver can be turned on slowly rather than sharply to reduce the size and slope of the current spikes.

Here the distributed RC constant of the polysilicon gate can be beneficial. If the transistors of output buffers are laid out with long polysilicon strips forming the gate electrode, as shown in Fig. 7.25, the beginning of the strip will turn on before its end, and the driver will not generate a sharp current spike. The distributed RC constant of the polysilicon gate structure can be significant because the gate capacitance per unit length is an order of magnitude greater than the wire capacitance per unit length. Polysilicon typically has a sheet resistance of 20 $\Omega/\Box$, and gate capacitance is $C_{gox} = 3$ fF per minimum feature size square for 1.3 μm CMOS technology; as a result, $RC = 0.06$ $psec/\Box^2$. Off-chip buffers usually have very large W/L ratios. If one of the fingers of the driver transistor has a ratio of $W/L = 100$, its RC constant becomes

$$R_{poly}C_{gate} = 100^2 \times 0.06 \ psec/\Box^2 = 0.6 \ nsec$$

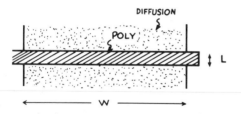

FIGURE 7.25   Large W/L ratio device with a long polysilicon finger

and the tip of the finger will turn on roughly 0.3 nsec later than its root. This RC delay tracks the process variations. If the polysilicon width is narrower than nominal, the buffer will be faster, but the resistance of the polysilicon line will increase more than the reduction in the gate capacitance, causing the RC constant to rise. If the gate oxide is thinner than the normal, the buffer will again be faster but gate capacitance will also increase and reduce the current transients. Silicided polysilicon gates have smaller RC constants, which may not be sufficient to reduce the transients. Then an off-chip driver can be partitioned into smaller buffers, and a delay line can be constructed to turn on the buffers one at a time rather than all simultaneously.

If the chip lies in the fast corner of the process, off-chip drivers may become actually slower than nominal parts due to power-supply-level collapse and ringing at the transmission lines. One way to avoid this is to use a voltage-controlled output buffer that equalizes its rise and fall times over process variations. Such circuits are reported to reduce the ΔI noise by a factor of two, and the chip-to-chip delays by 30 percent [7.24]. As illustrated in Fig. 7.26, the key feature is an on-chip process-dependent voltage source that equalizes the charge/discharge rate of a series transistor in the output buffer. For example, if nMOS transistors are very fast, VCP will go above its nominal value, and the rise time of the lower NOR gate in the output buffer will degrade; consequently the nMOS output device will turn on more slowly, compensating for its low impedance.

7.6   Comparison of ΔI Noise in CMOS and Bipolar Circuits

Figure 7.27 compares the power-supply-level fluctuations in CMOS and bipolar ECL circuits. In a CMOS chip, the portion of the circuit that is not switching at a given system cycle (represented by R_1 and C_1) helps the switching portion of the chip (represented by R_2 and C_2). This is because C_1 shares its charge with C_2. If only 10–20 percent of the chip capacitance switches at a

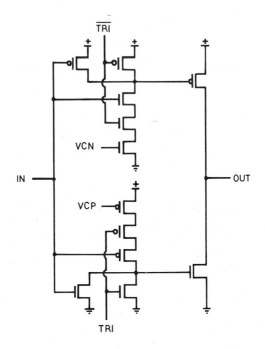

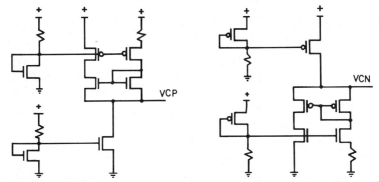

FIGURE 7.26   Off-chip drivers that track the process variations. The buffer is shown at the top. When the signal *TRI* is a "1," both output transistors are disabled, and the buffer is tristated. Voltage control signals *VCN* and *VCP* for *n*- and *p*-channel devices are generated by the circuits at the bottom.

given cycle, the voltage fluctuation will be less than 10–20 percent because

$$V_{DD} + \Delta V = \frac{C_1}{C_1 + C_2} V_{DD}, \tag{7.52}$$

which yields

$$\Delta V = -\frac{C_2}{C_1 + C_2}V_{DD}. \qquad (7.53)$$

Bipolar circuits function differently because a bipolar transistor is not a bidirectional switch like an MOS transistor. Because of the diode structures in a bipolar device, the current can flow in only one direction. This prevents switching bipolar gates from taking advantage of the capacitance tied to the outputs of nonswitching gates. In bipolar ECL chips there is another mechanism that helps to curb power-supply-level fluctuations: the dc current of the ECL gates. As can be seen in Fig. 7.27(b), when a group of ECL gates (represented by R_2) switches on, the potential V_X goes down, and a current proportional to the reduction in V_X switches over from R_1 to R_2. As a result a portion of the dc current can support the required charging current with little drop in the supply level if R_1 is much smaller than R_2.

When R_2 is off, the total current I_1 is

$$I_1 = \frac{V_{DD}}{R_1}. \qquad (7.54)$$

Immediately after R_2 is turned on, the current will remain constant because

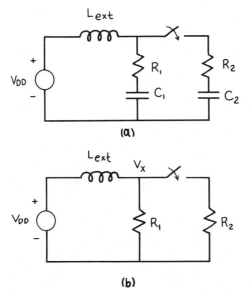

(a)

(b)

FIGURE 7.27   Comparison of power-supply-level fluctuations in CMOS and bipolar circuits: (a) simplified model for the internal circuitry of a CMOS chip, (b) simplified model for the internal circuitry of a bipolar ECL chip

of the pin inductance L_{ext}, and a voltage fluctuation ΔV will appear at the power-supply level as

$$
\begin{aligned}
V_{DD} + \Delta V &= I_1(R_1 \| R_2) \\
&= \frac{V_{DD}}{R_1} \frac{R_1 R_2}{R_1 + R_2},
\end{aligned}
\tag{7.55}
$$

which yields

$$
\Delta V = -\frac{R_1}{R_1 + R_2} V_{DD}.
\tag{7.56}
$$

Similarly when R_2 is turned off, a voltage fluctuation

$$
\Delta V = \frac{R_1}{R_2} V_{DD}
\tag{7.57}
$$

will occur. As a result, if $R_1 < 0.1 R_2$, the power-supply level will not fluctuate more than 10 percent.

Off-chip drivers are a source of concern in both technologies because their only current return path is the large inductance power/ground pins. When many drivers switch simultaneously, V_{DD} dips down, V_{SS} bumps up, and noise pulses are injected at the outputs of quiet drivers. In addition, steady clock lines can be perceived by the receivers as if they are switching, and active driver delays increase because of the reduction in power-supply level and current drive capabilities of output buffers.

7.7   Off-Chip Driver Simultaneous Switching Noise Calculations

Consider a situation in which 32 low-impedance CMOS buffers ($R_S \ll Z_0$) are switched simultaneously. In addition, the line impedance is 50 Ω, rise time is 2 nsec, output swing is 5 V, and the allowed power-supply-level fluctuation is 0.25 V. The rate of change of the output voltage can be calculated from the voltage swing and rise time as

$$
\frac{dV}{dt} = \frac{80\% \times V_{swing}}{t_r} = \frac{80\% \times 5V}{2nsec} = 2\, V/nsec.
$$

The current driven into the transmission line is $I = V/Z_0$, and its rate of change is

$$
\frac{dI}{dt} = \frac{1}{Z_0}\frac{dV}{dt} = \frac{2\,V/nsec}{50\,\Omega} = 0.04\, A/nsec.
$$

TABLE 7.1   Voltage and Current Transients at Output Buffers

Technology	Rise Time (nsec)	Voltage Swing (V)	dV_{OUT}/dt (V/nsec)	dI_{OUT}/dt (mA/nsec)
CMOS	3	5	1.3	27
Bipolar	0.4	1	2.0	40
GaAs	0.2	1	4.0	80

With 32 drivers, the total current transient is obtained as

$$\frac{dI_{TOT}}{dt} = N_{drv}\frac{dI}{dt} = 1.28\ A/nsec.$$

Through a 1 nH inductance, this gives rise to a voltage drop of

$$\Delta V = L\frac{dI}{dt} = 1.28\ V.$$

To guarantee a maximum of 0.25 V voltage fluctuation, the effective inductance of the power connections must be

$$L = \frac{\Delta V}{dI/dt} = 0.2\ nH.$$

which can be achieved by 10 pairs of power/ground connections with an effective inductance of 2 nH per pair.

Typical output voltage and current slopes for various technologies are listed in Table 7.1. They are calculated in a similar way as the above derivation but with different rise times.

7.8   Advanced Packages

The structures in Figs. 7.28 and 7.29, and the cover of this book incorporate a number of concepts from existing state-of-the art packages that were described in Chap. 3 [7.13],[7.22].

Chips are flip mounted on a polyimide/ceramic substrate. Ceramic forms a stable base, and thick film decoupling capacitors are integrated on it. This thick film metal can be molybdenum or tungsten, which have melting points compatible with ceramic processing, and their coefficients of thermal expansion match well with alumina. It is important that the decoupling capacitor has a small inductance; neglecting its inductance can be disastrous. A ceramic substrate ensures high capacitance because of its large dielectric constant ($\epsilon_r = 10$). Chip-to-chip connections are made with densely packed thin film copper signal lines fabricated on top of the ceramic substrate and insulated with polyimide. Copper has low resistivity ($\rho = 1.7 \ \mu\Omega\text{-cm}$), which provides low loss transmission lines, and polyimide has a small dielectric constant ($\epsilon_r = 2.5$), which yields low capacitance and fast propagation speed ($v = c_0/\sqrt{\epsilon_r}$). The dice are flip mounted on this high-performance substrate to minimize the parasitic capacitances and inductances of electrical connections to the chip and to provide a large number of connections. Engineering change pads surround the chips. These can be used to accommodate late changes to fix design errors with minimal effect on product schedule. Chips may also have decoupling capacitors that are integrated on them. To maximize the heat removal rate, cooling fins are fabricated on the back surfaces of the dice. In this way heat densities up to 1000 W/cm^2 can be cooled [7.19]. The packaging concepts in Figs. 7.28 and 7.29 can also be applied to silicon-on-silicon hybrids and wafer-scale integrated circuits.

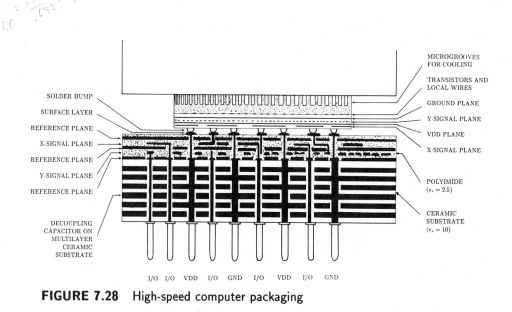

FIGURE 7.28 High-speed computer packaging

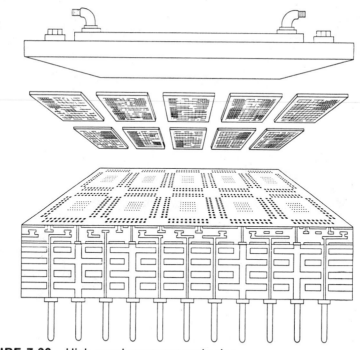

FIGURE 7.29   High-speed computer packaging

7.9   Effects of Scaling on Power-Supply-Level Fluctuations

This section analyzes scaling of power-supply-level fluctuations caused by the self-inductance of package and chip power-distribution lines. Voltage fluctuation ΔV is expressed as

$$\Delta V = L\frac{dI}{dt}, \tag{7.58}$$

where L is the inductance of the line, and I is the total chip current in off-chip power lines or the current supplied to a subblock in on-chip calculations. The difference between the two becomes indistinguishable in wafer-scale integration. In the following subsections, individual components of this noise phenomenon are analyzed.

7.9.1   Rise Time
The intrinsic gate delay τ_g of MOS transistors improves by the device-scaling factor S. Because of the increasing chip size, clock frequency may not improve

by an equal amount; nevertheless, through the use of creative architectures and circuit techniques, a continuous improvement in clock frequency is expected. As a result, the rise times t_r of all local and global signals will decrease as the minimum feature size is reduced:

$$dt \; \propto \; t_r \; \propto \; \frac{1}{S}. \tag{7.59}$$

7.9.2   Total Chip Current

Chip current increases very rapidly as the minimum feature size is reduced and the chip size becomes larger. In CMOS circuits, the chip current can be thought of as "limited" by the transistors and "required" by the total capacitive load on the chip (gates and interconnections).

Current Drive Capability of the Transistors

To determine the maximum current drive capability, assume a transistor packing-efficiency-limited chip. The number of gates N_g increases with larger chips and smaller gate dimensions:

$$N_g \; \propto \; S^2 S_C^2. \tag{7.60}$$

The on-resistance of a minimum-sized transistor remains approximately constant:

$$R_{tr} \; \propto \; 1. \tag{7.61}$$

Because the power supply is scaled down, current drive per device decreases by $1/S$,

$$V_{DD} \; \propto \; \frac{1}{S}$$

$$I \; \propto \; \frac{1}{S}. \tag{7.62}$$

As a result the total current drive capability of the transistors on the chip increases by SS_C^2,

$$I_{TOT} = f_d N_g I \; \propto \; SS_C^2, \tag{7.63}$$

where f_d is the portion of the devices that switches at any given time.

Current Demand of the Chip

The length of interconnections on the chip is

$$l_{intTOT} = \frac{D}{p_w} D n_w e_w, \tag{7.64}$$

where D is die size, p_w is average interconnection pitch, n_w is the number of interconnection levels, and e_w is wiring efficiency (see Chapter 9 for a more detailed derivation). Die size increases by S_C, p_w scales down by $1/S$, and n_w normally increases by S. Capacitance per unit length of interconnections remains approximately constant:

$$C \propto 1. \tag{7.65}$$

As a result the interconnection capacitance on the chip

$$C_{intTOT} = l_{intTOT} C \tag{7.66}$$

scales as

$$C_{intTOT} \quad \propto \quad \frac{S_C}{1/S} \times S_C \times S \times 1 \times 1$$

$$= \quad S^2 S_C^2. \tag{7.67}$$

Total gate capacitance is

$$C_{gateTOT} = N_g k_g F^2 C_{gox}, \tag{7.68}$$

where k_g is the size of the gate, F is minimum feature size, and C_{gox} is gate oxide capacitance per unit area; all scale by 1, $1/S$, and S, respectively. As a result the gate capacitance scales by

$$C_{gateTOT} \propto S^2 S_C^2 \times 1 \times \frac{1}{S^2} \times S = S S_C^2. \tag{7.69}$$

Consequently the average current drive requirement for the chip becomes

$$I_{TOT} \quad = \quad \frac{C_{TOT} \Delta V}{\tau_g}$$

$$\propto \quad \frac{S^2 S_C^2 \times 1/S}{1/S} \tag{7.70}$$

$$= S^2 S_C^2.$$

Because C_{TOT} is assumed to scale as the larger of the interconnection and gate capacitances, the average current drive requirement increases faster than the total current drive capability as a result of a higher number of interconnection levels. In practice, this does not cause a problem because logic chips are usually interconnection-capacity limited, and the total drive capability of transistors is never fully utilized. (There is always some empty silicon area where more or larger transistors can be placed; however, it is not always where it is needed.)

In conclusion, the total current in the chip rises by $S^2 S_C^2$:

$$dI \propto I_{TOT} \propto S^2 S_C^2 \tag{7.71}$$

TABLE 7.2   Scaling of Power-Supply-Level Fluctuations

Parameter	Ideal Scaling	Improved Scaling
Rise time (t_r)	$1/S$	$1/S$
Total current requirement of the chip (I_{TOT})	$S^2 S_C^2$	$S^2 S_C^2$
Rate of change of current $(dI/dt \propto I_{TOT}/t_r)$	$S^3 S_C^2$	$S^3 S_C^2$
Number of power connections (N)	1	$S S_C^2$
Effective inductance per connection (L)	S_C	$1/S^2$
Voltage fluctuation $[\frac{L}{N}(dI/dt)]$	$S^3 S_C^3$	1
Signal levels and noise margins (V_{DD})	$1/S$	$1/S$
Signal-to-noise ratio $[V_{DD}/L(dI/dt)]$	$1/S^4 S_C^3$	$1/S$

S: Scaling factor for device dimensions
S_C: Scaling factor for chip size

Table 7.2 summarizes the effects of scaling on dI/dt noise. In the following discussion, it is assumed that decoupling capacitors surround the chip carrier to eliminate fluctuations caused by the inductance of the board wires, and as a result only the inductance of the connectors between the die and the substrate and the inductance of on-chip power lines are considered. The goal is to give an idea of how fast the delta-I noise is increasing and to investigate the effectiveness of some of the proposed methods rather than to obtain precise measures.

In simple ideal scaling, feature size is scaled by $1/S$, and the rise time of the signals improves by $1/S$. Total switching current increases by $S^2 S_C^2$, and the time rate change of the current grows as $S^3 S_C^2$. Self-inductance of on-chip power lines rises by S_C because of larger chip dimensions. If the package technology is not improved, the inductance of the connections between the die and the substrate (bonding wire, lead frame, TAB, solder balls, power distribution networks on ceramic substrates) also increases by S_C. As a result the voltage fluctuation rises by $S^3 S_C^3$. Because the signal levels along with all other voltages are scaled by $1/S$, the signal-to-noise ratio is reduced by $S^4 S_C^3$, an alarmingly large ratio. Two factors reduce the severity of the situation. First, off-chip power lines (board wires, pins, lead frame, bonding wires) have much larger inductances than the on-chip wires, and their current demand does not increase as fast as on-chip circuitry. Because of their large sizes, they are also easier to decouple using capacitors. Second, on-chip current transients may be increasing very fast, but they have not yet reached significant levels except in high-speed DRAMs and SRAMs that generate large current spikes at the beginning of their sense and precharge cycles.

The improved scheme in Table 7.2 tries to curb the voltage fluctuations by increasing the number of power connections and by reducing the inductance per connection. A matrix of power connections proportional to SS_C^2 is spread over the chip, similar to the improved scaling that was suggested to minimize the IR voltage drops along the power lines in Chapter 5. In addition, inductance per connection and the powering tree on the chip is reduced by $1/S^2$. This can be accomplished by using TAB or flip-chip technology and by integrating decoupling capacitors on the chip. As a result the effective inductance of the power network drops by $1/S^3 S_C^2$. This yields a signal-to-noise degradation factor of $1/S$ rather than $1/S^4 S_C^3$.

7.10   Summary

Cross talk between signal wires and inductive voltage fluctuations along the power lines can increase circuit delays or even cause latches to switch to a wrong value. Cross talk is a result of mutual capacitances and inductances between neighboring lines. If the rise time t_r of signals is longer than the round-trip delay $2t_f$ along the line, inductive effects are negligible, and the coupling can be calculated based on mutual capacitances. For long wires or fast rise times, wires must be modeled as coupled microstrip lines. Using even- and odd-mode calculations, the worst-case coupling coefficient is obtained as

$$K_V = \frac{Z_{0e} - Z_{0o}}{Z_{0e} + Z_{0o}}. \tag{7.72}$$

When a pair of coupled lines is in a homogeneous medium, the electrical and magnetic couplings are described by the relation

$$\frac{\mathcal{C}_m}{\mathcal{C}} = \frac{\mathcal{L}_m}{\mathcal{L}}, \tag{7.73}$$

and the coupling coefficient is obtained as

$$K_V = \frac{\mathcal{C}_m}{\mathcal{C}} = \frac{\mathcal{L}_m}{\mathcal{L}}, \tag{7.74}$$

where $\mathcal{C}_m$ and $\mathcal{L}_m$ are mutual capacitance and inductance per unit length. In a homogeneous dielectric medium, the forward capacitive and inductive couplings are equal in magnitude and opposite in sign, and they travel at the same speed. As a result there would be no far-end cross talk if the lines were in a homogeneous medium and were terminated with matched resistors at the source end.

In a homogeneous medium, the backward capacitive and inductive couplings have the same sign, and the near-end noise is an attenuated version of the driving function. When the transmission lines are terminated with matched resistors at the source and receiving ends, the pulse width is equal to the round-trip along the line $2t_f$ and the amplitude is proportional to the amplitude of the driving function. The near- and far-end coupling coefficients are obtained as

$$K_{ne} = \frac{C_m}{2C}$$

$$K_{fe} = 0. \tag{7.75}$$

In an inhomogeneous medium, the proportionality described by Eq. 7.73 does not hold, and the inductive and capacitive couplings are not equal; noise waveforms are observed at the far end proportional to the line length, slope of the driving function, and

$$K_{fe} = \frac{1}{2}\left[Z_0 C_m - \frac{\mathcal{L}_m}{Z_0} \right]. \tag{7.76}$$

The couplings can be minimized by making sure that no two lines lie next to each other for longer than a certain length, by keeping the spacings between lines large, by placing reference planes under and above the lines, and by putting grounded wires between any two signal lines. The area, yield, and monetary costs of these methods are traded off against the performance requirements of the system.

Delta-I noise is a result of current transients generated by the circuits and the inductance of the power-distribution network. Simultaneous switching of output buffers and on-chip circuitry may result in large current surges, and the resulting power-supply-level fluctuations can interfere with on-chip circuitry and with active as well as quiet off-chip drivers. In CMOS circuits, non-switching chip capacitance helps the switching portion of the chip by charge sharing. In bipolar ECL circuits, any steady current dissipated by the chip helps to reduce ΔI noise because when a group of circuits switches, the power-supply level falls slightly and the current demand of the already conducting circuits decreases. The freed-up current is then channeled to the newly turned on circuits, and their current demand is satisfied with little fluctuation at the power-supply level. Off-chip drivers are a source of concern in both technologies because their only current return path is the high-inductance power/ground pins.

Multiple power and ground connections and a hierarchy of decoupling capacitors built on the module and chips can effectively control the power-supply-level fluctuations that result from dI/dt noise. Decoupling capacitors placed close to the chip store charge and energy during the steady state and then assume the role of power supply during current switching. Multiple power and ground connections reduce the effective inductance and can support a larger current transient than a single pair of connections can on their own. In addition, turn-on characteristics of off-chip drivers can be tailored such that large current spikes are avoided.

It is important to take the full current path into account while calculating the inductance that produces ΔI noise. For example, on-chip decoupling capacitors help to curb the noise generated by on-chip circuitry but are not as effective in reducing the simultaneous switching noise generated by off-chip drivers because the current return loop must be closed through package power/ground pins in any off-chip path. On-chip capacitor provides a stable voltage difference between on-chip power lines, but the chip power-supply levels may move up and down with respect to the board power-supply levels when current transients are generated. As a result it is best to keep the power lines and pins of on-chip and off-chip circuitry separate. The above argument also applies to signals that leave modules and boards.

In addition to transient noise, resonance effects at the board and chip levels must be taken into account. For example, the tank circuit formed by the module pin inductance and chip capacitance and resistance form a resonance circuit. A large voltage fluctuation can build up at the power lines if the resonance frequency is near the operation frequency or one of its harmonics. Similar resonance phenomena exist at the board level due to the inductance of the power connectors to the board and the lead inductance of the decoupling capacitors on the board. These resonance frequencies must be kept either well below or well above the system operating frequency. The impedance as seen by the switching circuits must also be kept small at these resonance frequencies.

8

CLOCKING OF HIGH-SPEED SYSTEMS

Clocking is a major topic in high-speed digital system design. In CPUs, the clock frequency determines the rate of data processing. (The instruction execution rate is determined by the ratio of the clock frequency to the average number of cycles required to execute an instruction.) In I/O and memory buses, the clock frequency determines the rate of data transmission. (The data transmission rate is determined by the product of the clock frequency and the bus width.) Consequently digital system designers strive to maximize the clock frequency in order to achieve high system performance.

The first step to attain a high clock frequency is to employ a fast circuit family. With faster circuits, a given amount of logical function can be performed in a shorter time. Standard functions, such as arithmetic logic units, multiplexors, RAMs, and off-chip drivers, have shorter delays when implemented in a faster technology. Of course, to harness the potential speed of a technology, the system must be carefully partitioned, critical paths must be finely tuned, and circuits must be carefully designed and optimized. Usually bipolar ECL systems have clock frequencies two to three times faster than CMOS systems, and GaAs systems have clock frequencies two to three times faster than bipolar ECL systems. In 1990, typical clock frequencies are

16–40 MHz for CMOS systems, 60–125 MHz for bipolar ECL systems, and 200–500 MHz for GaAs systems.

The second step to attain a high clock frequency is to provide a fast storage element (latch, flip-flop, register) and a robust clocking scheme that is free from race conditions and adds minimal "dead" time to the machine cycle. Usually dead time, such as nonoverlap between two clocks, is added to prevent race conditions, which also increases the cycle time. High-speed systems usually use clocking schemes that require careful design but also introduce minimal cycle time penalty. In general, simple clocking schemes work better than more complicated ones. It is best to keep the number of clock signals and the number of clock phases to a minimum. This reduces the negative effects of clock skew and simplifies the routing of the clock signals and the wiring requirements.

Race conditions are a major concern in synchronous systems. It is important that a signal does not inadvertently go through more than one storage element during a single clock cycle. This can happen due to clock skew and fast paths that violate the hold-time requirements. To avoid race conditions, one needs a robust storage element, a good clocking scheme, a well-laid-out clock distribution tree with a small skew, and careful delay analysis that takes into account not only long delay paths but also the fast paths.

The third step to attain a high clock frequency is to construct a clock distribution system with a small skew. If two circuits clocked by the same signal are not equidistant from the clock driver (such as points A and B in Fig. 8.20), they receive the clock signal at different times. This is one form of clock skew. Skew is also introduced by the variations in the delay of clock buffers throughout the system due to process-dependent transistor and interconnection parameter variations and capacitive loading variations. The most troublesome aspect of the skew is the uncertainty of the arrival of the clock signal at a given storage element. If it can be guaranteed that the clock signal always arrives at a given storage element a predetermined amount of time earlier than it arrives at another storage element, design techniques can be employed to compensate or even take advantage of this fact. But the nature of clock skew is such that the designer does not know which storage elements will receive the clock early and which storage elements will receive it late.

There are two reasons for this uncertainty. First, the logic design is usually done before the chips are laid out, so the relative positions of storage elements with respect to clock buffers are not known to the designer. Second, the variations in the clock buffer delays due to device parameter-variations are random in nature. The parameter-variation related clock skew is even more serious in chip-to-chip paths because circuits on two separate chips, which may be processed at different times in a manufacturing line or even at different manufacturing lines, have much more variation than circuits on the same die. Due to this variation of the arrival time of the clock signal, one needs to add the worst-case skew to the path delays and consequently to the machine cycle time in order to guarantee that the circuits will function properly. Controlling the skew becomes harder as circuits get faster, chips become larger, and minimum feature size is scaled down. For high-speed systems, a large amount of design effort is spent to minimize the clock skew and prevent it from becoming a significant portion of the cycle time.

In this chapter, an overview of clocking methods is presented. Single-phase latch and edge-triggered flip-flop machines, as well as two-phase single- and double-latch machines, are described. Cycle time and race calculations for these configurations are derived, and advantages and disadvantages of various approaches are pointed out. Tuning of clock signals and timing problems at the asynchronous interfaces are described. Self-timed design methods are presented that offer an alternative approach to synchronous designs. Later, a symmetric clock distribution tree is described that is suitable for VLSI circuits, as well as for the module- and board-level clock distribution of high-speed systems. In this technique, the clock signal is delayed by an equal amount before it reaches the subblocks in the system. Design considerations include transmission line properties of the clock lines, reflections at the discontinuities, cross talk, termination, driving of the lines, distributed interconnection RC delay, and skin effect. A design example is presented in which the clock signal is distributed over a 3-inch wafer with less than 0.5 nsec rise time and less than 0.5 nsec skew (suitable for a 250 MHz clock signal and 4 ns cycle time).

8.1   Clocking Schemes and Storage Elements

A variety of clocking schemes and associated storage elements are used in digital systems. The possibilities include one-phase clocking with latches or edge-triggered flip-flops and two-phase clocking with single or double latches.

A *D-latch* is a storage element that has data and clock inputs of D and CLK and a data output of Q. While CLK is high, Q follows D, changing whenever D changes. While CLK is low, Q remains constant, holding the last value of D (Fig. 8.1(a)).

An *edge-triggered D-flip-flop* has the same inputs and outputs as the D-latch, but Q changes only on the rising or falling edge of the CLK (Fig. 8.1(b)).

Two-phase systems usually have two nonoverlapping clocks. The storage element may be a double latch where one latch directly feeds the next and the two stages are clocked separately. These two stages are referred to as master and slave or L1 and L2 latches (Fig. 8.1(c)). Two-phase systems may also partition the logic between latches that are clocked by nonoverlapping clocks, in effect, placing logic between the master (L1) and slave (L2) stages, in addition to between L2 and L1 stages. Two-phase systems with single and double latches are shown in Fig. 8.2(b) and 8.2(c).

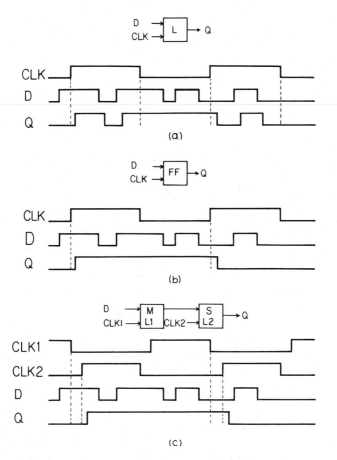

FIGURE 8.1   Input, output, and clock waveforms for various storage elements: (a) D-latch, (b) positive edge-triggered D-flip-flop, (c) two-phase double latch

The behaviors of latches, edge-triggered flip-flops, and two-phase double latches are summarized in Fig. 8.1. A latch is transparent when the clock is high; output follows the input. Edge-triggered flip-flops and two-phase double latches do not have the transparency property of a single-phase latch; they change states only during certain clock edges and only once during a cycle. Various storage elements, associated clocking schemes, and their optimization are described in reference [8.17].

Generic single-phase and two-phase finite state machines are shown in Fig. 8.2. Single-phase latch machines have the fewest elements, require minimal clock wiring, and offer high performance, but they also demand more

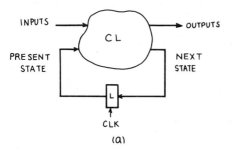

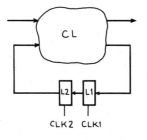

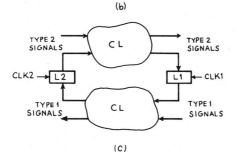

FIGURE 8.2   Generic structures of finite state machines: (a) single-phase system, (b) two-phase system with double latches, (c) two-phase system with single latches

careful control of minimum/maximum delays and clock pulse width because of the transparency property of the latch. The main disadvantage is the complexity of timing requirements due to the two-sided timing constraints, which require careful control of latch-to-latch delays and clock pulse width. Two-sided timing constraints can be avoided by using an edge-triggered flip-flop or a two-phase clocking scheme at the expense of a more complicated storage element and loss of some performance. Single-phase edge-triggered flip-flop machines have less severe race conditions because a flip-flop is never transparent. Two-phase machines shown in Fig. 8.2(b) and 8.2(c) employ nonoverlapping

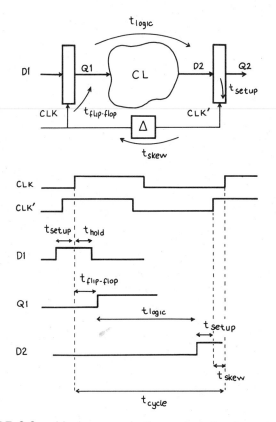

FIGURE 8.3   Maximum cycle time calculation in a single-phase system with edge-triggered flip-flops

clocks; as a result they have the best control of race conditions. In Fig. 8.2(b), double latches are used, and the system behaves similarly to an edge-triggered flip-flop machine. In Fig. 8.2(c), the logic is divided between two groups of latches clocked by two nonoverlapping clocks. Here it is important that no latch has its input connected to a signal that is driven by a latch controlled by the same clock phase. Latches clocked by $CLK1$ should be fed by only type 2 signals, and latches clocked by $CLK2$ should be fed only by type 1 signals. A signal changes its type as it goes through a latch.

Good storage element design and clock distribution are essential to achieve high performance. It is important that the sum of storage-element delay, setup time, and clock skew is kept to a minimum, and race conditions are avoided with a good storage element design and a carefully laid out clock distribution network.

8.1.1   Single-Phase Clocking with Flip-Flops

Figure 8.3 illustrates the factors that contribute to the cycle time in a single-phase clock system based on edge-triggered flip-flops.[1] The minimum cycle time is given by

$$t_{cycle,min} > t_{flip\text{-}flop,max} + t_{logic,max} + t_{setup,max} + t_{skew,max}. \qquad (8.1)$$

Here $t_{flip\text{-}flop}$ is the flip-flop delay: time from the clock edge that captures the data to the time that the data is available at the output of the flip-flop. The second term, t_{logic}, is the maximum delay through any logic block between two flip-flops, including array accesses and chip crossings. Setup time, t_{setup}, is the amount of time the inputs of a flip-flop should be stable prior to the clock edge. Finally, t_{skew} is the worst-case skew between clocks; the maximum amount of time the clock of the receiving flip-flop can be earlier than the clock of the sending flip-flop.

Hold time, t_{hold}, is the amount of time the input must stay stable after the clock edge to guarantee capturing the correct data. It is not uncommon to have a zero or negative hold time. To guarantee that the data is captured, the clock width must be greater than the hold time:

$$t_{hold} < t_{clk\text{-}width}. \qquad (8.2)$$

In addition, to prevent a race condition in which the data may go through two flip-flops at the same cycle, minimum flip-flop-to-flip-flop delay must be greater than the maximum hold time:

$$t_{hold,max} < t_{flip\text{-}flop,min} + t_{logic,min} - t_{skew,max}. \qquad (8.3)$$

If this inequality does not hold, the output of a flip-flop may reach the next flip-flop while the second flip-flop is still trying to capture the data from the previous cycle. Even with 0 ns hold time, there can be problems due to clock skew. To prevent a race condition, when a flip-flop directly feeds another flip-flop, buffers must be inserted to add delay or the minimum flip-flop delay may have to be guaranteed to be larger than the sum of clock skew and hold time.

8.1.2   Single-Phase Clocking with Latches

Edge-triggered flip-flops have a sharp cycle boundary—the clock edge that activates the flip-flops. In a latch-based design, the latch is transparent while the clock is high and the boundary is not sharply defined. A latch accepts the data during a time period rather than being triggered by a clock edge. The latch opens when the clock goes high; data is accepted continuously while the clock is high; and latch closes when the clock goes down. While

---

[1]As shown in Fig. 8.3, any cycle time or critical path delay calculation must close a *complete loop* through various delay subblocks.

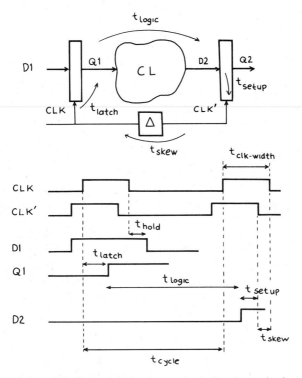

FIGURE 8.4   Maximum cycle time calculation in a single-phase system with latches

the clock is high, any change in the input is reflected at the output after a latch delay. This is useful when the delays of two succeeding cycles are not equal because delay from one cycle can be transferred to the next cycle by an amount that can be as much as a clock width. Since the cycles of a pipeline do not have the exact same delay, the flexibility in allocation of the delay is advantageous. This optimizes the cycle time but also introduces race conditions in fast paths. Single-phase latch-based designs were common in early computers. Currently they are used in some high-end designs, but due to complicated timing requirements, single-phase latch designs are not commonly used in VLSI circuits. The design requirements are too complicated to be handled by most of the existing CAD tools for VLSI.

The cycle time in a latch-based system cannot be defined as clearly as in a flip-flop-based system because of the interfusion of two succeeding cycles while the clock is high. For example, in Fig. 8.4, both the preceding and succeeding stages are assumed to have delays less than the stage depicted in the figure. The data from the preceding stage $D1$ arrives before the *rising* edge of a clock, and the data $D2$ does not have to get to the next latch until near the *falling* edge of the following clock. If this is the critical path, the cycle time is defined as

$$t_{cycle,min} > t_{latch,max} + t_{logic,max} + t_{setup,max} + t_{skew,max} - t_{clk\text{-}width,min}.$$
$$(8.4)$$

Here the cycle time is shorter than a flip-flop-based system by $t_{clk\text{-}width}$ because some of the delay from the critical stage is transferred to the preceding and succeeding stages.

The penalty for the additional flexibility is the complexity of the race conditions in latch-based single-phase systems. To prevent race conditions, the *clock width* has to satisfy a *two-sided* relationship:

$$t_{setup,max} < t_{clk\text{-}width} < t_{latch,min} + t_{logic,min} - t_{hold,max} - t_{skew,max}. \quad (8.5)$$

Clock width must be greater than the maximum setup time to guarantee the latching of the data. Clock width must also be less than the sum of minimum latch and logic delays *minus* the maximum hold time and skew. This is to prevent a race condition in which the clock goes high, and fast data from an earlier stage propagates through a minimum logic path and arrives at the next latch before its clock closes, violating the hold-time condition of the previous data. As a result the circuits must be designed such that the slowest signal can propagate to a latch within a clock cycle and at the same time the fastest signal does not go through more than one latch while the clock is high. This is challenging in latch-based systems because when the clock is high, all the latches are transparent. Only careful design can prevent the data from zipping through more than one latch.

Two-sided constraints, such as the condition in Eq. 8.5, are hard to meet, especially in VLSI circuits where circuit delays can vary greatly from one wafer lot to the next, as well as across the same die. To guarantee that all the race conditions are prevented, detailed timing analysis of maximum and minimum delays involving large amounts of circuits may be required. The validity of these calculations will depend on the accuracy of the delay equations. One of the worst features of fast path races is that they cannot be resolved by reducing the clock frequency unlike long critical path delays, which can be resolved by running the clock slower. In addition, some fast path races are intermittent; they show up only during certain conditions. These differences become important during debugging. A race condition can severely hinder the bring-up process.

8.1.3 Two-Phase Clocking

As described above, single-phase latch- and edge-triggered flip-flop-based systems have race conditions. Latches have more serious race conditions due to two-sided requirements on clock width, minimum logic delays, and clock skew. Edge-triggered flip-flops have relatively less serious race conditions due to

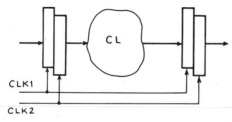

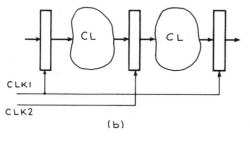

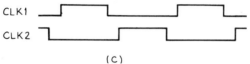

FIGURE 8.5   Two-phase nonoverlapping clocking: (a) double-latch design, (b) single-latch design

clock skew. Two-sided constraints of latches can be avoided by using an edge-triggered flip-flop or a two-phase clock. A two-phase clock requires more clock lines, and its double latches have more transistors and a longer delay, but it eliminates fast path races. To avoid race conditions, most nMOS and CMOS VLSI circuits are designed with nonoverlapping two-phase clocks.

Two-phase double latches are *master-slave* or *L1-L2* type where the data is first captured by the *master (L1)* latch and then passed to the *slave (L2)* latch. The nonoverlapping nature of the clocks prevents both latches from being turned on simultaneously; as a result, the double latch is never transparent, and no two-sided requirements exist. The designer mainly has to worry about the longest delays. Even in a nonoverlapping two-phase clocking scheme, however, some fast path analysis is required to avoid race conditions that may result from clock skew. It should be ensured that hold-time requirements are met under even minimum logic delays and maximum clock skew.

Two-phase nonoverlapping clocking schemes are shown in Fig. 8.5. Double-latch design is not the only way to build a two-phase system; in many designs, the logic is divided into two halves, and the halves are placed between two

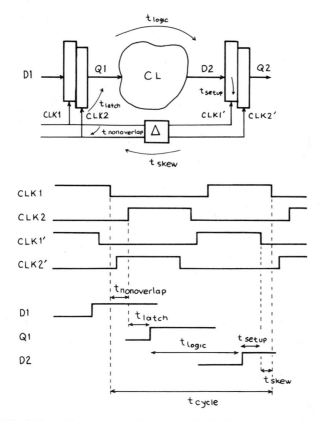

FIGURE 8.6   Maximum cycle time calculation in a two-phase system with nonoverlapping clocks and double latches

latches, each clocked by one of the clock phases (Fig. 8.5(b)). In some cases, four or more clock phases are used to divide the cycle time into many subcycles. The best approach to designing high-speed systems, however, is to keep the clocking as simple as possible. This minimizes the area overhead due to clock wires and the cycle time overhead due to clock skew.

Figure 8.6 illustrates the factors that contribute to the cycle time in a two-phase system with nonoverlapping clocks and double latches. The minimum cycle time is given by

$$t_{cycle,min} > t_{nonoverlap,max} + t_{latch,max} + t_{logic,max} + t_{setup,max} + t_{skew,max}.$$
$$(8.6)$$

To guarantee that the data is captured, the clock width must be greater than the setup time:

$$t_{setup} < t_{clk-width}.$$
$$(8.7)$$

In addition, to prevent a race condition, minimum latch-to-latch delay must be greater than the maximum hold time:

$$t_{hold,max} < t_{nonoverlap,min} + t_{latch,min} + t_{logic,min} - t_{skew,max}. \qquad (8.8)$$

If this inequality does not hold, the output of a double latch may reach the next latch too fast and disrupt the capturing of the data from the previous cycle.

It is important to note that the only difference between the equations of edge-triggered flip-flop and double latch with nonoverlapping clocks is the addition of the $t_{nonoverlap}$ to the delay and race equations. In a two-phase nonoverlapping clocking scheme, a performance penalty is paid due to the addition of $t_{nonoverlap}$ in the cycle time described in Eq. 8.6. But this also gives an independent control for the race condition described by Eq. 8.8. The nonoverlap time can be increased to guarantee that the hold-time requirement will be satisfied under even the worst-case situation. It may be necessary to insert buffers to add delay only in severe cases involving chip-to-chip signals and where chip-to-chip clock skew is too large to be covered by the clock nonoverlap.

Two-phase double latches also have distinct advantages in testing of the circuits. When double latches are used, a simple multiplexing circuit can be added to the input latch, which enables test data to be scanned in and out of a chip by a tester. If a strict two-phase clocking discipline is followed, this scanning operation can be made free from any race conditions. This provides a structured approach to testing with good observability and controllability. Automatic test generation programs compatible with this design method simplify the test pattern generation task. The penalty is the additional complexity of the storage element, increased wiring requirements, and possibly a slightly longer cycle time.

8.2   Storage Element Implementations

Various CMOS latch implementations are shown in Fig. 8.7. A dynamic latch is shown in Fig. 8.7(a). Here a full CMOS transmission gate is used to clock the data input. This is more expensive than an nMOS dynamic latch with only an nMOS pass transistor because a CMOS transmission gate requires a pMOS-nMOS transistor pair, and, more importantly, complementary clocks CLK and $\overline{CLK}$ have to be routed to the latch. The full transmission gate is used because it passes a good logic "1" and "0" values providing high noise margins and eliminating any dc power dissipation. If a single nMOS transistor is used as a transmission gate in a CMOS latch, the logic "1" level would degrade due to the threshold drop across the transistor and cause the inverter to dissipate some dc power.

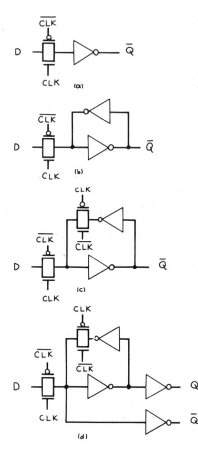

FIGURE 8.7 CMOS latch circuits: (a) dynamic CMOS latch, (b) static CMOS latch with cross-coupled inverters, (c) static CMOS latch with a clocked feedback, (d) buffered version of the circuit in (c)

In many applications dynamic storage is undesirable. The circuit in Fig. 8.7(b) provides a feedback path via a second inverter and makes the latch static. If the charge at the storage node leaks or is removed due to noise or other interference, the feedback inverter would supply the charge to restore the correct value. The two inverters and the pass transistors are ratioed to make sure that the latch will switch properly when data is being written to it. The feedback inverter has to be sufficiently weak so that the circuit that drives the latch can overpower the feedback inverter and change the latch state. The problem is that the circuit that drives the latch is part of the ratioed circuit. Furthermore, if one circuit drives more than one latch, it becomes harder to switch the latches because one driver has to fight against multiple feedback

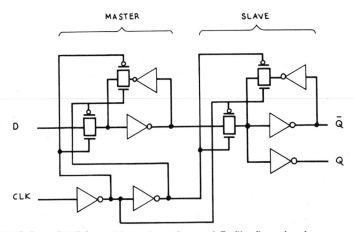

FIGURE 8.8   CMOS positive edge-triggered D-flip-flop circuit

inverters. The feedback inverter in the latch should be weak enough to guarantee that the weakest circuit that can possibly drive latches can switch them under the maximum fan-out condition, taking into account the interconnection resistance effects.

The circuit in Fig. 8.7(c) also provides a feedback loop, but it avoids the need for ratioed design by placing a transmission gate at the feedback loop. Either the data input or the feedback loop is connected to the storage node but never both simultaneously.

The circuits in Fig. 8.7(b) and 8.7(c) are prone to charge-sharing problems because, directly or indirectly, their storage nodes are tied to their outputs. Due to charge sharing, the output can be pulled up or down, causing the latch to change its state and the stored value to be lost. For example, the latch output may be tied to a pass transistor multiplexor that feeds a large capacitive load. When the pass transistors are turned on, the charge sharing between the capacitance at the input of the multiplexor and the capacitance at the output of the multiplexor may cause the latch to change its state. If the latch is designed for a custom application feeding a well-characterized node, there may be no problems. For a more generic building block, however, the circuit in Fig. 8.7(d) may be more appropriate where the outputs are buffered, isolating them from the storage node.

A CMOS positive-edge-triggered D-flip-flop is shown in Fig. 8.8. Here, two latches similar to the one in Fig. 8.7(c) are tied in a master-slave fashion. A complementary clock pair is generated by two inverters that are part of the flip-flop circuit.

A fully static CMOS master-slave D-flip-flop that does not use transmission gates is shown in Fig. 8.9. It is built using AND-OR-INVERT (AOI) and

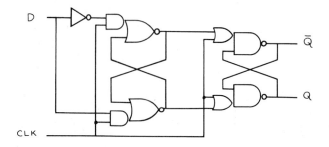

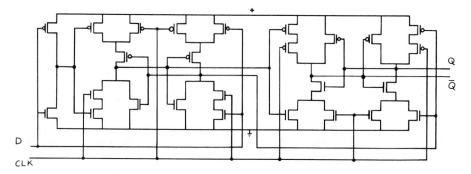

FIGURE 8.9   Fully static CMOS negative-edge-triggered D-flip-flop circuit

OR-AND-INVERT (OAI) gates. Using an AOI in the master and an OAI in the slave eliminates the need for a clock inverter. The flip-flop can be converted to a positive-edge-triggered one by using an OAI in the master and an AIO in the slave.

Two-phase double-latch (L1-L2) circuits are shown in Fig. 8.10. A dynamic double-latch design is in Fig. 8.10(a), and static circuits are in Fig. 8.10(b) and 8.10(c). The static double latch in Fig. 8.10(b) is similar to the D-flip-flop in Fig. 8.8 except the clocks are nonoverlapping, and they are provided by a clock generator external to the latch. In this scheme, the nonoverlap between $CLK1$ and $CLK2$ can be adjusted to avoid any critical race conditions.

An ECL latch circuit is shown in Fig. 8.11. Here *series gating* is used, and as a result the clock signal levels are one diode drop below the data input. In addition, a differential clock is used. The gate at the left brings the data in when the clock is high, and the gate at the right holds the data when the clock goes low. As in CMOS, ECL D-flip-flops and two-phase double latches can be constructed by putting two latches together in a master-slave fashion.

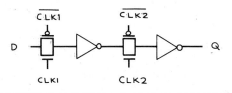

(a)

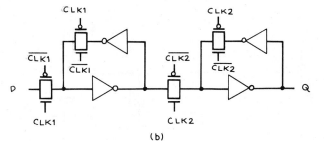

(b)

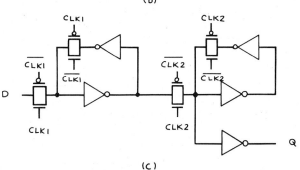

(c)

FIGURE 8.10   CMOS two-phase double-latch circuits: (a) dynamic, (b) static unbuffered, (c) static buffered

8.3   Clock Skew

As described in the previous sections, clock skew directly adds to the cycle time. In a high-speed bipolar mainframe computer with a single-chip-module-on-a-planar-board packaging technology, a typical on-board critical path can be allocated in this way [8.1]:

- Silicon delay, 60 percent

- Package delay, 30 percent

- Clock skew, 10 percent

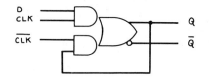

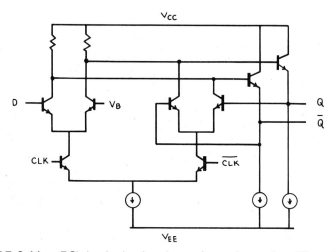

FIGURE 8.11   ECL latch circuit using series gating and a differential clock

For a board-to-board path, package delay may dominate, and the delay break-down may become as follows:

- Silicon delay, 20 percent

- Package delay, 70 percent

- Clock skew, 10 percent

In both cases, clock skew accounts for 10 percent of the total cycle time.

On-chip skew problem becomes more severe as die size increases (because of longer interconnection lengths) and as minimum feature size is scaled down (because of a higher RC constant per unit length of interconnection) [8.2]–[8.5]. As circuits get faster and cycle time is reduced, keeping the clock skew a small portion of the cycle time becomes harder. In addition, new architectures, such as heavily pipelined systolic arrays, may eliminate long signal wires and may make the clock distribution lines the only global signals. Then clock skew may become a major cycle time limiter [8.6].

These developments introduce more stringent requirements for the clock signal, and as a consequence the clock distribution network must be carefully designed. This applies to on-chip clock distribution in VLSI circuits, as well as system-level clock distribution on the modules and boards of high-performance computers.

8.4   "Useful" Clock Skew

Clock skew is not always bad; controlled clock skew can be used constructively to improve the performance. This is referred to as *tuning* the clock signal or the clock net. If the delays in stages of a pipeline are not balanced, cycle time can be optimized by delaying the clock that captures the data from the pipe stage with the longer delay or by bringing the clock earlier to latches that feed the longer pipe. This is illustrated in Fig. 8.12. Here it is important to control the skew precisely, which is even more challenging than its elimination. Using this technique, one can achieve the same optimization effect in edge-triggered flip-flop or two-phase double-latch-based systems as in single-phase latch-based systems, namely transferring delay from a critical cycle to the preceding or succeeding cycles. The main difference is that by tuning some clock paths, the delay is shifted only in a few select critical paths, and possible new race conditions are introduced only in these paths. The rest of the system benefits from the robustness and simplicity of the two-phase nonoverlapping clocking scheme, which guarantees that there are no race conditions without requiring any complicated delay analysis. Only the nets with the adjusted clocks need to be analyzed for the new types of race conditions.

When moving clocks with respect to each other, minimum and maximum delays through the logic blocks must be taken into account to prevent race conditions. For example, if CLK' in Fig. 8.12 is moved too much to the left, data launched by a CLK' flip-flop can go through a minimum delay path and arrive at a CLK'' flip-flop before the data from the previous cycle is latched. As a result the output of the CLK' flip-flop can go through the CLK'' flip-flop and arrive at CLK''' flip-flop one cycle too early, causing a timing failure. It should be noted that this race condition arises only when a storage element with an early clock feeds a storage element with a late clock. Data can be properly transferred in one direction with late storage elements feeding early ones. In most cases, however, data needs to be transferred in both directions.

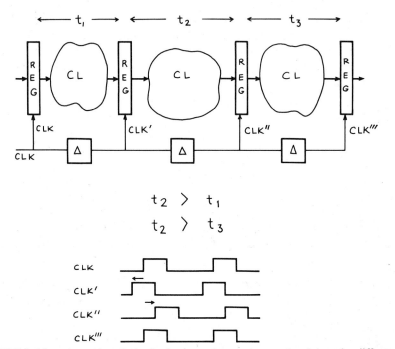

FIGURE 8.12   Intentional clock skew used to balance the delays in different stages of a pipeline. The stage in the middle has more delay than the stages before and after it ($t_2 > t_1$ and $t_2 > t_3$). To minimize the cycle time, the designer can "steal" some time from the stages with less delay. As illustrated by the clock waveforms at the bottom, this can be accomplished by bringing CLK' earlier (which steals time from the first stage) and/or by delaying CLK" (which steals time from the third stage).

8.5   Asynchronous Interfaces and Synchronizer Failure

The communication among systems not sharing a common clock gives rise to synchronization problems [8.22]–[8.28]. Asynchronous signals incoming to a synchronous system must be synchronized with the rest of the system. This is usually done by feeding the asynchronous signal to a latch or flip-flop referred to as the *synchronizer*. A serious concern in the asynchronous interfaces of synchronous systems is the prolonged indecision of the synchronizing flip-flop when it enters a *metastable* state, as illustrated in Fig. 8.13. If the incoming asynchronous signal switches at the same time as the clock signal that is supposed to synchronize it, the setup time requirement may be violated, and the flip-flop can go into a metastable state and stay there for many cycles (*synchronizer failure*).

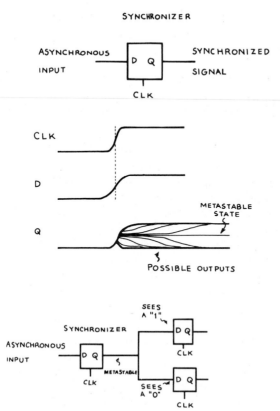

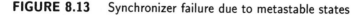

FIGURE 8.13   Synchronizer failure due to metastable states

The synchronizer bugs are hard to trace because they are probabilistic in nature and cannot be predictably repeated. As shown in Fig. 8.13, the metastable output of the synchronizer latch may be interpreted as a "1" by part of the system and "0" by the rest, further complicating the debugging process.

If the asynchronous signal is not performance critical, the output of the first flip-flop is fed to another flip-flop. This minimizes the possibility of injecting undefined states into the system because the system is not exposed to the asynchronous signal until it goes through two flip-flops. The metastability problem is confined to these few synchronizer flip-flops. The larger the number of synchronizer flip-flops the signal goes through, the smaller the possibility of injecting an undefined state into the system.

All methods that minimize the probability of a system crash due to synchronization failure cost time. If one waits long enough before using the sampled data, the probability of synchronizer failure will be small. Even when the failure possibility is very small, however, a machine with a 20 nsec cycle time samples an asynchronous input 50 million times a second, making it hard to achieve a very long mean-time-to-failure.

8.6 Phase-Locked Loops

A phase-locked loop (PLL) can be used to dynamically adjust the phase of a signal to match it with the phase of another signal [8.29], [8.33]–[8.38]. Phase-locked loops can be used to compensate for the clock skew that results from chip-to-chip variations in the delays associated with on-chip clock generators and buffers. Clock generator/buffer delay may vary from one chip to another because of the differences in the loading at the clock buffer outputs and chip-to-chip variations in process-dependent device parameters and junction temperature. Device parameter variations can be especially problematic because the chips may be manufactured by two different silicon foundries and may have significantly different device characteristics, both meeting the cycle time requirements. Even on the same fabrication line, the process can vary greatly. Junction temperature is another variable. The temperature of the chips can vary due to differences in their power dissipation, as well as differences in their location on the board with respect to the air flow and differences in the chip carriers and heat sinks.

The PLL eliminates only the skew due to clock generator/buffer. The chip-to-chip clock skew and on-chip clock distribution skew cannot be eliminated by a PLL. The PLL compares the phase of the waveform at the output of the on-chip clock buffer with the external clock signal and aligns the edges of the two waveforms by shifting the phase of the on-chip clock buffer output. As a result the clock signals on two chips will be matched only as well as the external clocks arriving at the chips. It is important to note that a PLL cannot compensate for the skew due to board wiring and packaging because the clocks on both chips are matched with the external clock at the chip pins. The skew due to board wiring and package can be minimized by making all the clock traces of equal length on the board and by placing the clock pins at identical positions on the chip carriers, trying to keep the lead length and capacitive loadings as identical as possible. In addition, the portion of on-chip skew is not eliminated that is due to distributed RC delays of the on-chip clock distribution lines and due to one circuit's being closer to the on-chip clock buffer than another circuit. This can be minimized by using a clock distribution tree that keeps all circuit blocks at an equal distance from the clock buffer, such as the H-clock tree described later in this chapter.

The general block diagram and a CMOS implementation of a phase-locked loop circuit is shown in Fig. 8.14. The circuits shown are from reference [8.38]; a different circuit implementation of a similar concept is also presented in [8.37]. The second reference uses a voltage-controlled oscillator

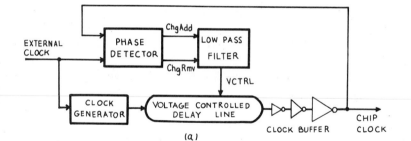

(a)

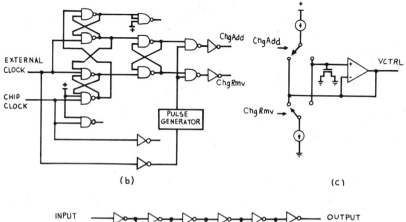

(b)

(c)

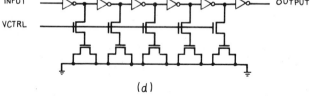

(d)

FIGURE 8.14   Block diagram of a phased locked loop and its CMOS implementation: (a) block diagram, (b) phase detector, (c) low-pass filter, (d) voltage-controlled delay line

(VCO) instead of a voltage-controlled delay line (VCDL). In Fig. 8.14, the phase of the clock buffer output is compared with the phase of the external clock by the *phase detector*, which is based on an edge-triggered D-flip-flop. The phase detector generates two signals, *ChgAdd* and *ChgRmv*. The first will be high if the chip clock is ahead, and the second one will be high if the external clock is ahead. Note the dummy loads that ensure symmetry in the phase detector circuit and increase its accuracy. The *low-pass filter* smooths out the output of the phase detector. The filter consists of a large nMOS capacitor and an op amp connected as a unity gain buffer providing a low-impedance output. The capacitor is charged or discharged by a pair of matched current sources under the control of the phase detector outputs $ChgAdd$ and $ChgRmv$. The phase detector outputs are gated by a pulse so that at every cycle, a fixed amount of charge is placed on or removed from the capacitor independent from the clock frequency. If $ChgAdd$ is high, charge is added to the capacitor, and the control voltage $VCTRL$ slightly goes up. If $ChgRmv$ is high, charge is removed, and $VCTRL$ slightly goes down. This control voltage is fed to a *voltage-controlled delay line* (VCDL), which is a series of capacitors tied to the outputs of inverters. The higher the $VCTRL$, the larger a capacitance the inverters see and the more the chip clock is delayed. As a result, if the chip clock is ahead, the phase detector brings $ChgAdd$ high, charge is added to the capacitor in the low-pass filter, $VCTRL$ goes up, and the capacitive loading in the VCDL increases and slows down the chip clock. If the chip clock is behind, it will be pushed ahead in a similar fashion. In other words, the clock generator and buffer are placed in a negative feedback loop, and the effect of the process, temperature, and loading variations is tracked and removed by the PLL. One needs to make sure that there is sufficient adjustment range and sufficient delay from the external clock to chip clock because, in this configuration, chip clock may need to be delayed by half or one full cycle with respect to the external clock to match their phases.

Phase-locked loops can be used to synchronize signals in many applications. For example, reference [8.40] describes how a PLL is used to synchronize the enable signals of multiple tristated off-chip buffers all tied to the same bus. Using a PLL, enable signals on the chips are synchronized, and it is guaranteed that no two drivers will be on simultaneously, even for a short time. This is required in order to avoid any bus contention. In addition, this is done with minimal cycle time and area penalty relative to other alternatives, such as disabling all drivers for a period of time analogous to the nonoverlap period in two-phase systems, which would hurt the cycle time. As will be described in the next section, PLLs also can be used to synchronize the communication between systems with different clocks in order to avoid metastable states and synchronizer failures.

8.7   Self-Timed and Asynchronous Systems

To avoid some of the timing problems associated with a synchronous system, self-timed asynchronous circuit design techniques can be used [8.7], [8.29]–[8.34]. There are three major types of asynchronous design applications:

1. Locally synchronous regions of logic communicating via asynchronous connections or via a bus clocked at a slower rate than the regions.

2. Self-timed logic circuits that communicate through a set of asynchronous protocols.

3. Memory and array circuits that use self-timed signals to optimize the speed and area.

Here the main goal is to eliminate the dependence on a global clock and to avoid the associated timing problems. Most of the self-timed circuit techniques are still at the development stage and have not yet gained widespread acceptance by the design community except in memory design.

8.7.1 Locally Synchronous Systems

Here the system (which can be a VLSI chip, a board, or an entire computer) is divided into synchronous islands with independent clocks [8.31], [8.32]. Actually this technique is fairly common in general-purpose computers where the CPU has its own clock and is tied to an asynchronous I/O bus with adapters on it. Each I/O adapter card may have its own clock. This is necessary because distributing a single clock signal over the entire computer with a reasonable skew would be very hard; it would limit the clock frequency and compromise the overall performance.

As the number of transistors on a VLSI chip increases and the circuit speed improves, clock skew becomes harder to manage, mainly because of the distributed RC delays of wires. Then it may become attractive to divide even a single VLSI chip into locally synchronous regions analogous to a large computer system.

Once the system is divided into locally synchronous regions, the clock skew problem is transferred to the communication mechanism among the regions. The communication among the regions can be in one of two ways [8.32]:

1. Asynchronously using a self-timed discipline.

2. Synchronously using a global clock slower than the local clocks.

Self-timed circuit techniques (method 1) are described in the next section. A general scheme that uses a synchronous global bus (method 2) is shown in Fig. 8.15. This structure can be a single chip that has independent synchronous regions, or it may be a board with chips that are synchronous but use a slower global clock for chip-to-chip communication, or it may be a large computer where the synchronous regions are separate boards tied to the system bus.

Here the synchronous regions use the faster clock, and the global communication is established by a bus that operates with a clock N times slower than the local clocks [8.31], [8.32]. It is important that the communication requirement among the regions is satisfied by the slow global bus; otherwise system performance will suffer. One would first determine the clock frequency that the bus can support and then select a bus width that satisfies the system communication requirements.

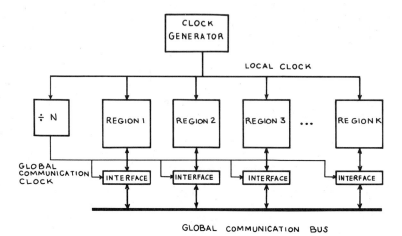

FIGURE 8.15   System with fast local clocks and a slower global communication clock and bus

The global clock frequency is an exact fraction $(1/N)$ of the local clock frequency; accordingly every edge of the slower global clock occurs simultaneously with an edge of the faster local clock. Due to skew, however, the edges can shift such that metastable states may be generated at the interface circuitry. This problem can be avoided by maintaining a correct relationship between the local clocks and the global clock. As shown in Fig. 8.16, phase-locked loop circuits placed in each module can be used to adjust dynamically the phase of the local clock to match it with the phase of the global communication clock to avoid metastability at the interface circuitry [8.31]. This is similar to the use of local synchronization in mainframe computers to ensure synchronous behavior between the frames.

8.7.2   Self-Timed Logic Circuits

A self-timed system is built from speed-independent modules that communicate via a set of asynchronous protocols that do not require a global clock [8.7], [8.29], [8.30], [8.34]. The system changes its state only when all modules signal completion. The lack of a global clock signal, by definition, eliminates the clock skew problem. In addition, in a synchronous system, cycle time is determined by the worst-case path. In essence, the rest of the logic circuitry has to wait for the longest path *every* machine cycle. In a self-timed system, on the other hand, initiation of a computational step depends on the completion signals generated by its sequential predecessors. The total delay is the aggregate delay of the computational steps; as such, it reflects the *average* de-

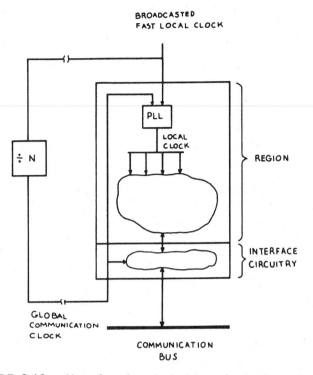

FIGURE 8.16   Use of a phase-locked loop for local synchronization of the clocks

lay rather than the *worst-case* delay [8.7]. If the worst-case path is activated rarely, a self-timed design will have a better performance than a synchronous implementation. On the negative side, self-timed systems require considerably more hardware than the synchronous systems because of the extra logic required for communication. They are also harder to design.

In a self-timed system, it is important that the events occur in the right order, but the exact timing is not critical. The goal is to ensure that a logically correct system cannot fail due to any timing problem. Self-timed systems are built from *elements* that perform computational steps initiated by the activation of an input called *Request* or *Go* and signal the completion of a step by an output referred to as *Acknowledge* or *Done* (Fig. 8.17). The sequence of the computational steps is determined by the connection of the elements, and the time needed to do the computation is given by the sum of the element delays between initiation and completion. All of the timing issues are confined into the elements, and interelement communication is asynchronous.

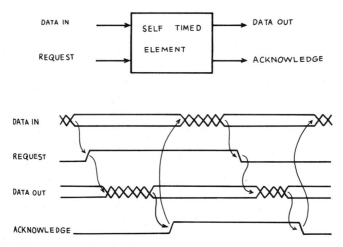

FIGURE 8.17 A self-timed element and a two-cycle signaling scheme [8.7]. Reprinted by permission of Addison-Wesley Publishing Co.

The sequence illustrated in Fig. 8.17 is a two-cycle signaling scheme [8.7]. The sequence is as follows. When the data inputs become stable, the *Request* line changes its state. The element then starts working on the data, and its outputs are changing. Once the work is finished and outputs are stable, the element switches the *Acknowledge* line to activate the other elements that may be waiting for its outputs. Then the other elements start working; their outputs start changing. This may effect the inputs of the first element. When the element that feeds the first element is finished, it switches the *Request* line, prompting the first element to start working on the data again. The entire sequence keeps repeating—the elements passing data to each other and using *Request* and *Acknowledge* lines to control the sequence in an asynchronous fashion. A more detailed treatment of self-timed logic circuits can be found in [8.7].

8.7.3 Self-Timed Memory and Array Circuits

Self-timed circuit techniques are used widely in memory circuitry. A majority of the SRAMs and DRAMs described in Chapter 4 use self-timed signals to generate many clocks such as precharge (PC) and sense amplifier enable (SAE).

Because of the differences in the physical structures of self-timed logic and memory circuits, their implementations differ significantly. Self-timed logic circuits are usually fully asynchronous and require large amounts of circuitry for signaling. The element detects the completion using combinational logic and signals it to the next element. Memory circuits, on the other hand, generate the completion signal by using an *analog* delay that mimics the function that is supposed to be completed [8.9]. This requires more careful design but takes much less area. Here the analog delay used should be longer than the longest functional delay through the block. The challenge is to make the delay long enough to ensure correct operation but not so long that the performance suffers. In addition, the delay should track the delay of the block it is mimicking under process, temperature, and power-supply-level variations. Usually a dummy circuit that is almost an exact copy of the functional block is used, such as a dummy word line or a dummy bit line in a RAM.

Analog features require special care and careful design; as a result they are hard to use in complex logic designs with tens of thousands of gates. The regular nature of the memory and array structures, however, makes it easier to control the timing and allows for extreme customization because a memory chip has better characterized and fewer unique paths than a logic chip. The relatively few self-timed signals in a RAM can be hand honed and carefully analyzed using a circuit simulator to ensure correct operation.

CMOS PLAs also can take advantage of self-timed signals to improve their speed and density. A self-timed precharged PLA is shown in Fig. 8.18. The circuit works as follows. When $\overline{PC}$ goes low, the AND plane is precharged. In the meantime, once the signal $DONE$ goes low, the OR plane also starts precharging. The pMOS precharge transistors of the OR plane are clocked by $DONE$ to prevent any dc paths during precharge. Because of the nMOS transistors between the planes and the ground, all the paths to ground are turned off during the precharge, and no dc paths are activated. The PLA inputs must be stable when $\overline{PC}$ goes high. Then the ground connection of the AND plane is enabled, and the AND plane starts to evaluate. Meanwhile the OR plane is still disabled because $DONE$ is low. Extra parasitic capacitance is placed on the dummy line that generates the $DONE$ signal to ensure that it is slower than any of the functional lines. As a result, when $DONE$ goes high, all the outputs of the AND plane are already stable, and the OR plane starts to evaluate.

This structure enables one to construct a NOR-NOR PLA, which has only parallel pull-down transistors. Alternatively a domino PLA would have to implement its AND plane as a series chain because domino logic requires an inverter between the AND and the OR plane (Fig. 8.19). The series chain is undesirable because it has a high resistance and a longer delay than a parallel pull-down. In addition, a large chain can suck up a lot of charge and discharge the precharged node inadvertently. This happens when the nMOS transistor closest to the ground has a low input and all the others above it have high inputs, and the transistor at the top of the chain goes high right before the precharge clock is disabled. To prevent this charge-sharing problem, the setup time of the PLA inputs with respect to the precharge clock must be specified longer than the setup time of a comparable-sized PLA implemented with a NOR-NOR structure. This will give more time for the drains of the

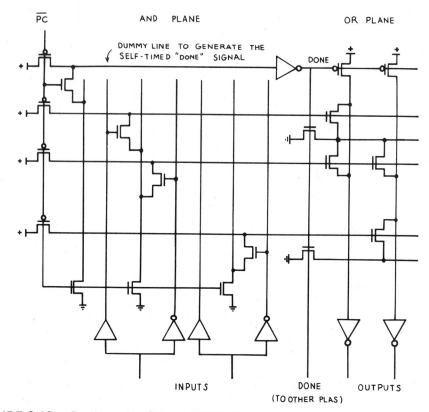

FIGURE 8.18   Precharged self-timed CMOS PLA

transistors in the chain to be precharged through the precharge transistor rather than stealing charge from the precharged node after the precharge clock is deactivated. This guarantees correct operation but also gives away some performance.

Multiple self-timed PLAs can be cascaded by feeding the *DONE* line of the preceding PLA to the $\overline{PC}$ of the next PLA. Then the AND plane of the first PLA will precharge when $\overline{PC}$ is low. When *DONE* goes low, the OR plane of the first PLA and the AND plane of the second PLA will also precharge. The AND plane of the first PLA would evaluate when $\overline{PC}$ goes high. The OR plane of the first PLA will start to evaluate when *DONE* goes high. The AND plane of the second PLA is ready to evaluate when *DONE* goes high but cannot discharge inadvertently because the outputs of the first PLA are precharged low and will go high only if their correct final value is high. A small *p*MOS device with a grounded gate can be used as a trickle device to replenish any charge lost due to leakage or noise. This is similar in principle to domino logic (see Chapter 2). For two cascaded PLAs to function

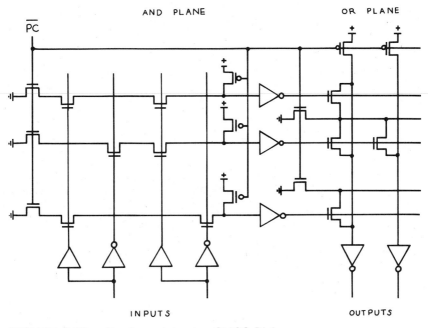

FIGURE 8.19   Precharged domino CMOS PLA

correctly, the second PLA cannot have any inverters at its input. The output of the first PLA is precharged low, and an inverter at the input of the second PLA will make it high, generating unintended discharge paths in the AND plane of the second PLA. The inverters shown at the left half of Fig. 8.18 must be taken away at the second and any succeeding PLAs. When both the true and complement of a signal are required, the PLAs must generate both signals separately and send this double-rail signal to the next PLA. A number of such PLAs can be connected one after another, providing a dense and fast precharged CMOS PLA implementation. A few simple rules must be followed. If a PLA accepts inputs from more than one self-timed PLA, it should use the $DONE$ signal from the latest PLA as its precharge clock. A PLA cannot accept inputs from another PLA that is later in the cascade chain than itself.

8.8   H-Clock Tree For Reduced Clock Skew

In the rest of this chapter, a regular clock distribution structure will be described that can reduce the clock skew at the chip, board, and system levels. In slow VLSI circuits, clock frequency is relatively low and minimum-width

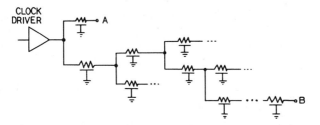

FIGURE 8.20   Clock skew as a result of different RC delays in branches A and B

lines are employed for clock distribution. Under these conditions, clock skew is dominated by the difference in the RC time constants of interconnections (such as the lines connecting the clock driver to points A and B in Fig. 8.20). In some circuits, a powering tree is used to drive the clock lines (Fig. 8.21). Here the important point is to use identical drivers and to equalize the load so that every driver sees the same capacitive load. Then the clock skew is mainly due to the mismatch among the drivers due to device parameter variations

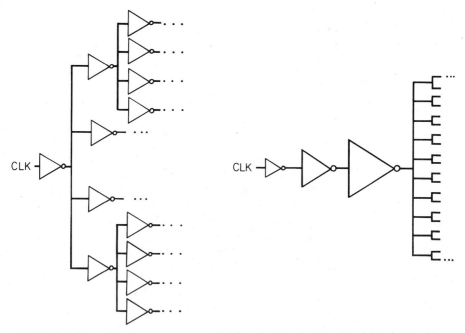

FIGURE 8.21   Clock power up tree (left) and a big clock buffer (right). The big buffer usually results in smaller skew and should be used instead of the powering tree when possible.

across the chip. Driver delays can be matched by using the identical layout for all the drivers and placing them next to each other and in the same orientation on the chip. Placing them in the same orientation guarantees that all are affected similarly by the orientation dependence of photolithography, etching, ion implantation, oxide growth, and diffusion steps. One big buffer is usually preferable to a powering structure.

In high-speed circuits with submicron feature sizes, the RC effects of long interconnections should be minimized by proper scaling of higher levels of wires as described in Chapter 5. This is true for PC board and multichip module wires, consequently they have very small distributed RC delays. Then the difference in the transmission line flight times of clock distribution lines sets a fundamental lower bound on the clock skew.

The skew can be minimized by distributing the clock signal in such a way that the interconnections carrying the clock signal to the functional subblocks are of equal length. If the clock signals are delayed equally before arriving at the subblocks, they will be perfectly synchronous. Because a skew proportional to the block size will be introduced by the clock distribution lines within the subblocks, the block size will be determined by the clock skew that can be tolerated. Using this basic concept, the clock distribution scheme in Fig. 8.22 minimizes clock skew by repeating an H-shaped structure recursively. In this distribution tree, all points labeled 4 are equidistant from the origin labeled 0. The H structure can be duplicated again and again until a small enough subblock is obtained such that the skew within the block is tolerable. The only additional clock skew is the result of variations in the logic thresholds of the receiver circuits at the end of the clock lines (if such circuits exist) and variations in the insulator thickness and interconnection width and height. This method is applicable to the layout of clock lines on the PC boards, multichip modules, and large VLSI and WSI chips. The definition of the subblock can be single-chip packages placed on a PC board, bare dice placed on a multichip module, or subsections of a circuit laid out on a large VLSI or WSI chip.

As shown in Fig. 8.23, based on the concepts described in the earlier sections, one can envision a computer system with a minimal skew. The clock is distributed to various chips with equal-length lines, which yields minimal skew. Chip-to-chip variations in clock generator/buffer delays are minimized by using phase-locked loops. The clock signal is again distributed within the chip using equal-length lines. The above description can be made more general by replacing chip with "region," where a region can be a board, a chip, a subsection on a large VLSI chip, or a subsection on a WSI circuit.

If the designer wants to introduce "intentional skew" to improve performance, the H-clock tree can be modified to accommodate such special needs. Intentional skew can be included by inserting an additional interconnection in the clock path that will introduce approximately 66 psec skew per 1 cm of line. Alternatively, 2 μm CMOS inverters inserted in the clock path can

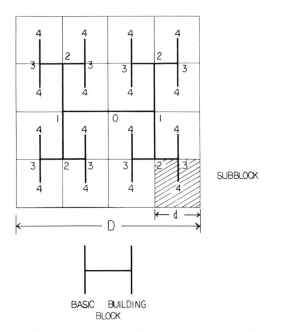

FIGURE 8.22   H-clock tree. Clock signal is delayed by an equal amount before it reaches the subblocks; as a result clock skew is reduced. Nodes labeled 4 are equidistant from the origin labeled 0.

produce a skew of 1 nsec per stage. Polysilicon lines are another option. A 0.25 μm thick and 2 μm wide polysilicon has a resistance of 10 kΩ/cm. If it also has a capacitance of 2 pF/cm, a skew of 20 nsec can be added per 1 cm of line. These three options offer a spectrum of delay lines with different controllabilities. For short ($<$ 100 psec) and rigidly controlled delays, metal transmission lines can be used, and for longer delays polysilicon lines or inverters are suitable. Here, control of the added skew is the main challenge because all these delay elements vary with process-dependent device parameters, junction temperature, power-supply level, and so forth. For example, it may be necessary to trim the polysilicon lines with a laser to ensure good control of their RC delays [8.41].

8.9   Design Considerations for the H-Tree

In this section, design considerations including reflections at the discontinuities, cross talk, termination and driving of the clock lines, interconnection resistance, skin effect, and the electrical properties of interconnections are

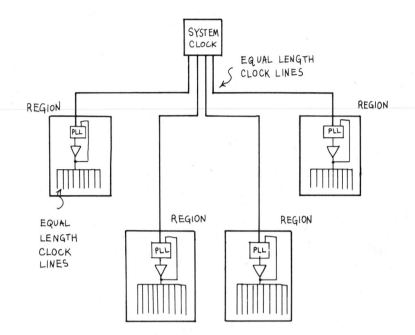

FIGURE 8.23   Minimal skew system. Clock lines to all regions are of equal length, minimizing the region-to-region skew. Phase-locked loops are used to minimize the skew due to region-to-region variations in clock generator/buffer delays. Clock distribution within a region is again minimized by using equal-length clock lines

discussed. The general principles described apply to clock distribution nets on PC boards, multichip modules, and VLSI and WSI circuits. Various design factors are treated in detail for two reasons. First, they are important in ensuring successful operation of the H-clock tree. Second, they demonstrate how some of the distributed RC delay and transmission line concepts described in the previous chapters can be applied to a practical design case.

8.9.1   Reflections at the Discontinuities and Cross Talk

In order to distribute a 250 MHz signal (4 nsec clock period) with a 0.5 nsec skew and 0.5 nsec rise time over a 3-inch wafer (an area of 5.4 × 5.4 cm), the interconnections must be treated as transmission lines because the target delays and rise times are comparable to the time-of-flight delay of the electromagnetic waves (the propagation speed of electromagnetic waves in SiO_2 or polyimide is approximately 15 cm/nsec).

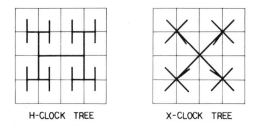

H-CLOCK TREE   X-CLOCK TREE

FIGURE 8.24   Symmetric clock trees

When the H-tree is used for high-speed clock distribution in the transmission line mode, reflections and cross talk must be minimized to ensure correct operation. For example, another scheme that yields equal-length interconnections is the X-clock tree (Fig. 8.24). The H-clock tree is superior to the X-tree for the following reasons:

- Because the clock lines do not produce corners sharper than 90°, inductive discontinuities are held to a minimum and, as a result, reflections are small. These discontinuities can be further reduced by tailoring the branch points (Fig. 8.25).

- The fan-out at the branching points is always 2, which simplifies matching the line impedance and minimizes reflections.

- No two clock lines are ever in close proximity, and therefore cross talk is small.

To avoid reflections at the branching points, the line separates into two branches with characteristic impedances twice the impedance of the incoming line. In parallel, they act like a single line with the same impedance as the incoming line. As illustrated in Fig. 8.26, the following condition must be

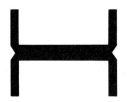

FIGURE 8.25   Reduction of inductive discontinuities at the corners of the H-clock tree

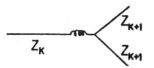

FIGURE   8.26 Matching   condition   at   the   branching   point: $(Z_k = Z_{k+1}\|Z_{k+1} = \frac{Z_{k+1}}{2})$. The inductive discontinuity caused by the corners is also shown.

satisfied to obtain perfect matching:

$$Z_k = Z_{k+1}\|Z_{k+1} = \frac{Z_{k+1}}{2}. \tag{8.9}$$

This can be accomplished by narrowing the line width at the branching points because the characteristic impedance of a line Z_0 is inversely proportional to its capacitance. An H-clock tree with lines that become narrower at the branching points is illustrated in Fig. 8.27.

8.9.2   Termination of the Clock Lines

In addition to the reflections from the branching points, reflections from the end points must also be considered. If the lines are not terminated properly, reflections may disturb the clock signal. Both the required on-resistance of the driver and the average power dissipation can be improved by the source-end termination method described in Chapter 6. In this scheme, the on-resistance of the driver is equal to the characteristic impedance of the line, which reduces the area required by the driver transistor because the aim is for an $R_S = Z_0$ source impedance rather than 0 Ω. In addition, the interaction between inci-

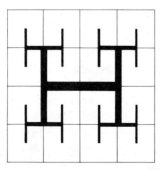

FIGURE 8.27   Tailored H-clock tree. Line width is reduced at the branching points for proper impedance matching.

dent and reflected waveforms eliminates continuous power dissipation. Power is consumed only during transitions.

Figure 8.28 shows the results of SPICE simulations of a transmission line terminated at the source end. As described in Chapter 5, a series of cascaded drivers that increase in size until the last one is large enough to drive the load can be used to minimize the overall delay in high-capacitance nodes such as I/O buffers [8.48]–[8.50]. In clock distribution, skew and rise time are the major concerns. Delay is not important if it is introduced equally for all the subblocks. Taking advantage of this fact, the inverters in the cascaded driver in Fig. 8.28 are increased in size up to the inverter before the final one. First, the size of the final inverter is calculated to satisfy the transmission line termination condition; then a cascaded chain of inverters is formed such that they increase in size until the size is somewhat larger than the inverter that drives the transmission line. In this way, the final inverter is driven by an inverter larger than itself. As a result the waveforms are sharper than they would have been if straightforward cascaded drivers were employed. Device parameters of Stanford's 2 μm CMOS process were used in the simulation. (See Chapter 4 for the SPICE parameters.)

8.9.3   Driving the Clock Lines

When the H-tree is used in a high-speed CMOS WSI environment in which the system may contain single or multiple wafers, the H-clock distribution tree may be driven by an on- or off-wafer clock driver. An off-wafer driver can be a bipolar buffer chip. Depending on the total load resulting from all the clock lines and the current drive capability of the clock driver circuit, the local clock lines in the subblocks may or may not be buffered from the global clock lines. Receiver circuits similar to Schmitt-trigger circuits can be used to "sharpen" the clock signal before it is distributed within the subblock. If the RC constant of the clock tree is not small enough, high-speed repeaters may be employed [8.4]–[8.5]. Source-end-terminated buffers between hierarchy levels, as illustrated in Fig. 8.29, can reduce the distributed RC delays and restore the "sharpness" of the clock signal from one level to the next. As a result the requirements for the RC constant of the interconnections are less stringent. Matching at the branching points is also made easier by the existence of the buffers, and the widths of the clock-tree lines can be made equal. Clock skew, however, is now determined by how well the repeaters are matched. If all the repeaters in a path are at the slow or fast end of the process window, a relatively large clock skew may occur. Because the repeaters are distributed on a very large die or, in the case of WSI, over a whole wafer (Fig. 8.29), the device parameters must have a very tight distribution for this approach to be successful. As a result these additional buffer circuits should be avoided if possible.

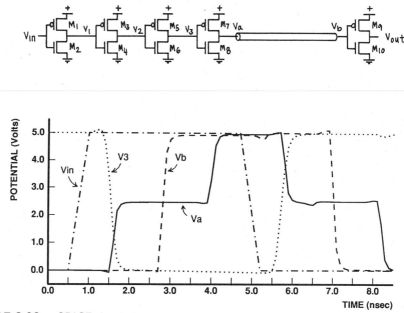

FIGURE 8.28   SPICE simulations of transmission line termination at the driver end. Transistor sizes are M1: 25/2 μm, M2: 10/2 μm, M3: 100/2 μm, M4: 40/2 μm, M5: 400/2 μm, M6: 200/2 μm, M7: 340/2 μm, M8: 176/2 μm. Note that the inverter before the final one is slightly larger than the final inverter to achieve a sharp waveform. The transmission line impedance Z_0 is 50 Ω, and its length is 18 cm ($t_f = 1.2$ nsec). Stanford's 2 μm CMOS process parameters were used in the simulation (Table 2.2).

8.9.4   Interconnection Resistance

When the H-clock tree is used, all subblocks will receive identical clock signals; however, if the interconnections have large RC constants, the waveforms will have long rise times, and a high-frequency clock signal will not be possible (Fig. 8.30(a)). If the rise time is too long, the clock waveform may not attain sufficient amplitude at high frequencies. This gives rise to an effect called *clock pulse-width narrowing.*

The RC effects in the clock lines must be minimized to obtain clock signals with short rise times (Fig. 8.30(b)). This can be accomplished by scaling the clock lines such that the interconnection and insulator are thick enough to produce low RC lines.

It is not generally possible to calculate the exact closed-form waveforms at the nodes of an RC tree; however, the symmetric structure of the H-clock tree simplifies the analysis. The RC network in Fig. 8.31 (a) represents the clock

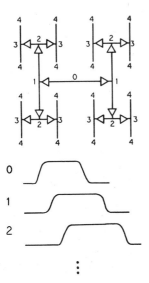

FIGURE 8.29 Buffered H-clock distribution network

tree in Fig. 8.22. Because all points labeled 1 are symmetric, the waveforms at these nodes are identical, and as a result they can be assumed to be virtually shorted together. The same assumption applies to nodes $2, 3, \ldots, n$. This simplification transforms the RC tree into the distributed RC line in Fig. 8.31(b), which makes it easier to analyze. Recall that the line width is approximately halved at the branching points to satisfy Eq. 8.9 because Z_0 is

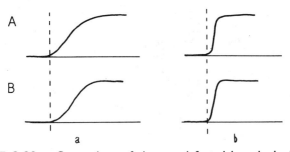

FIGURE 8.30 Comparison of slow- and fast-rising clock signals. Even if the subblocks receive identical signals, the RC constant of the interconnections must be sufficiently small to achieve high-frequency clock signals: (a) no clock skew and long rise time, (b) no clock skew and short rise time.

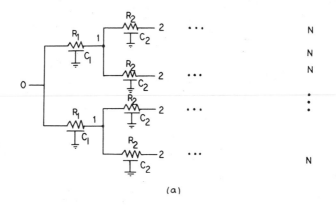

(a)

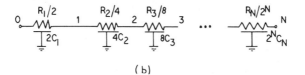

(b)

FIGURE 8.31   RC network representations of the H-clock tree: (a) distributed RC network, (b) simplified to a distributed RC line

inversely proportional to the unit length capacitance of the line (Eq. 8.18). As a consequence, the interconnection width is approximately halved from one hierarchy level to the next. The fringing field effects are taken into consideration in determining the width so that the interconnection capacitance per unit length is reduced by a factor of two. The resistance per unit length is increased approximately by a factor of two at each hierarchy level. This structure is called a binary H-clock tree, and the resulting network is equivalent to a uniformly distributed RC line.

The properties of the binary tree described above are summarized in Table 8.1. Here k is a hierarchy label for a class of lines ($k = 1$, for example, represents the lines that connect nodes 0 and 1 in Fig. 8.22), $\mathcal{R}_{int}$ and $\mathcal{C}_{int}$ are the resistance and capacitance per unit length of line, and r and c are arbitrary units. The length and characteristic impedance of a branch are denoted by l_{int} and Z_k. The number of branches in a given class and the subblock size related to the hierarchy level are also listed. As can be seen in Table 8.1, the length of the interconnections that join nodes $2k$ to $2k + 1$ and $2k + 1$ to $2k + 2$ is $D/2^{k+2}$ where D is the die size. The total length of the line from

TABLE 8.1   Properties of the Binary H-Clock Tree

Level (k)	Connected Nodes	$\mathcal{R}_{int}$ (Ω/cm)	$\mathcal{C}_{int}$ (pF/cm)	$\mathcal{R}_{int}\mathcal{C}_{int}$ (ns/cm^2)	l_{int} (cm)	Z_k (Ω)	No. of Branches	Block Size
1	0-1	r	c	rc	$\frac{D}{4}$	Z_0	2	—
2	1-2	$2r$	$c/2$	rc	$\frac{D}{4}$	$\frac{Z_0}{2}$	4	$D/2$
3	2-3	$4r$	$c/4$	rc	$\frac{D}{8}$	$\frac{Z_0}{4}$	8	—
4	3-4	$8r$	$c/8$	rc	$\frac{D}{8}$	$\frac{Z_0}{8}$	16	$D/4$
5	4-5	$16r$	$c/16$	rc	$\frac{D}{16}$	$\frac{Z_0}{16}$	32	—
6	5-6	$32r$	$c/32$	rc	$\frac{D}{16}$	$\frac{Z_0}{32}$	64	$D/8$
$\vdots$	$\vdots$	$\vdots$	$\vdots$	$\vdots$	$\vdots$	$\vdots$	$\vdots$	$\vdots$
(n-1)	(n-2)-(n-1)	$2^{(n-2)}r$	$c/2^{(n-2)}$	rc	$\frac{D}{2^{(\frac{n}{2}+1)}}$	$\frac{Z_0}{2^{(n-2)}}$	$2^{(n-1)}$	—
n	(n-1)-n	$2^{(n-1)}r$	$c/2^{(n-1)}$	rc	$\frac{D}{2^{(\frac{n}{2}+1)}}$	$\frac{Z_0}{2^{(n-1)}}$	2^n	$D/2^{\frac{n}{2}}$

the origin 0 to the end point n then becomes

$$
l_{TOT} = \frac{D}{4} + \frac{D}{4} + \frac{D}{8} + \frac{D}{8} + \dots + \frac{D}{2^{\frac{n}{2}+1}} + \frac{D}{2^{\frac{n}{2}+1}}
$$

$$
= D \sum_{k=1}^{n/2} \frac{1}{2^k}
$$

$$
= D \left[1 - \left(\frac{1}{2} \right)^{\frac{n}{2}} \right]
$$

$$
\approx D.
$$

(8.10)

If the resistance of the driver is neglected, the time required for the last node to reach 90 percent of its final value is

$$
T_{90\%} = R_{TOT}C_{TOT} = \mathcal{R}_{int}\mathcal{C}_{int}l^2{}_{TOT} = \mathcal{R}_{int}\mathcal{C}_{int}D^2, \qquad (8.11)
$$

where

$$
\mathcal{R}_{int} = \rho_{Al} \frac{1}{H_{int}W_{int}}
$$

$$
\mathcal{C}_{int} = \epsilon_{ox} \frac{W_{int}}{t_{ox}}
$$

(8.12)

$$
\mathcal{R}_{int}\mathcal{C}_{int} = \rho_{Al}\epsilon_{ox} \frac{1}{H_{int}t_{ox}},
$$

where H_{int} and W_{int} are the thickness and width and ρ_{Al} is the resistivity of the interconnection, and t_{ox} and ϵ_{ox} are the thickness and dielectric constant of the insulating layer. When the die size and clock signal requirements are specified, the interconnection structure should be designed such that

$$T_{90\%} = \rho_{Al}\epsilon_{ox}\frac{D^2}{H_{int}t_{ox}} \qquad (8.13)$$

is less than the required clock rise time. An ideal parallel plate approximation is used for the capacitance calculation, and the fringing fields are neglected because, for most of the length of the clock distribution tree, $W_{int} \gg H_{int} = t_{ox}$.

The above discussion considered the interconnection as a lumped distributed RC line. In a transmission line environment, the resistive losses must be regarded differently. The response of a uniform lossy line to a unit step input has two components. The first is a step function that travels down the line with the speed of light, and its magnitude is attenuated exponentially along the length of the line; the second is a slow-rising waveform that exhibits RC-like behavior (see Chapter 6). If $\mathcal{R}_{int}l_{int} \gg 2Z_0$, the fast-rising portion of the waveform is negligible, and the transmission line behaves like an RC line. The delays and rise times are then determined by the distributed RC constants rather than by the time-of-flight delays. To achieve high-speed signals, the fast-rising portion of the waveform

$$(1 - loss) = e^{-\mathcal{R}_{int}l_{int}/2Z_0} \qquad (8.14)$$

must be close to 1.0 (for example, $(1 - loss) = 0.8$). Neglecting fringing fields and substituting the following values of Z_0 and $\mathcal{R}_{int}$ in Eq. 8.14,

$$Z_0 = \frac{1}{v\mathcal{C}_{int}} = \sqrt{\frac{\mu_0}{\epsilon_{ox}}}\frac{t_{ox}}{W_{int}}$$

$$\mathcal{R}_{int} = \frac{\rho_{Al}}{W_{int}H_{int}},$$

the requirement for the cross-sectional dimensions of the interconnection is obtained as

$$H_{int}t_{ox} = \sqrt{\frac{\epsilon_{ox}}{\mu_0}}\frac{\rho_{Al}D}{2\ln[1/(1-loss)]}. \qquad (8.15)$$

In determining the cross-sectional dimensions, this design criterion must be considered in addition to Eq. 8.13.

In summary, at least one of the metal levels should be scaled such that its RC constant per unit length is not too high. The lower limit for the RC constant will be determined by how thick the insulators and metal conductors can be made. Thick insulators may crack because of large stresses, and they also increase the challenges associated with via openings for interlayer contacts. Thick metal layers may encounter problems with stress, planarization, and etching.

In calculating the resistance of thick interconnections for high-speed signals, the skin effect must also be taken into consideration (see Chapter 6). A steady current is distributed uniformly throughout the cross-section of a conductor through which it flows; however, time-varying currents concentrate near the surfaces of conductors. This is known as the skin effect [8.52]. At high frequencies, the effective resistance of an interconnection will not decrease by making it thicker than two to four times the skin depth. At 1 GHz for aluminum at room temperature, skin depth is 2.8 μm.

8.9.5   Electrical Design of Interconnections

Future high-speed ICs with very large die dimensions will have rigid requirements for the interconnections. Currently this problem is not too serious because in digital systems the existing wiring hierarchy (short interconnections on the chip level and longer ones on the various packaging levels) meets the different requirements for chip and system wires. As chips get larger or when the entire system is fabricated on a single wafer, a similar hierarchy of interconnections may need to be constructed on the silicon to meet the needs of the long global and short local lines. At the lower levels, densely packed lines with low capacitance and resistivity will be required; however, because of the relatively short interconnection lengths and large interconnection counts, density will be the major concern, and low capacitance and resistance will be secondary. On the other hand, there will be fewer long interconnections, and the overall performance of the system will be strongly dependent on their speed. Their electrical properties therefore will be more important than their packing density.

A major problem with interconnections fabricated on silicon is their high capacitance caused by the thinness of the insulating layers. This results in a low characteristic impedance because

$$v = \frac{c_0}{\sqrt{\epsilon_r \mu_r}} = \frac{1}{\sqrt{\mathcal{L}C}} \tag{8.16}$$

and

$$Z_0 = \sqrt{\frac{\mathcal{L}}{C}} \tag{8.17}$$

yield

$$Z_0 = \frac{1}{vC}, \tag{8.18}$$

where C and $\mathcal{L}$ are the capacitance and inductance per unit length of interconnection, and c_0 and v are the propagation speed of the electromagnetic waves in free space and in the transmission line medium with relative permittivity and permeability ϵ_r and μ_r. The low characteristic impedance in the transmission line requires a greater current drive capability from the clock driver to achieve transmission line speed.

TABLE 8.2 Comparison of Interconnection Geometries

Parameter	(a)	(b)	(c)	(d)	(e)
C_{int}	1.8 pF/cm	2.5 pF/cm	2.5 pF/cm	1.6 pF/cm	1.6 pF/cm
R_{int}	$\dfrac{140\Omega/cm}{[x(\mu m)]^2}$	$\dfrac{70\Omega/cm}{[x(\mu m)]^2}$	$\dfrac{93\Omega/cm}{[x(\mu m)]^2}$	$\dfrac{93\Omega/cm}{[x(\mu m)]^2}$	$\dfrac{70\Omega/cm}{[x(\mu m)]^2}$
$R_{int}C_{int}$	$\dfrac{250ps/cm^2}{[x(\mu m)]^2}$	$\dfrac{175ps/cm^2}{[x(\mu m)]^2}$	$\dfrac{230ps/cm^2}{[x(\mu m)]^2}$	$\dfrac{150ps/cm^2}{[x(\mu m)]^2}$	$\dfrac{112ps/cm^2}{[x(\mu m)]^2}$
Z_0	37 Ω	26 Ω	26 Ω	44 Ω	44 Ω
C_{int}	good	poor	poor	best	best
R_{int}	poor	best	good	good	best
$R_{int}C_{int}$	poor	good	poor	good	best
Z_0	good	poor	poor	best	best
Technology	poor	good	best	poor	worst
Etching	worst	poor	good	good	poor
Step coverage	good	good	good	poor	poor
Stress	good	good	good	poor	poor
Packing density	best	good	good	good	good

Another major concern is the large RC time constant per unit length of interconnection. No matter how strong the clock driver is made, high clock frequencies will not be possible if the line has a large RC constant. As demonstrated in Eq. 8.12, the time constant per unit length is determined only by the interconnection and insulator thicknesses; therefore RC delay will not improve with a wider interconnection because, neglecting fringing fields, capacitance will increase and resistance will be reduced approximately by the same factor. In other words, the performance of the interconnection can be enhanced only by adjusting the vertical dimensions. This places the responsibility on process design because no adjustments can be made at the mask level (with the exception of the possible use of repeaters and other circuit solutions) to increase the speed of interconnections.

Interconnections can be fabricated by one of many techniques and with

TABLE 8.3 Conductors for High-Speed Interconnections

Material	Bulk Resistivity ($\mu\Omega$-cm)
Al	2.74
Au	2.20
Cu	1.70
Ag	1.61

various aspect ratios. A number of them are illustrated and compared in Table 8.2. Fringing fields are taken into account in the capacitance calculations by using Eq. 8.19. In Table 8.2, x is the size of the rectangular grid in units of μm. As can be seen, thicker insulating layers and rectangular cross-sections have the best electrical properties, but they are also the most technologically challenging structures because of the problems associated with stress in thick insulator and metal layers, step coverage, and via openings and fillings, in addition to the challenges in developing dry-etching or lift-off techniques necessary to obtain rectangular cross-sections.

As described in Chapter 3, alternative interconnection materials can be employed to reduce resistance. As exhibited in Table 8.3, the resistivities of gold, copper, and silver are lower than in aluminum; however, replacing aluminum with any one of them would be difficult because they are not as compatible with IC processing as is aluminum. Failure to adhere to the insulator or substrate (Au, Cu), corrosion (Cu, Ag), degradation of MOS devices due to contamination (Cu), difficulty in dry etching (Cu), migration (Ag), reactions with other metals (Ag), incompatibility with high-temperature processing (Au, Ag), and high cost (Au, Ag) are some of the problems that must be resolved before aluminum can be replaced. Among the possible candidates, copper is the most likely successor because it has significantly lower resistivity than gold and fewer complications than silver. To avoid the problems related to copper, a thin layer of metal (such as nickel, titanium, or tungsten) must be used as an undercoat to ensure adhesion and to prevent diffusion into the insulator; a reliable lift-off technique must be developed to obtain interconnections with a high-aspect ratio, and copper interconnections must be goldplated to confine them in order to avoid corrosion and diffusion.

Alternative dielectric materials may be used to reduce interconnection capacitance. Low-dielectric constant polymers (such as polyimide $\epsilon_r < 3.5$) are possible candidates. They have good planarizing characteristics because they can be deposited as viscous liquid films. They can also tolerate more stress than SiO_2 and as a result form more reliable thick films. On the negative side, they deteriorate at temperatures as low as $500°C$ and have inferior heat-conduction properties.

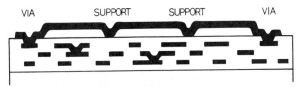

FIGURE 8.32   Air bridges for high-speed interconnections. Periodic supports enhance the strength of the interconnections.

The capacitance and transmission line properties of the interconnections can be improved by air bridges at the top-most level (Fig. 8.32). These structures can be fabricated through lift-off techniques; a thick layer of photoresist can support the bridge during deposition, and another layer of resist can confine the deposition to the desired area. Periodic supports can be placed to enhance the strength of the interconnection as necessary. Because interconnections are surrounded by air ($\epsilon_r = 1$) instead of SiO_2 ($\epsilon_r = 3.9$), their capacitance is approximately four times smaller. Air bridges are used in microwave integrated circuits [8.53]; however, potential yield, packaging, and hermeticity problems must be resolved before they can be applied successfully to VLSI circuits because these ICs require much higher packing densities and integration levels than do microwave circuits, and the volume and cost specifications necessitate relatively high yields.

As described in Chapter 2, cooling the chips to liquid nitrogen temperature reduces interconnection resistance significantly, improves the skin effect limits, and enhances transistor performance (increases transconductance) and reliability (eliminates latch-up, reduces electromigration). The supply voltage should be lowered to avoid hot-electron-induced device degradation, and ohmic contact problems encountered at low temperatures must be solved. The low characteristic impedance of the interconnections is not affected by reducing chip temperature. With or without liquid nitrogen cooling, an interconnection scheme similar to the one illustrated in Fig. 8.33, where lower levels are densely packed and higher levels have thicker interconnections and insulators, will be desirable for high-speed VLSI and WSI circuits.

In fast circuits, noise is also a major problem. As the number of interconnection levels increases, the capacitance and mutual inductance between neighboring lines become important. In the absence of a proper ground plane, the neighboring signal lines become current return paths, and excessive noise and cross talk may result; furthermore the rise time of signals degrades because, in a multiconductor environment, non-TEM modes of propagation may become excited, which causes dispersion because each mode propagates at a different speed. The silicon substrate also introduces problems because it acts like a conductor for capacitive effects and like an insulator for inductive effects.

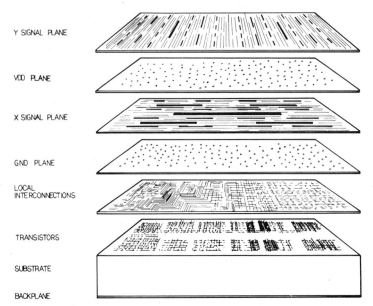

Y SIGNAL PLANE

VDD PLANE

X SIGNAL PLANE

GND PLANE

LOCAL INTERCONNECTIONS

TRANSISTORS

SUBSTRATE

BACKPLANE

FIGURE 8.33   Hierarchically designed interconnection system.   Upper levels have sparsely packed lines with small RC constants and large Z_0 for global communication. Lower levels have densely packed short local lines. The power and ground reference planes between the X and Y distribution levels solve noise problems and provide a good transmission line medium.

This gives rise to so-called slow wave mode [8.54] in which the propagation speed is much less than $v = c_0/\sqrt{\epsilon_r \mu_r}$ because the geometries that define $\mathcal{L}$ and $\mathcal{C}$ are not identical, and as a result $\mathcal{L}\mathcal{C}\neq\epsilon\mu$. If the geometries that determine the capacitance and inductance were identical (as they normally are), $v = 1/\sqrt{\mathcal{L}\mathcal{C}} = c_0/\sqrt{\epsilon_r \mu_r}$. When the substrate acts as an insulator for the inductive effect, $\mathcal{L}$ is larger than it typically would be, and as a result $v = 1/\sqrt{\mathcal{L}\mathcal{C}}$ is smaller than $v = c_0/\sqrt{\epsilon_r \mu_r}$. It is best therefore not to rely on the silicon substrate as the current return path but to supply a nearby metal reference plane. Alternating ground and V_{DD} planes between signal planes or coupled transmission lines placed on the same level may be possible solutions to avoid non-TEM modes and the adverse effects of a semiconductor substrate.

An interconnection scheme as illustrated in Fig. 8.33 would be very desirable. Here the two top-most interconnection layers are reserved for long lines in the X and Y directions for ease of routing, and these planes are separated by continuous (with the exception of via holes) V_{DD} and ground planes that distribute power to the system. These planes serve multiple purposes. They supply a nearby low-resistivity return path for interconnections to ensure good transmission line properties, eliminate cross talk between the X and Y planes, and reduce coupling between the lines on the same level. They also provide a low-resistance (small IR drops) and low-noise (small LdI/dt power-supply-level fluctuations) power-distribution scheme. Under these global wiring and ground and power planes, scaled-down regular layers for local interconnections are available as needed. Successful manufacturing of this complicated structure with acceptable yields, however, is a major challenge. The possible formation of shorts between the ground and power planes is of special concern. Using a mesh structure instead of continuous planes would improve the yield but at the expense of degraded electrical properties. Despite all the challenges, a structure similar to the one in Fig. 8.33 may be the best way to take full advantage of the performance potentials of submicron electronics.

8.10   Design Example

This section discusses a design example in which a 250 MHz clock signal (4 nsec cycle time) is distributed over a 3-inch wafer ($D = 5.4$ cm). The specified rise time ($T_{90\%}$) of the signal distributed to the subblocks is 0.5 nsec, and the maximum allowable clock skew within a block is 0.5 nsec.

The first step is to determine the subblock size. It is assumed that the local interconnections that route the clock signal within the subblocks have the following dimensions: $W_{int} = 2$ μm, $H_{int} = 1$ μm, and $t_{ox} = 1$ μm. Equation 8.19 is used to calculate the resistance and capacitance per unit length for this level [8.55].

$$\mathcal{R}_{int} = \rho_{Al} \frac{1}{W_{int} H_{int}}$$

$$\frac{C_{int}}{\epsilon_{ox}} = 1.15 \left(\frac{W_{int}}{t_{ox}} \right) + 2.80 \left(\frac{H}{t_{ox}} \right)^{0.222}$$

$$+ \left[0.06 \left(\frac{W_{int}}{t_{ox}} \right) + 1.66 \left(\frac{H}{t_{ox}} \right) - 0.14 \left(\frac{H}{t_{ox}} \right)^{0.222} \right] \left(\frac{t_{ox}}{W_{sp}} \right)^{1.34} .$$

$$(8.19)$$

With the above assumptions, interconnection parameters are obtained as $\mathcal{R}_{int} = 150$ Ω/cm, $C_{int} = 2$ pF/cm, and $\mathcal{R}_{int} C_{int} = 0.3$ nsec/cm^2.

Because the maximum allowable clock skew within a subblock is 0.5 nsec, based on the delay expression for distributed RC lines, the largest subblock size d_{max} is calculated as

$$T_{90\%} = \mathcal{R}_{int} C_{int} l_{int}^2 = \mathcal{R}_{int} C_{int} \left(\frac{d_{max}}{\sqrt{2}} \right)^2$$

$$d_{max} = 1.83 \ cm.$$

$$(8.20)$$

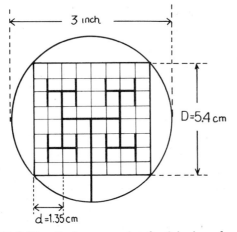

FIGURE 8.34 Design example of a 3 inch wafer

Because $d = D/2^{n/2}$ and $D = 5.4$ cm, n is chosen to be 4, which yields $d = 1.35$ cm (Fig. 8.34).

The second step is to determine the number of levels that can remain unmatched at the branching points. This is important because if the final branch is a 50 Ω line and all the levels are matched, the input impedance of the clock tree becomes 3.12 Ω. It is difficult to drive such small impedances with CMOS circuits. SPICE simulations of the equivalent transmission line structures in Figs. 8.35 and 8.36 demonstrate that the last two branches can remain unmatched and a rise time of less than 0.5 nsec can still be obtained. Because this does not include the lossy nature of the lines and the safety margin necessary to achieve high yield, only the final set of branches is left unmatched in the actual design. For the other branches, the interconnection width is increased to double the capacitance per unit interconnection length to satisfy the matching conditions. For an interconnection technology with the ratios in option (a) in Table 8.2 and the subblock size as calculated above, Table 8.4 lists the design parameters of the 3-inch wafer. As can be seen, the interconnection width must be increased by a factor larger than two to account for the reduced fringing field effects as the width is increased. As a result, $\mathcal{R}_{int}C_{int}$ improves at the higher hierarchy levels. As demonstrated in Eqs. 8.16, 8.17, and 8.18, C_{int} determines all other transmission line parameters. Fringing fields are taken into account by using Eq. 8.19.

In the third step, the vertical dimensions of the clock lines that can yield a fast-rising clock signal must be obtained. In Table 8.4, the dimension x (representing the interconnection and insulator thicknesses) is a variable to be determined by the loss requirements (Eq. 8.14).

There are two subtleties in the following calculation: the last branch is

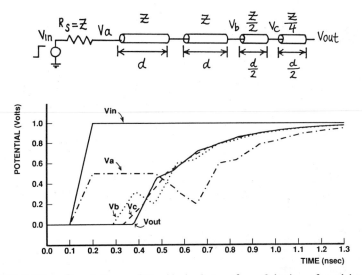

FIGURE 8.35 Simulations of the H-clock tree for a 3-inch wafer with the last two stages unmatched. Transmission lines with no loss are assumed, and $d = 1.35$ cm.

TABLE 8.4 Design Parameters of the 3 Inch Wafer

$(D = 5.4$ cm, $d = 1.35$ cm, $W_{int} = 2x$ μm, $H_{int} = x$ μm, $t_{ox} = x$ $\mu m)$

k	l_{int} (cm)	W_{int} (μm)	R_{int} (Ω/cm)	C_{int} (pF/cm)	$R_{int}C_{int}$ (ns/cm^2)	Z_0 (Ω)
4	0.68	$2x$	$\dfrac{140}{x^2}$	1.8	$\dfrac{0.25}{x^2}$	37
3	0.68	$2x$	$\dfrac{140}{x^2}$	1.8	$\dfrac{0.25}{x^2}$	37
2	1.4	$7x$	$\dfrac{40}{x^2}$	3.6	$\dfrac{0.14}{x^2}$	18.5
1	1.4	$17.5x$	$\dfrac{16}{x^2}$	7.2	$\dfrac{0.12}{x^2}$	9.3
0	—	$38x$	$\dfrac{7.4}{x^2}$	14.4	$\dfrac{0.11}{x^2}$	4.6

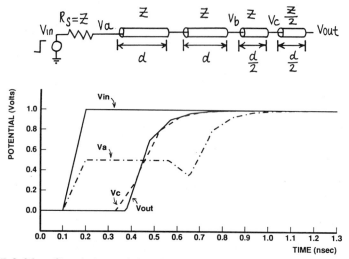

FIGURE 8.36 Simulations of the H-clock tree for a 3-inch wafer with only the last stage unmatched. Transmission lines with no loss are assumed, and $d = 1.35$ cm.

not matched, and because of the fringing field effects, the clock tree will not act like a uniformly lossy line.

Because the final branch is unmatched, the reflections in Fig. 8.37 must be taken into account. When all the components are summed,

$$
\begin{aligned}
V_{OUT} &= 2x_1x_2(1+\Gamma) - 2x_1x_2^3(1+\Gamma)\Gamma + 2x_1x_2^5(1+\Gamma)\Gamma^2 \\
&\quad -2x_1x_2^7(1+\Gamma)\Gamma^3 + ... \\
\\
&= 2x_1x_2(1+\Gamma)\left[1 + (-x_2^2\Gamma) + (-x_2^2\Gamma)^2 + ...\right] \qquad (8.21) \\
\\
&= 2x_1x_2(1+\Gamma)\left[\frac{1 - (-x_2^2\Gamma)^{N+1}}{1 + x_2^2\Gamma}\right].
\end{aligned}
$$

Here x_1 is the portion of the step input that arrives at the final set of branches after losses, and x_2 is the amount that goes through the final branch. Both x_1 and x_2 are less than 1 and are defined by the $(1 - loss)$ term in Eq. 8.14. The parameter Γ is the reflection coefficient at the discontinuity between the last set of branches and the lines that feed them. The time element is included in N ($N = f(t)$). As also illustrated by the waveforms in Fig. 8.36, $(-x_2^2\Gamma)^{N+1} \approx 0$ for the frequencies and dimensions of concern; therefore,

$$
V_{OUT} \approx \frac{2x_1x_2(1+\Gamma)}{1 + x_2^2\Gamma}. \qquad (8.22)
$$

Because of the fringing fields, the H-clock tree does not act like a uniformly lossy line; the lines at the end have more loss. As a result, total loss must be

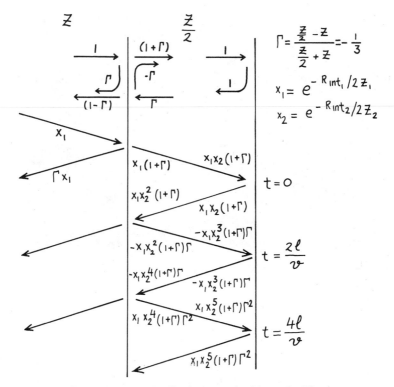

FIGURE 8.37   Reflections in the final unmatched branch. The lossy nature of the line is taken into account.

calculated as a product of the losses at individual levels. From the information in Table 8.4 and using Eq. 8.14, x_1, x_2, and Γ are derived as

$$x_1 = exp\left(-\sum_{k=1}^{3} \frac{l_{int} R_{int}}{2 Z_k}\right) = e^{-3.9/x^2}$$

$$x_2 = exp\left(-\frac{l_4 R_4}{2 Z_4}\right) = e^{-1.28/x^2} \tag{8.23}$$

$$\Gamma = \frac{\frac{Z_k}{2} - Z_k}{\frac{Z_k}{2} + Z_k} = -\frac{1}{3}.$$

Using Eqs. 8.22 and 8.23 and requiring that V_{OUT} is 70 percent of the full swing, $x = H_{int} = t_{ox}$ is determined to be 4 μm. Because the slow-rising portion of the waveform is neglected in this calculation, the actual signal will have a rise time even better than 0.5 nsec. The skin effect limitations must also

FIGURE 8.38   Subsection of the H-clock tree on an upscaled model. The module measures 15×10 cm. The ground plane and transmission line are copper, and the insulator is Teflon.

be checked. Because skin depth δ is 2.8 μm for aluminum at room temperature at 1 GHz, the designed interconnection thickness is within the limits of the skin effect.

An on-wafer CMOS clock driver reduces the burden on the system-level driver that distributes the signal to the wafers. An on-wafer driver terminated at the source end is the most suitable method because of the difficulty in driving transmission lines with low characteristic impedance using CMOS. Because Z_0 of the initial line is 4.6 Ω, and following the design steps described in Chapter 6, the transistor sizes of the driver are found to be $W/L_{NMOS} = 2456$ and $W/L_{PMOS} = 6140$ for a 2 μm CMOS process ($k_n = \mu_n C_{ox} = 30\ \mu A/V^2$, $k_p = \mu_p C_{ox} = 12\ \mu A/V^2$). To achieve a high current drive, parasitic diffusion resistances and polysilicon RC delays must be carefully considered.

The feasibility of the H-clock tree was first investigated by building upscaled models (Fig. 8.38), and the results indicated that the signal can be distributed over a 15×15 cm area with a 30 psec skew and 100 psec rise time (Fig. 8.39). The H-tree was then fabricated on a 3 inch silicon wafer (Fig. 8.40). In the process sequence, a 7400 A pad oxide was first grown on the substrate, and a 5 μm aluminum ground plane was deposited via electron beam sputtering. A 5 μm SiO_2 layer was then grown by plasma-enhanced chemical vapor deposition (PECVD), and another 5 μm aluminum layer was deposited. After all layers were placed on the wafer, the top aluminum was wet etched to form the transmission lines, and then the SiO_2 layer was dry etched to expose the aluminum ground plane.

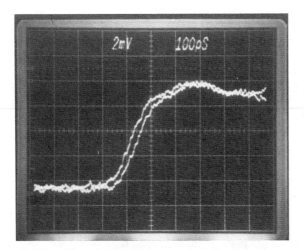

FIGURE 8.39   Waveforms of the transmission measurements of the upscaled model as observed on the oscilloscope screen. As can be seen, the H-tree can distribute a clock signal over a 15×15 cm area with a 30 psec skew and 100 psec rise time.

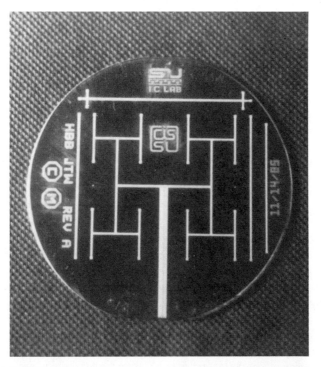

FIGURE 8.40   Photograph of the H-clock tree on a 3-inch wafer

8.11 Summary

Clocking is very important in high-speed systems. The choice and design of the storage element are intimately tied to the clocking scheme. Single-phase latch and flip-flop designs and two-phase double-latch designs are some of the possibilities. Single-phase latch machines require the least amount of circuitry and offer the best performance, but timing requirements are hard to meet because a latch is transparent when the clock is high, which gives rise to race conditions. In single-phase latch designs, minimum latch-to-latch delays, as well as maximum delays, must be carefully tuned. Flip-flops are edge triggered and do not have the transparency property of a latch; as a result timing requirements are much easier to meet. Two-phase machines with nonoverlapping clocks are easiest to design because race conditions can be eliminated by increasing the nonoverlap period. In addition, they make the testing of the chips easier.

Asynchronous interfaces have a problem called the *synchronizer failure*. This happens when an asynchronous signal changes at the same time as the clock that is sampling it. Then the flip-flop can enter a metastable state and cannot decide if it should go to zero or one. Undefined states that can be interpreted as ones or zeros are injected into the system, which then usually crashes. This problem is hard to trace because it may happen seldom and may not repeat itself in the exact same way.

Asynchronous self-timed design techniques attempt to solve some of the problems associated with synchronous systems by eliminating the need for a global clock. In one approach the design is partitioned into locally synchronous regions that operate with a fast clock. The regions communicate either via an asynchronous protocol or use a slower clock for global signals. A second approach is to use fully asynchronous self-timed modules that signal each other when their outputs are ready. Some of these techniques have become popular in regular array structures such as memories and PLAs. Self-timed logic, on the other hand, requires too much circuitry for signaling. In addition, there is little design tool support and limited experience base in self-timed logic design.

As the clock frequency improves, it becomes harder to keep the clock skew a small portion of the cycle time. Usually the clock lines on the board are kept the same length to minimize the skew between the chips. The skew due to chip-to-chip variations in clock generator/buffer delay can be reduced by using phase-locked loops. The skew due to clock lines can be handled in a systematic way by using a regular structure that keeps the clock lines equal length. This can be used in boards, modules, and chips. For example, an H-clock distribution tree has equal delay to all the subblocks and as a result minimizes the skew. The following factors must be taken into account in the design of the H-tree distribution network:

- To obtain "sharp" clock signals, the RC constant of the lines must be sufficiently small, which requires proper scaling of the vertical dimensions (interconnection and insulator thicknesses). This indicates that the electrical properties of the interconnections must be carefully considered during the design of a process because interconnection performance is closely related to vertical dimensions, and little improvement is achieved by adjusting the horizontal dimensions at the mask level. The skin effect must be taken into consideration in distributed-RC delay calculations.

- The characteristic impedance of the lines should be tailored to avoid reflections at the branching points. This can be achieved by a small fan-out structure that does not produce corners sharper than 45°or 90°and by reducing the interconnection impedance by a factor of two at each level in the hierarchy.

- Nearby low-resistivity current return paths should be provided to ensure good transmission line properties. This reduces the coupling between the lines, shields the adverse effects of the semiconductor substrate, minimizes the relative contribution of non-TEM modes, and enhances transmission line behavior.

- The clock drive and termination methods should not require an excessively small source resistance (necessitating very large area transistors) and should not dissipate dc power.

9

SYSTEM-LEVEL
PERFORMANCE
MODELING

This chapter introduces a system-level circuit model for central processing units (CPUs) referred to as the SUSPENS model (Stanford University System Performance Simulator) [9.1]. The model emphasizes the interactions among devices, circuits, logic, packaging, and architecture. In this chapter, first, the derivation of the SUSPENS model is presented in detail, and then its results are shown to agree with the clock frequency, power dissipation, and chip and module sizes of existing microprocessors, gate arrays, and mini- and mainframe computers. Later the model is used to predict the performance of future microprocessors and to compare CMOS, bipolar, and GaAs technologies in a system environment.

The major performance-limiting factors identified and included in the SUSPENS model are:

- *Device properties:* current drive capability and input capacitance of transistors.

- *On-chip interconnections:* electrical properties (resistance and capacitance), packing density, and number of levels of on-chip wires.

- *Chip-to-chip interconnections:* electrical properties (resistance, capacitance, and characteristic impedance), packing density, and number of levels of module-level wires.

- *Cooling capability:* power-dissipation density of the module.

- *Machine architecture, organization, and implementation:* these determine the interconnection requirements of the system, average interconnection lengths, and the logic-depth and fan-out of the circuits. They also determine the number of machine cycles required per executed instruction—the cycles-per-instruction (CPI) ratio of the machine.

The SUSPENS model can support various chip and packaging technologies. The chip technology can be silicon *n*MOS, CMOS, or BJT, or GaAs MESFET, HEMT, or HBT. Differences due to basic transistor properties and on-chip wiring densities of these technologies can be compared using the SUSPENS model. The module, as defined in the SUSPENS model, can be a PC board, a ceramic multilayer hybrid, a silicon-on-silicon package, or global connections in a wafer-scale integrated system. The major differences among these packages are the electrical properties and packing densities of their interconnections, electrical properties of the connections between the module and chips (capacitive and inductive discontinuities), and the power-dissipation capability of the module. The basic principles of the model are general enough to apply to various packaging schemes.

Using the SUSPENS model, the optimal chip integration level for a multichip CPU can be calculated for various packaging options. The results indicate that as minimum feature size is scaled down, interconnections and packaging become major performance limiters. One of the goals of this chapter is to model the hierarchical nature of interconnecting a digital system and to determine the best way of partitioning the interconnections between the package and silicon. This can be used to establish the optimal level of integration (number of logic gates per chip). The effects of technology parameters, such as minimum feature size, number of interconnection levels, and interconnection pitch, are also examined by varying them independently. The overall performances of systems built from different technologies (CMOS, *n*MOS, BiCMOS, cyrogenic CMOS, ECL, GaAs) are compared for various minimum feature sizes. The relative importance of chip- and package-level limitations and their interrelation are also studied.

The SUSPENS model is very general and is structured such that it can be easily modified. For example, one can replace the SUSPENS gate delay equations with a set of library-specific delay equations to describe a particular circuit library more accurately. Similarly, chip-to-chip delay equations and packaging parameters can be replaced by more detailed models. As such, the SUSPENS model can be tailored and tuned to reflect a particular design style or product line accurately. The important aspect is the approach taken in this model, which ties the material, device, circuit, logic design, packaging, and system architecture parameters.

The SUSPENS model is basically an analytical model that takes some material, device, circuit, logic, package, and architecture parameters as inputs and using them calculates the clock frequency of the system. The SUSPENS model has to be complemented by a system-level architectural model to calculate the organizational efficiency of the system—the CPI ratio. The cycle time and cycles-per-instruction ratio together yield the system performance. Other information that can be obtained using the SUSPENS model are the chip sizes, power dissipation, and package dimensions.

9.1 Optimal Integration Density

Optimal integration density (number of logic gates per chip) depends on such interrelated design goals as *cycle time, cycles-per-instruction ratio, power dissipation, cost,* and *physical size* of the computing machine. High-performance systems, such as mainframe computers, are built usually with bipolar chips that do not contain many gates but consume a large amount of power. Large numbers of these chips are housed in modules with many layers of signal lines that have controlled characteristic impedances. These modules also have reference planes placed between the signal layers to shield the noise generated at high speeds. Sophisticated cooling mechanisms are used to remove the large amount of heat produced by the closely packed chips. These systems are costly. Entry-level systems, such as personal computers, are built from very highly integrated and usually single-chip CPUs (microprocessors) that dissipate orders of magnitude less power and cost only a small fraction of the price of high-end systems. Minicomputers and workstations that cover the price range between personal computers and mainframes are usually based on a CPU chip set that consists of a few custom or semicustom chips.

The factors that determine the optimal chip integration level for different design goals are described below.

Yield. In most cases, the number of circuits integrated on a CMOS chip is limited by the die yield. If more circuits are placed on a chip, the yield would go down, and the cost would become unacceptably high. Chip integration level can be limited by factors other than yield, as will be described below. The following factors will gain importance as performance levels improve and larger chips become economically feasible. Then one could choose not to increase the integration level for performance reasons rather than yield.

Power-dissipation density. Fast bipolar circuits (especially ECL) consume large amounts of power. As a result, in high-performance ECL chips, integration density is usually determined by the ratio of maximum allowable power dissipation per chip P_c to average power dissipation per gate P_g,

$$N_g = \frac{P_c}{P_g},$$

where N_g is the number of logic gates on the chip. To add more gates on a chip, the current and power consumption of a gate must be decreased, which in turn reduces the circuit speed. Since CMOS does not dissipate any dc power, this is not a major concern in CMOS.

Availability of on-chip interconnections. High-speed systems employ greater levels of parallelism and pipelining that demand a large number of simultaneous information transfers and increase the wiring requirements. Consequently, the number of logic gates that can be interconnected may be limited by the availability of on-chip signal lines. High-performance bipolar gate arrays usually have up to three or four levels of wiring to provide high wireability. A lower-performance CMOS chip with many gates can be implemented by sharing on-chip interconnections among the gates or subblocks, as is done routinely in microprocessors that rely heavily on precharged or tristated buses for on-chip information transfers. This saves wires, as well as multiplexors that would have been required to select the data from a number of possible sources.

Distributed-RC delays of on-chip interconnections. The interconnections on the chip are more densely packed than the interconnections on the module. Chip wires also have larger RC constants and inferior transmission line properties. In addition, they generate more coupled noise because of insufficient shielding and small spacings. When chip size is varied, the relative distribution of the chip- and module-level interconnections changes. By reducing the number of gates integrated on a chip, more interconnections are transferred to the module, and as integration level increases, more wires are transferred from the board to the chips. If the chip is very large, there will be few module-level wires, but system performance may degrade because of the large RC delays of the global chip-level interconnections. Secondary problems that are not significant for current chips but may surface at the 1 GHz range include the reflections caused by poor on-chip transmission lines, degradation of waveform edges because of dispersion, and excessive cross talk resulting from insufficient shielding of on-chip lines. In the past and at the present, increasing the integration level has been and continues to be beneficial in nMOS and CMOS chips. As separate components are integrated on a chip the performance improves because chip-crossing delays are eliminated, and a larger number of parallel interconnections are available among the components integrated on a die. But it is conceivable that, for example, in wafer-scale integration the performance may be worse than it would be if the system were partitioned into multiple chips because of the distributed-RC delays of thin film wires. This would depend on the electrical characteristics of the WSI interconnections.

Module and/or board size. If the chips have very few logic gates on them, the function will have to be partitioned into many chips, and the demand for module and/or board interconnections will increase. Because the package-level interconnections are not densely packed, the footprint of the chips will be much greater than the actual chip size. Performance may degrade[1] because of the large overall system size for the following reasons:

- Speed-of-light limit: The signals cannot travel faster than the propagation speed of electromagnetic waves; larger size translates into longer time-of-flight delay.

- Capacitive loading: Longer lines result in bigger capacitive loads, which increases the delay and dynamic power dissipation.[2]

Both of these factors encourage a higher chip integration level because wires are shorter and capacitive loadings are smaller on the chips.

Noise considerations. Circuit speed can be also limited by cross-talk noise between the signal wires and dI/dt noise at the power lines. Intricate and expensive packaging techniques are required to reduce the noise generated at high speeds. Cross coupling between signal lines can be shielded by reference planes between the signal layers or by grounded lines between the wires at the same level. In addition, dI/dt noise can be minimized by packages that avoid long leads with large inductances and by decoupling capacitors placed very close to or fabricated on the chips. (See Chapter 7.) Because of smaller distances and dimensions, noise is usually easier to handle on the chips than at the board level.

---

[1]Lower integration density does not necessarily yield lower performance. As will be described later, Cray supercomputers are designed with chips that carry very few gates. Here, high performance is achieved by measuring the delay of every individual chip and accounting for variations in delays by adjusting the wires (a fast chip will get a longer wire in front of it, and a slower chip will get a shorter wire). In addition, the package and wiring delays are carefully tailored. This requires complicated packaging schemes and fine tuning, which yield an expensive system.

[2]The relation between the capacitive loading and power dissipation actually depends on the drive and termination conditions. In a typical bipolar transmission line driving scheme, the resistance of the driver is much less than the impedance of the line, and the line is terminated by a resistor at the receiving end; as a result, line length does not affect power dissipation. For example, when a 10 Ω bipolar buffer drives a 50 Ω transmission line terminated by a 50 Ω resistor at the receiving end, power dissipation is $V_{supply}^2/60\Omega$, independent of line length. Low-power driving schemes, however, are common and often preferable. As described in Chapter 6, CMOS drivers that terminate a transmission line at the driver end eliminate dc power dissipation and require smaller driver transistors. Here the dynamic component of power dissipation is dominant, and it is proportional to interconnection capacitance. Similarly, in low-end systems, no termination is necessary because the rise time of signals is longer than the time-of-flight delay. Therefore in the unterminated lines and in those terminated at the source end, power dissipation of the off-chip drivers increases linearly with chip-to-chip interconnection length.

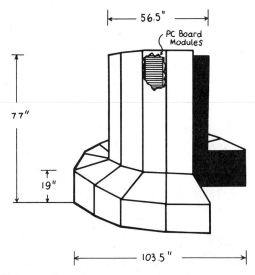

FIGURE 9.1   Cray 1 computer tower

9.2   Examples of Central Processing Units

Systems at both ends of the integration spectrum have been developed. A set of samples that includes supercomputers, a mainframe, a CMOS supermini-computer, and a microprocessor is presented in Table 9.1. Because of the fast pace of change in the computer field, the definition of the "state-of-the-art computer" in a given class of computers constantly changes. The following examples of commercial computers are presented in order to give the reader a sampling of various design styles and design objectives. These are a collection of implementations that were defined within the technology limits of their time. Of course, the parameters such as cycle time and price ranges change and the information about them can become obsolete so rapidly that they may be misleading. Here they are presented to give the reader a feel for the order of magnitude of various parameters for different classes of computers. They should be studied with that in mind rather than taken as absolute quantities.

Cray series of computers are some of the most popular supercomputers. The Cray-1 (Cray Research Inc., 1976) uses only four chip types: 16×4 bit bipolar register chips (6 nsec cycle time), 1024×1 bipolar memory chips (50 nsec cycle time), and bipolar ECL logic chips with 5/4 NAND gates with subnanosecond propagation times [9.2]. All integrated circuits are packed in 16-pin flat packs, and 288 packages can be mounted on a module. The module is a five-layer PC board with both surfaces populated by chips. The outer wiring planes on both surfaces are used for signal lines, and the three inner

TABLE 9.1   Examples of Central Processing Units

Parameter	CRAY-1	CRAY-2	CRAY-3
Year	1976	1985	1990
Logic technology	Silicon ECL	Silicon ECL	GaAs
Integration level (N_g)	2	16	200–400
Memory technology	1Kbit Bipolar	256Kbit MOS	
Module technology	PC board both	Three-d	16-chip
	sides populated	8-board module	multichip modu
Module size	$6in \times 8in$	$1in \times 4in \times 8in$	$4in \times 4in$
$N_{Wiring\ Planes}$	5	24	8
P_{MODULE}		500 W	16 W
P_{SYSTEM}	115 kW	150 kW	150 kW
Clock cycle ($nsec$)	12.5	4.1	2
Number of processors	1	5	
System performance	80–250 MFlops	0.25–1.2 GFlops	10 GFlops
Approximate price ($)	5–10 M	17 M	

Parameter	IBM 3081	Convex C-1	Intel 80386
Year	1981	1984	1985
Logic technology	TTL	CMOS	CMOS
Integration level (N_g)	700	8000	50,000
Module technology	Ceramic hybrid	PC board	132 pin PGA
Module size	$9cm \times 9cm$	$20in \times 20in$	$3cm \times 3cm$
$N_{Wiring\ Planes}$	33		2
P_{MODULE}	300 W		2 W
P_{SYSTEM}		3 kW	2 W
Clock cycle ($nsec$)		50–100	62
Number of processors	2	5	1
System performance		60 MFlops	3–4 Mips
Approximate price ($)	1–5 M	0.5 M	300

planes are used for −5.2 V, −2.0 V, and 0 V power supplies. The boards are 6 inches wide and 8 inches long. As many as 72 modules can be inserted into a 28-inch-high chassis, and two chassis form a 58-inch-high tower (Fig. 9.1). The Cray-1 CPU contains 2.5 million transistors and fits into four towers. Its clock cycle is 12.5 nsec (80 MHz), and it can execute 80 to 250 million floating-point operations (MFlops) per second. It dissipates 115 kW with 1 Mwords (2.25 Mbytes) of memory. Cray-1 costs about $ 5 million to $ 10 million.

The Cray-2 (announced by Cray Research Inc. in June 1985) follows the tradition of Cray-1 [9.3]. It has a clock cycle of 4.1 nsec (250 MHz), contains one foreground and four background processors, and provides 0.25–1.2 GFlops of processing power. Cray-2 is built from 16-gate ECL logic arrays and 256

FIGURE 9.2   Three-dimensional module of Cray-2. The module contains eight layers of chips that are interconnected in x-, y- , and z-directions (Photo courtesy of Cray Research Inc.)

Kbit MOS memory chips. The Cray-2 system contains 165,000 logic chips and 75,000 memory chips mounted in three-dimensional eight-board modules with circuit connections in all three dimensions. A photo of this module is shown in Fig. 9.2. The module has 750 packages arranged in an $8 \times 8 \times 12$ configuration, measures $1 \times 4 \times 8$ inches, has a circuit density of nearly 40 percent by volume, and consumes 300–500 W of power. The modules plug into 14 towers arranged in a $300°$ arc.

Cray-3 (Cray Computer Corp.) uses GaAs chips and is expected to be introduced in 1990 [9.4]–[9.5]. It is projected to have a 2 nsec clock cycle time (500 MHz) and 10 GFlops at peak performance.

The packaging of Cray-3 consists of direct dice mounting on multichip modules. This is the first multichip module technology that uses direct chip attach in Cray. The multichip module measures 1×1 inch, has eight layers of wiring, and can carry 16 dice. Sixteen of these multichip modules are mounted onto 4×4 inch printed circuit boards. Four printed circuit boards form an assembly. The packaging hierarchy of Cray-3 is illustrated in Fig. 9.3. In a Cray-3, there are about 200 of these assemblies, each of which holds up to 1,024 chips. The system has 50,000 GaAs chips, and the total chip count is 90,000 (the balance is mainly CMOS RAM chips). The average power dissipation of a chip is 0.8 W. Total system power is approximately 150 kW. Like Cray-2, Cray-3 is also liquid cooled.

The packaging hierarchy of Cray-3 can be summarized as follows:

- 200–400 gates per GaAs die

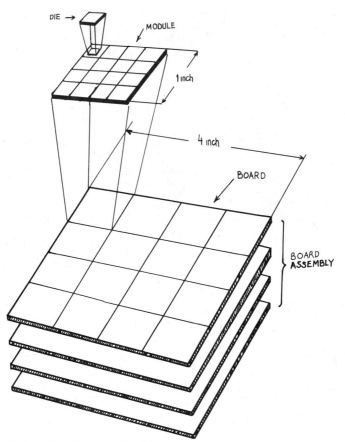

FIGURE 9.3   Cray-3 packaging hierachy

- 16 dice per module
- 16 modules per board
- 4 boards per assembly
- 4 boards per CPU
- a total of 200 boards in the system

This yields an integration density of

- 200–400 gates per die
- 3K–6K gates per module
- 50K–100K gates per board
- 200K–400K gates per CPU

The use of GaAs chips and dense multichip modules are the most significant technological factors that make it possible to achieve the performance levels projected for Cray-3. Packaging, and especially package manufacturing and chip/package assembly, is reported to be one of the most challenging aspects of Cray-3. The chips are bonded to substrates with 3-mil wires (3-mil diameter gold posts) that are so tiny that they require special robotic gear. The design has apperently gotten to a point where the robotics needed to build the dense modules are as important as the design itself [9.5]. This packaging approach is essential in trimming the size of Cray-3 to about one-half that of Cray-2 (32-inch octagon/34-inches high for Cray-3 versus 53-inch diameter/45-inches high for Cray-2). Small size is important in a machine where the cycle time is 2 nsec because light travels in common dielectrics only one foot in 2 nsec. Consequently, even with the dense packaging technology of Cray-3, it takes a good portion of a machine cycle just to send a signal across a board.

Clock skew was also a major challenge in the design of Cray-3. Cray-2 has a single-phase clock. Cray-3 has moved to a two-phase clock scheme. Two-phase clocking requires twice as many latches but also has broader clock tolerances, which helps the clock skew. The logic depths (number of logic gates traversed in a machine cycle) of Cray-1, Cray-2, and Cray-3 are reported to be respectively 8, 4, and 6 levels.

Cray-4 is also expected to be based on GaAs, and its projected cycle time is 1 nsec (1 GHz). It is expected to have a two-phase clock scheme that will require a latch-to-latch delay of 500 psec if logic is placed between L1 and L2 latches as well as between L2 and L1 latches.

The IBM 3081 (1981) uses LSI chips. The packaging of 3081 was described in Chapter 3, and a photograph of the system is shown in Fig. 9.4. Its CPU was developed from bipolar gate arrays that contain up to 704 circuits with a 1.1 nsec gate delay. One hundred to 133 of these chips are placed on a thermal conduction module (TCM) that measures 9×9 cm [9.6], [9.7]. A TCM contains 33 layers of molybdenum conductors (bulk resistivity is three times that of copper), and aluminum oxide (a ceramic) is used as the insulator ($\epsilon_r = 9.4$). At the top, there is a bonding level on which the chips are mounted with solder bumps. The next five layers are for redistribution and have a 0.25 mm pitch that matches the chip contact pad spacing. Sixteen of the following layers are x- and y-signal planes with a controlled impedance of 55 Ω and a line pitch of 0.5 mm, which yields 320 cm of signal wiring per cm^2 of TCM. Typically there are 130 meters of signal wiring per TCM. Between each pair of signal planes is a voltage reference plane (total of eight). Three power-distribution planes (for two supply levels and ground) bring the number of layers up to 33 ($1+5+16+8+3 = 33$). A TCM can support 300 W of power dissipation and 4 W per chip, which corresponds to 4 W/cm^2 at the module level and 20 W/cm^2 at the chip level. A typical TCM contains 52 logic chips (25,000 logic circuits), 34 memory chips for cache (65,000 memory bits), and five terminator chips (500 terminating resistors). It makes 1800 connections to the next level (PC board), 500 of which are power connections.

FIGURE 9.4   IBM 3081 system (Courtesy of International Business Machines Corparation.)

Nine TCMs are placed on a PC board that measures 60×70 cm [9.8], [9.9]. The PC board has 20 layers of copper lines, six of which are signal planes and 12 are power reference planes. The signal wires are 80 ± 10 Ω transmission lines. The PC board can supply 600 A of total current with a 15 mV power fluctuation. It contains one of the dual CPUs of the IBM 3081. A full-featured 3081 has 26 modules that require three PC boards; two are CPUs and one is memory.

An example of a 64-bit superminicomputer is the Convex C-1 system (1984) based on CMOS gate arrays [9.10]. It runs at 60 MFlops and costs approximately $ 0.5 million. The C-1 is built around five asymmetrical parallel processors; the physical-cache unit, address translation unit, instruction processing unit, address and scalar unit, and the vector-processing system. Each processor fits on a 20×20 inch board except the vector-processing system, which occupies three boards. The entire CPU fits on seven boards. The vector-processsing unit uses the CMOS gate arrays. Without the 24 CMOS 8000-gate logic arrays, 12 additional boards would have been necessary. The gate arrays are 2.5 micron Fujitsu CMOS arrays with 179 pins, and the remainder of the CPU logic is built with standard off-the-shelf Fairchild Advanced Schottky TTL chips.

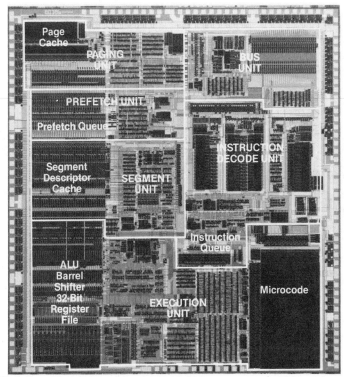

FIGURE 9.5   Die photo of Intel 80386 (Photo courtesy of Intel Corp.)

Because of the CMOS gate arrays, a fully loaded C-1 consumes only 3200 W and runs with a 100 nsec major clock cycle; it also has a 50 nsec minor clock. The C-1 can have up to 128 megabytes of DRAM memory (16 megabytes per card).

Intel 80386 is a 32-bit microprocessor (1985). The die photo of Intel 80386 is shown in Fig. 9.5, and the packaged chip is shown in Fig. 9.6. Intel 80386 is built using a 1.5 μm CMOS process with two layers of metal wiring. It contains 275,000 transistor locations (including the unused positions in microcode ROM) and 180,000 actual transistors. The die size is 9.5 mm.

The initial 80386 parts in 1985 were 12 to 16 MHz. The clock frequency was later improved to 25 MHz in 1988 and to 33 MHz in 1989. The average processing rate of 80386 pipeline is 4.4 cycles per instruction, giving an overall execution rate of 3 to 4 Mips at 16 MHz clock frequency [9.22]. The 33 MHz part is capable of operating at 8 MIPS. The chip is housed in a 132-pin PGA, dissipates in the order of 1 W, and costs a few hundred dollars.

80386 is organized as eight logical units: bus interface unit, prefetch unit, instruction decode unit, control unit, data unit, protection test unit, segmen-

FIGURE 9.6 Intel 80386 die mounted on a pin grid array (Photo courtesy of Intel Corp.)

tation unit, and paging unit [9.22]. The 32-bit *bus interface unit* provides communication to I/O and memory. It controls all address, data, and control signals to and from the CPU. Instructions are fetched from the memory by the *prefetch unit* in advance and stored in a temporary instruction queue. This queue also acts as an instruction buffer between the prefetch unit and the instruction decode unit. The *instruction decode unit* decodes the instructions from the queue and passes them to the execution unit. The *execution unit* operates on the decoded instruction, performing the required steps for execution. The ALU, barrel shifter, and 32-bit register file are some of the main data flow elements in the execution unit. The *control unit* includes the microcode ROM shown at the lower right corner, which provides the main control for the execution unit. The 80386 supports both segmentation and paging. A memory reference goes first to the segment unit, then to the paging unit, and finally to the bus unit, which puts the address on the external bus. The *segment unit* calculates the logical address and checks for segment protection violations. The *paging unit* translates the logical address into a physical address and checks for paging violations. Segment and paging units contain subsets of the segment and page tables in on-chip segment descriptor and page caches for speedy translation and protection. As can be seen in the photo, the chip floorplan closely follows the pipeline of the machine. Control is concentrated at the middle of the chip, and the data path elements surround

FIGURE 9.7   Intel 80486 die mounted on a pin grid array (Photo courtesy of Intel Corp.)

the control logic. In this way, the wires are kept short among various control sections, as well as between the control logic and the data path. This prevents critical paths that cross the chip back and forth multiple times.

Intel 80486 (announced in 1989) integrates the functions of the 80386 CPU and 80387 floating-point math coprocessor on a single chip and also features an on-chip cache that contains recently used data and instructions [9.23]. Intel 80486 is built using a 1 μm CMOS technology and integrates 1.2 million transistors on a single chip. The die area is 518.3 mils square. The die size 10.5 × 15.7 mm, which is equivalent to 13 × 13 mm. It achieves 15 MIPS with 25 MHz clock frequency and 20 MIPS with 33 MHz clock frequency.

Intel 80486 uses pipelining and RISC design techniques to execute frequently used instructions, such as register-register operations, loads, and stores, in a single clock cycle. Execution of these instructions takes multiple cycles in 80386 (load 4 cycles, store 2 cycles, register-register operations 2 cycles)— the predecessor of 80486. Intel 80486 has a four-way set-associative 8 KByte on-chip cache to minimize the memory access delays and to reduce the system bus traffic. It also has a 32-entry four-way set-associative on-chip translation lookaside buffer (TLB) to translate virtual addresses to real addresses. A burst data transfer mechanism allows four 32-bit words to be read sequentially from

FIGURE 9.8   Die photo of Intel 80486 (Photo courtesy of Intel Corp.)

TABLE 9.2 Critical Parameters of Some Microprocessors

Microprocessor	Year	Tech.	F [μm]	n_w	p_w [μm]	N_{tr}	N_p	D_c [cm]
Intel 4004	1971	pMOS		1+1		2,300	18	0.3
Intel 8008	1972	pMOS		1+1		3,500	18	0.35
Intel 8080	1974	nMOS		1+1		5,000	40	0.45
Intel 8085	1976	nMOS		1+1		6,500	40	0.5
Intel 8048	1977	nMOS		1+1		17,000	40	0.5
Intel 8086	1978	nMOS		1+1		20,000	40	0.6
iAPX-43201	1981	HMOS		1+1		110,000	64	0.8
iAPX-43202	1981	HMOS		1+1		49,000	64	0.8
iAPX-43203	1981	HMOS		1+1		60,000	64	0.8
Intel 80286	1982	HMOS		1+1		130,000	68	0.8
Intel 80386	1985	CMOS	1.5	1+2	6	275,000	132	0.95
Intel 80486	1989	CMOS	1			1,200,000	168	1.3
Motorola 6800	1974	nMOS		1+1		5,000	38	0.5
Motorola 68000	1979	HMOS	3	1+1	7.2	68,000	64	0.66
Motorola 68010	1982					69,000		
Motorola 68020	1984	CMOS	2	1+1	5.4	180,000	98	0.92
Motorola 68030	1987	CMOS				300,000		
Motorola 68040	1990	CMOS				1,200,000		
Zilog Z80	1976	nMOS		1+1			40	
Zilog Z8000	1979	nMOS		1+1		17,500	48	0.6
NS 16032	1982	nMOS	3.5	1+1		60,000	48	0.74
NS 32332	1985	nMOS	2.5				84	
μVAX Chip Set								
I-Fetch Execute	1983	nMOS	2	1+2	9	60,210	132	0.9
Memory Subsys	1983	nMOS	2	1+2	9	53,720	132	0.83
Floating Point	1983	nMOS	2	1+2	9	42,520	132	0.73
Control Store	1983	nMOS	2	1+2	9	207,950	44	0.75
Bus Interface	1983	nMOS	2	1+2	9	24,330	132	0.67
μVAX 32720	1984	nMOS	2	1+2	9	125,000	68	0.86
Bellmac-32A	1982	CMOS	2.5			146,000	84	1.0
HP Focus	1982	nMOS	1.5	1+2	2.5	450,000	83	0.56
Berkeley RISCI	1981	nMOS	4	1+1	12	44,500	54	0.9
Stanford MIPS	1984	nMOS	3	1+1	9	24,000	84	0.58
Fairchild Clipper	1985	CMOS	2	1+2	6.5	132,000	132	1.0

memory to keep on-chip cache and instruction prefetch queue full. These features are said to enable 80486 run the same software two to four times as fast as 80386. Figures 9.7 and 9.8 show the die and package of Intel 80486.

Low-cost systems such as personal computers use microprocessors as their CPUs. Tables 9.2–9.4 list the critical parameters of various microprocessors [9.11]–[9.21]. Because microprocessors are driven by low-cost requirements, they are densely packed single-chip CPUs that dissipate little power and are

TABLE 9.3 Critical Parameters of Some Microprocessors

Microprocessor	Year	Tech.	N_{tr}	N_p	D_c [cm]	P [W]	f_c [MHz]	Mips Rate
Intel 4004	1971	pMOS	2,300	18	0.3			
Intel 8008	1972	pMOS	3,500	18	0.35		0.5–0.8	0.03
Intel 8080	1974	nMOS	5,000	40	0.45		2–3	0.5
Intel 8085	1976	nMOS	6,500	40	0.5		3–5	
Intel 8048	1977	nMOS	17,000	40	0.5			
Intel 8086	1978	nMOS	20,000	40	0.6	2.5	5–8	
iAPX-43201	1981	HMOS	110,000	64	0.8	2.5	8	
iAPX-43202	1981	HMOS	49,000	64	0.8	2.5	8	
iAPX-43203	1981	HMOS	60,000	64	0.8	2.5	8	
Intel 80286	1982	HMOS	130,000	68	0.8	3	6–10	0.6–0.9
Intel 80386	1985	CMOS	275,000	132	0.95	1–2	12–16	3–4
Intel 80486	1989	CMOS	1,200,000	168	1.3		25–33	15–19
Motorola 6800	1974	nMOS	5,000	38	0.5		1	
Motorola 68000	1979	HMOS	68,000	64	0.66	1.5	4–12.5	0.4–0.6
Motorola 68010	1982		69,000				12.5	
Motorola 68020	1984	CMOS	180,000	98	0.92	0.75	16.7–25	2–3
Motorola 68030	1987	CMOS	300,000				16.7–33	
Motorola 68040	1990	CMOS	1,200,000					
Zilog Z80	1976	nMOS		40		1.5	2–6	
Zilog Z8000	1979	nMOS	17,500	48	0.6		2.6–6	
NS 16032	1982	nMOS	60,000	48	0.74	1.25	10	0.5
NS 32332	1985	nMOS		84		2–3	15	
μVAX Chip Set								
I-Fetch Execute	1983	nMOS	60,210	132	0.9	4	10	
Memory Subsys	1983	nMOS	53,720	132	0.83	2.5	10	
Floating Point	1983	nMOS	42,520	132	0.73	2.0	10	
Control Store	1983	nMOS	207,950	44	0.75	1.2	5	
Bus Interface	1983	nMOS	24,330	132	0.67	3.5	10	
μVAX 32720	1984	nMOS	125,000	68	0.86	3	10	
Bellmac-32A	1982	CMOS	146,000	84	1.0	0.7	8–10	1
HP Focus	1982	nMOS	450,000	83	0.56	4	18	0.7
Berkeley RISCI	1981	nMOS	44,500	54	0.9		4	
Stanford MIPS	1984	nMOS	24,000	84	0.58	2	4	
Fairchild Clipper	1985	CMOS	132,000	132	1.0	0.5	33	5

housed in simple and inexpensive packages with a modest number of pins. The package pin count limitations may require multiplexed data and address pins (Table 9.4). In addition, to fit all the circuitry and interconnections on a single chip, most microprocessors rely on buses for on-chip information transfer. Sharing of pins and on-chip interconnections for multiple functions limits the performance of microprocessors because of resource conflicts. If two operations need the same on-chip or chip-to-chip bus, only one of them can use it at a given time, causing the other one to wait for one or more cycles.

TABLE 9.4 Bus Structures and Pin Counts of Some Microprocessors

Microprocessor	Year	Architecture (Data path width)	Address Bus Width	Data Bus Width	Address and Data Multiplexed	N_p
Intel 4004	1971	4				18
Intel 8008	1972	8	8	8	YES	18
Intel 8080	1974	8	16	8	NO	40
Intel 8085	1976	8	16	8	NO	40
Intel 8048	1977	8				40
Intel 8086	1978	16	20	16	YES	40
iAPX-43201	1981	32	32	32	YES	64
iAPX-43202	1981	32	32	32	YES	64
iAPX-43203	1981	32	32	32	YES	64
Intel 80286	1982	16				68
Intel 80386	1985	32	32	32	NO	132
Motorola 6800	1974	8	16	8	NO	38
Motorola 68000	1980	16	24	16	NO	64
Motorola 68020	1985	32	32	32	NO	98
Zilog Z80	1976	8				40
Zilog Z8000	1979	16	16	16	YES	48
NS 16032	1982	16	24	16	YES	48
NS 32332	1985	32	32	32	YES	84
μVAX Chip Set						
I-Fetch Execute	1983	32	2×32	2×32	YES	132
Memory Subsys	1983	32	2×32	2×32	YES	132
Floating Point	1983	32	2×32	2×32	YES	132
Control Store	1983	32	32	32	YES	44
Bus Interface	1983	32	2×32	2×32	YES	132
μVAX 32720	1984	32	32	32	YES	68
Bellmac-32A	1982	32				84
HP Focus	1982	32	32	32	YES	83
Berkeley RISCI	1981	32				
Stanford MIPS	1984	32				84
Fairchild Clipper	1985	32	2×32	2×32	YES	132

The abbreviations and symbols used in the tables are F minimum feature size, n_w number of wiring levels (1+2 means one level of polysilicon and two levels of metal), p_w wiring pitch (pitch is wire width plus spacing or, in other words, the distance between the midpoint of a wire to the midpoint of its neighbor), N_{tr} number of transistors on the chip, N_p number of signal pins, P power dissipation, and f_c clock frequency.

9.3   High-Performance Logic Design Techniques

In this section two common logic design techniques classified under the general term of *concurrent operation* will be described: *pipelining* and *parallelism* [9.24]. These important implementation techiques will be revisited many times in the rest of this chapter because they strongly affect the cycles-per-instruction ratio, cycle time, and wiring and hardware requirements of processors. (See Fig. 9.9.)

9.3.1   Pipelining

Pipelining is a technique commonly used to improve the execution rate of a CPU where the logic function is divided into smaller pieces, called *stages,* that operate concurrently [9.24]. The important aspect of a pipeline is that all the stages are active simultaneously, working on different instructions. Instructions are fed into the pipeline at a rate determined by the slowest stage of the pipeline. The time required for a function from its beginning to the end is not reduced by pipelining, but the rate at which instructions are executed improves because many operations are performed concurrently. As a result pipelining improves the performance by overlapping the execution of several instructions because instructions begin to be processed before previous ones are completed.

A typical CPU pipeline may be as follows:

```
| IF  | DEC | EX  | WB  |
      | IF  | DEC | EX  | WB  |
            | IF  | DEC | EX  | WB  |
                  | IF  | DEC | EX  | WB  |
                        | IF  | DEC | EX  | WB  |
            time
      ----------->
```

IF stands for instruction fetch, DEC is the decode stage, EX is the cycle in which the instruction is executed, and WB is the stage when the result of the operation is written back into the register file. The progression of time is represented at the horizontal axis. An instruction is fetched during one machine cycle, and in the next cycle it progresses to the decode stage. The next instruction is fetched when the first one is in the decode stage. The important point is that the start of an instruction does not wait for the previous ones to be completed. In this way, all portions of the hardware (instruction fetch unit, decode unit, execute unit, etc.) are utilized every cycle, making best use of the available resources and increasing the throughput of the machine.

Dependencies between instructions prevent the pipeline from achieving its maximum performance. To achieve a high efficiency in the pipeline, the

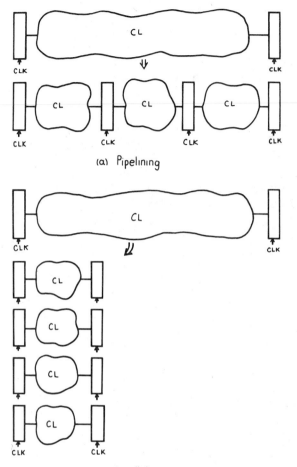

(a) Pipelining

(b) Parallelism

FIGURE 9.9   Improving the execution rate by (a) pipelining and (b) parallelism

implementation of instructions that break the regular flow of the instruction stream, such as branches, must be carefully optimized [9.25]. In a pipeline with sequential prefetching, the instructions that are in the pipeline must be flushed after a successful branch because they were incorrectly prefetched and loaded. The processor must then fetch instructions from the new path. Reference [9.25] reviews some of the methods that minimize the branch penalty in pipelined processors.

Because an instruction can start before a previous one is completed, a pipelined architecture has many dependencies besides branches. An instruction may need some data that has not yet been generated by the previous instruction, or the results of certain computations at the advanced stages of the pipeline may affect the order of the instructions that are still in the initial stages of the pipeline. For example, an arithmetic instruction that needs the data that is supposed to be brought from the memory by a load instruction cannot proceed before the data arrives. As another example, assume that in a pipelined CPU, registers are read during the decode cycle, and the result of an arithmetic operation is written back into the register file at the write-back cycle, as in the pipeline shown above. When an instruction is supposed to write a value to a register, the correct information will not be there on time if the following instruction needs to read this data from the register file. This requires "forwarding" or "bypassing" the information around the register file to the ALU inputs so that the pipeline will function properly. In many cases similar to these, pipelining creates dependencies and requires additional hardware and interconnections to evaluate and pass information between sections of the CPU.

Mainly due to branches and other dependencies and memory latency, the overall efficiency of a typical pipeline is less than one instruction per cycle. The key design trade-off is to minimize the average number of cycles required per instruction while keeping the design simple so that the resulting logic depths and interconnection requirements could support a short cycle time. The execution rate of the machine is given by the ratio of the clock frequency and the average number of cycles required per instruction:

$$\text{Instruction execution rate (Mips)} = \frac{\text{Clock frequency (MHz)}}{\text{Cycles per instruction}}$$

As an example, with a typical instruction mix, Intel 80386 requires 4.4 cycles per instruction, giving an overall execution rate of 3 to 4 Mips at 16 MHz clock frequency [9.22]. Reduced instruction set computer (RISC) processors usually have better cycles-per-instruction ratios than complex instruction set computer (CISC) processors.

9.3.2   Parallelism

Parallelism is also very effective in reducing the logic depth and increasing the execution rate. Parallelism achives concurrency by replicating a hardware structure several times. Performance improves because all structures execute simultaneously on various parts of the problem to be solved. Pipelining and parallelism are compared in detail in reference [9.24].

Figure 9.10 illustrates, at a very low level, how parallelism can reduce the logic depth at the cost of increased hardware. At the left, the function is implemented in four logic levels using only six gates. At the right, the logic depth is reduced to two, but the gate count goes up to nine. The first implementation is "short and fat," and the second implementation is "tall and thin." It is important to note that, in a tall and thin implementation, not

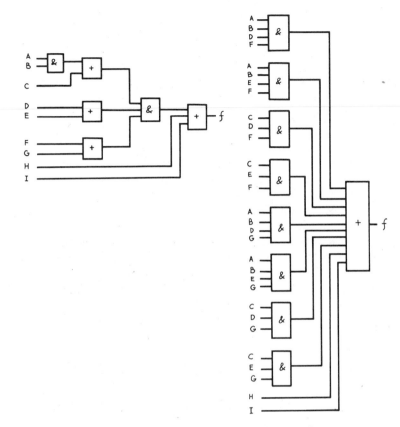

FIGURE 9.10   Reducing the logic depth by parallelism

only the number of gates increases, but the gate fan-in (number of inputs) also goes up dramatically. The short and fat implementation uses five two-input gates and one three-input gate. The tall and thin implementation uses mostly four- and three- input AND gates and a ten-input OR gate. The number of transistors in a gate is directly proportional to its fan-in. In static CMOS, the short and fat implementation requires 14 pMOS/nMOS transistor pairs, and the tall and thin implementation requires 38 pMOS/nMOS transistor pairs—almost a factor of three difference. One needs to be careful when using this technique to reduce the logic depth in CMOS chips. In CMOS circuits increased fan-in degrades the performance because more transistors are placed in series either in the pull-up or pull-down path. Consequently, overall delay may increase even when logic depth is reduced. Even if the overall delay does not increase, improvement may be miniscule. Bipolar circuits are less sensitive to increased fan-in. Consequently, they usually have logic gates that perform more functions than CMOS gates, and bipolar chips usually have shallower logic depths than CMOS chips.

The essence of this example also applies to more global system level implementation issues. At a higher level, to achieve high-speed operation, multiport register files and many data buses are necessary. If one can read and write to a register file at the same time, the amount of lost cycles for load/store operations would be reduced, and the access to registers during address calculations would not conflict with data manipulation. At the system level, vector processors are the ultimate example of how parallel hardware can be used to improve performance. The efficiency of vector processors depends on the type of problem. The efficiency is high when there are many operations that can be performed in parallel independently, such as multiplication of two vectors or inversion of a matrix, which are very common in scientific problems that involve numerical analysis. To solve such problems, vector units can be employed for processing more than one piece of data at a time via parallel vector processing.

Another area where parallelism (therefore performance) is traded off with cost is the chip pin count. The more pins a chip has, the wider a bandwidth it will have to and from the outside world, making it possible to perform multiple data transfers in the same cycle. As a result, to ensure the highest possible execution rate, multiplexed I/O pins and serial transmission of data are avoided, and this results in higher pin counts.

Parallel designs are faster, but they require more hardware and pins than less parallel implementations that execute in a more serial fashion.

9.4   Machine Organization and Rent's Rule

E.F. Rent of IBM published two internal memoranda in 1960 that contained the log plots of "number of pins" versus "number of circuits" in a logic design [9.26]. These data tend to form a straight line in a log-log plot and yield the relationship

$$N_p = K_p N_g{}^\beta. \tag{9.1}$$

Here, N_p is the number of pins or the number of external signal connections to a logic block, N_g is the number of logic gates in the block, β is the Rent's constant, and K_p is a proportionality constant. The values of K_p and β for the IBM computers were reported to be 2.5 and 0.6, respectively.

Rent's rule was later applied by IBM researchers to derive the average interconnection length in a gate array as a function of number of gates and gate pitch [9.27]–[9.32]. This was obtained by hierarchically dividing the logic design into quadrants and calculating the number of connections to a block from Rent's rule. By setting the upper limit on wire length equal to the size of the block that forms the next level of the hierarchy, a series sum of interconnection lengths was formed that establishes an upper bound on total wire length. (This will be discussed in Section 9.8.1.)

Because Rent's rule is an empirical result obtained by observing existing designs, it is useful in predicting the pin requirements and average interconnection lengths of well-studied architectures and follow-on computers that have designs similar to current systems, but it may give misleading results if extended to dissimilar architectures and too far into the future. The limitations of Rent's rule must be recognized. For accurate predictions, it is necessary to understand the design from which the initial Rent's data were obtained and to ensure that the architecture and implementation of that design are similar to the system being studied. Like any empirical (or even fundamental) relationship, Rent's rule can be misleading if it is applied without adequate understanding.

The key factors that affect Rent's constants include machine architecture, organization, and implementation. The design philosophy and methodology also affect Rent's constants. If the machine from which the initial data were obtained and the computer design for which the predictions are sought have similar machine organizations and implementations, the model will be successful. On the other hand, if the predictions are made for a system with an entirely different design philosophy from the one from which Rent's data were obtained, the results will have little meaning.

Many architectural and implementation factors affect the pin count and interconnection requirements and, as a result, the Rent's constant. For example, to ensure the highest possible speed, multiplexed I/O pins and serial transmission of data are avoided, and this results in higher pin counts. On the other hand, to achieve low-cost packaging in commercial microprocessors and memories, bidirectional and multiplexed I/O pins and partially serial data transformation are employed (Table 9.4).

It is important to note that two chips with exactly the same number of gate counts may have different pin counts due to package cost considerations. As a result, drawing conclusions concerning on-chip interconnection lengths from pin count data may be misleading because two chips with identical layouts and on-chip interconnection lengths may have drastically different pin counts if multiplexed or bidirectional pins and serial I/O ports are employed in one of the chips. For example, chip vendors usually offer a version of their 32-bit microprocessors that is basically identical to the original one except the external interface is 16 bits. The concerns are similar when the average interconnection calculations derived for a design methodology (for example, gate arrays) are applied to a chip designed by another methodology (for example, custom-designed microprocessors).

Another design area that illustrates the relation between average interconnection length and architectural design choices is pipelining. As described in Section 9.3, throughput of a design can be improved by pipelining, but this also increases the communication requirements (and therefore wiring) between the subunits because of the dependencies between the instructions that are executed simultaneously.

TABLE 9.5   Rent's Constants for Various System Types

System or Chip Type	Exponent β	Multiplier K_p
Static memory	0.12	6
Microprocessor	0.45	0.82
Gate array	0.50	1.9
High-speed computer		
Chip and module level	0.63	1.4
Board and system level	0.25	82

The following considerations relate to supercomputers and mainframes:

- Because of their high level of complexity and the limited design times between new generations and models, supercomputers and mainframes are usually implemented in gate arrays.

- These high-performance systems require a parallel I/O and avoid multiplexed or bidirectional pins.

- High performance also requires concurrency (pipelining and parallelism), which increases the communication demands among the subsections of the CPU. As a result, even if the chips are implemented as custom designs, their interconnection requirements may still be comparable to gate arrays.

Trends in high-performance systems are important for lower-cost designs because, historically, advanced design techniques such as pipelining were first implemented in mainframes and eventually found their way into microprocessors.

This chapter considers a wide variety of systems. Rent's data obtained from super- and mainframe computers and several commercial integrated circuits will be used to predict the pin requirements and average interconnection lengths of different architectures and implementations. Figure 9.11 shows the pin counts of several chips and packages, and Fig 9.12 plots the same data classified according to the type of system instead of product identification. The data are grouped into distinctive classes, and the slopes of the lines in these plots determine Rent's constant β in Eq. 9.1 for the listed system types.

Expectedly, memory devices have small pin counts. Faster static RAMs have higher pin counts than DRAMs because SRAMs usually have separate pins for all address bits. In DRAM chips, half the address (row address) is sent during one clock cycle and the other half during the following cycle (column address), and they share the same pins.

Commercial microprocessors form the next group. In microprocessors, data input, data output, and occasionally address share the same pins. In

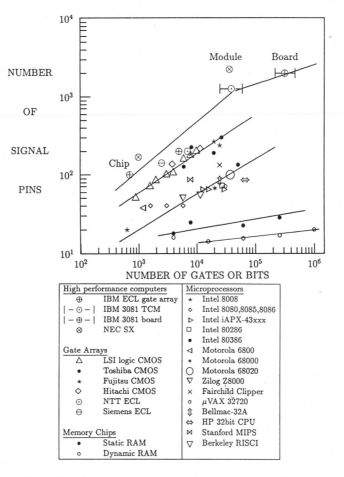

FIGURE 9.11   Rent's curves for various digital systems. Data points are classified according to product identification.

addition, microprocessors are *self-contained* systems that constitute a functionally independent portion of a larger system. This functional completeness reduces the pin requirements.

In gate arrays, the placement of the gates and circuit design is done long before the logic is designed. In this way, most of the semiconductor processing can be done ahead of time, and only the metal wiring steps are left to personalize the design. Processed wafers are stockpiled and quickly turned around after chip definition. Much of the development cost is shared among many designs, reducing the cost. This makes them very attractive for low-volume parts. Because of the rigidity of placement, however, gate arrays are

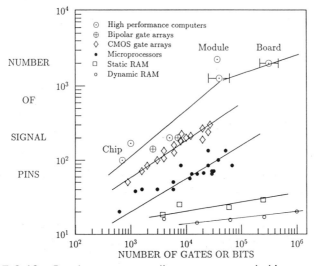

FIGURE 9.12   Rent's curves according to system and chip types

not as efficient as custom designs. Because they have to meet the requirements of a wide range of designs, for any given design, gate arrays are not as dense and fast as a custom implementation would be. In addition, gate arrays require more wiring space because the placement is done before the logic design. Wiring is more efficient in custom designs because placement can be optimized to minimize interconnection lengths. In the pin count plot, gate arrays are immediately above microprocessors because, due to their limited integration density, gate arrays are not as functionally self-contained as microprocessors and, as such, require more pins.

High-performance computers have the largest pin counts because they use more parallelism and pipelining, which, in turn, require a greater number of simultaneous information transfers. They normally avoid multiplexed and bidirectional I/O when possible. Module and board levels are also shown on the plot as if they were individual chips. This symbolizes a design in which one single module or board is mapped to a chip with an identical logic structure and the same pin count. When the "unit" (chip, module, or board) contains the totality of a CPU, the rate of increase in pin count drops; however, this does not occur until the late stages of completion. In the IBM 3081, one board contains a CPU, and each board consists of nine modules; as a result a module contains one-ninth of a CPU but still appears on the same Rent curve as the chips. When the gate count of the unit increases and contains more of the entire CPU, the pin count will move to another curve with a less steep slope (smaller Rent's constant) as illustrated in Figs. 9.11 and 9.12.

Rent's constants obtained from Fig. 9.11 are listed in Table 9.5. The quantitative results support the earlier observation that these constants are determined by the architecture and implementation of the system and that pin and interconnection requirements are more demanding in high-performance systems with large amounts of parallelism and pipelining.

The model described in the remainder of this chapter is very flexible; the architecture-dependent aspects can be changed by using a different Rent constant, and the results can be compared.

9.5   CMOS Technology

CMOS technology is a strong contender in all areas of integrated-circuit applications. With its low power dissipation and superior noise margins, CMOS is taking over the SRAM, DRAM, and microprocessor markets long dominated by nMOS. By enabling the integration of digital and analog functions on the same die, CMOS is becoming preferable to bipolar technology in many analog applications. Because reducing the minimum feature size benefits CMOS more than it does bipolar transistors, CMOS is competing with bipolar technology even in the high-performance digital circuit area. Cooling CMOS chips to liquid nitrogen temperature improves carrier mobilities, reduces parasitic junction capacitances, solves the latch-up problem, and decreases the resistance of metal lines by almost an order of magnitude. On the other hand, the gain of bipolar transistors degrades at low temperatures. As a result the same phenomenon that solves the latch-up problem for cyrogenic CMOS cripples bipolar chips if they are cooled to liquid nitrogen temperature. With parasitic interconnection resistance becoming a major performance concern in scaled technologies, the possibility of cyrogenic operation is a great advantage for CMOS.

For these reasons, CMOS is selected as the principal technology for studying the limits of integration here. Later in this chapter, the performance of CMOS integrated circuits is compared to those of silicon bipolar and GaAs circuits as a function of integration level.

Table 9.6 lists the electrical parameters of nMOS and CMOS transistors [9.33]–[9.42] and chip interconnections [9.43]–[9.45] at various feature sizes. As can be seen, interconnection resistance increases significantly as minimum feature size is reduced.

9.6   Packaging Technologies

Packaging is as important as, and often even more critical than, transistors in determining the overall performance of a system. This is the reason for

TABLE 9.6 *n*MOS and CMOS Technology Parameters

Parameter	7 μm nMOS	3 μm nMOS	2 μm CMOS	1.3 μm CMOS	0.7 μm CMOS	0.35 μm CMOS
F (μm)	7.0	3.0	2.0	1.3	0.7	0.35
L_{eff} (μm)	5.0	2.0	1.4	1.0	0.5	0.25
L_D (μm)	1.0	0.5	0.3	0.15	0.1	0.05
X_j (μm)	1.7	0.8	0.5	0.25	0.2	0.1
t_{gox} (Å)	1000	700	400	250	100	50
V_{DD} (V)	12.5	5.0	5.0	3.3	2.0	1.0
V_T (V)	2.5	1.0	0.8	0.65	0.4	0.2
R_{tr}(NENH) (Ω)	15000	15000	10000	8000	8000	8000
R_{tr}(NDEP) (Ω)	25000	25000	—	—	—	—
R_{tr}(pMOS) (Ω)	—	—	20000	16000	16000	16000
C_{tr} (fF)	20	6	4	3	2	1
W_{int} (μm)	14	5	3	2	1	0.5
W_{sp} (μm)	14	5	3	2	1	0.5
H_{int} (μm)	1	0.5	0.5	0.4	0.3	0.25
p_w (μm)	28	10	6	4	2	1
n_w	1	1	2	3	4	5
$\mathcal{R}_{int}$ (Ω/cm)	21	120	200	375	1000	2400
C_{int} (pF/cm)	5	3.5	2.0	2.0	2.0	2.0
$\mathcal{R}_{int}C_{int}$ (ps/cm^2)	105	420	400	750	2000	4800

the detailed analysis of the package in addition to the chip analysis in the SUSPENS model. The high-performance packaging alternatives considered here are wafer-scale integration, thin film and ceramic multilayer hybrids, and printed wiring board [9.46]. Table 9.7 summarizes these packaging technologies and their assumed properties, and Fig. 9.13 illustrates their basic physical structures. All these packages are treated in detail in Chapter 3.

In wafer-scale integration (WSI), "building blocks" are fabricated on a wafer and interconnected by additional global interconnections. These building blocks correspond to the chips in a hybrid technology. In the calculations, it is assumed that the building blocks in WSI are in all aspects identical to the chips in the alternative packaging options and that these blocks are interconnected by the two top-most aluminum layers that are in addition to the "chip-level" interconnections. The circuits within the blocks are connected with the same interconnections as the chips in the corresponding technology, and the wafer-level global lines are thicker and wider for reduced distributed-RC delays. For example, a WSI circuit based on 1.3 μm CMOS technology will have the three levels of wiring shown in Table 9.6, and it will also have the two additional levels of wafer-scale interconnections shown in Table. 9.7. Silicon dioxide or polyimide is the insulator [9.47], [9.48].

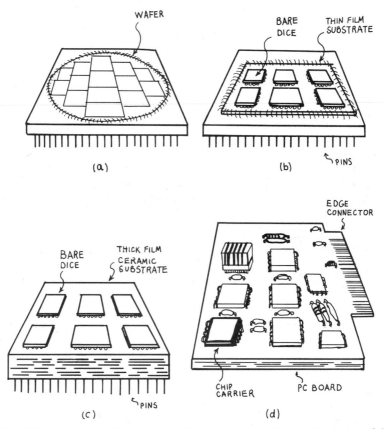

FIGURE 9.13 Conceptual illustration of various packaging technologies: (a) wafer-scale integration, (b) thin film hybrid, (c) ceramic hybrid, (d) printed wiring board with single chip carriers

The thin film multilayer hybrid (for example, silicon-on-silicon packaging) is similar to WSI in that it supplies two layers of copper interconnections on a silicon substrate on which the chips are flip-mounted or wire bonded. Polyimide is the dielectric. The thin film hybrid enables many chips to share the same package without requiring a 100 percent yield simultaneously on the chips and global interconnections on the substrate because they are fabricated separately. In addition, a mixture of CMOS, bipolar, and GaAs chips can be placed on the same substrate, and this is not easily achievable with WSI. The silicon substrate minimizes the thermal stress and facilitates conventional fabrication techniques. Unlike WSI, redundancy is not necessary, and engineering changes and repairs are relatively simple [9.49]–[9.51]. The substrate of the thin film hybrid can be another material besides silicon.

Ceramic multilayer hybrids are widely used in mainframe computers.

TABLE 9.7 Packaging Technologies

Parameter	Wafer-Scale Integration	Thin Film Hybrid	Ceramic Hybrid	Printed Wiring Board
P_W (μm)	20	50	500	160
N_W	2	2	16	4
W_{INT} (μm)	10	25	250	80
W_{SP} (μm)	10	25	250	80
H_{INT} (μm)	2	2	10	50
δ @$1GHz$ (μm)	2.8	2	4.8	2
ρ_{INT} $(\mu\Omega\text{-}cm)$	3	1.7	10	1.7
$\mathcal{R}_{INT}$ (Ω/cm)	15	3.4	0.4	0.5
ϵ_r	3.9	3.4	9.4	3.2
v_M $(cm/nsec)$	15	16	10	17
C_{INT} (pF/cm)	1.5	1.0	1.9	0.8
Z_0 (Ω)	44	60	55	75
C_{PAD} (pF)	0	0.25	0.5	10
P_P (μm)	40	100	250	1000
Q (W/cm^2)	40	20	4	0.5
Typical size $(cm \times cm)$	5×5	10×10	10×10	60×60

Here, thick film refractory metal lines are laid down on unfired ceramic sheets with already punched via holes. These metalized sheets are then dried and inspected, and as many as 33 of them are stacked and fired to form the substrate on which the chips are flip-mounted [9.6]–[9.7].

In a high-performance printed wiring board, dice housed in chip carriers are mounted on a multilayer board with four to eight conductor planes. The principal disadvantages of this packaging alternative are the area wasted by the chip carriers and the relatively limited power-dissipation density.

Combinations of these techniques are also possible. The structure in Fig. 9.14, for example, integrates a ceramic substrate and thin film signal lines insulated with polyimide. The ceramic substrate forms a stable base and provides large-value decoupling capacitors. The polyimide ensures densely packed low-capacitance signal lines.

The skin effect is taken into account in the resistance calculations in Table. 9.7. A steady current is distributed uniformly throughout the cross-section of a conductor through which it flows; however, time-varying currents concentrate near the surfaces of the conductors. This is known as the skin effect (see Chapter 6). At high frequencies, the effective resistance of an interconnection will not decrease when its thickness is increased beyond a critical value. As listed in Table 9.7, the skin depths in WSI and thin film hybrid packages at

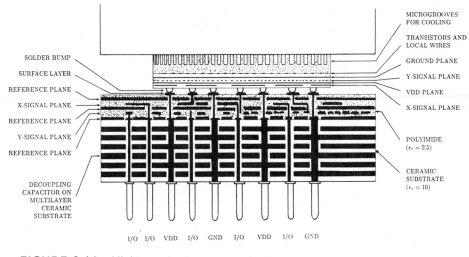

FIGURE 9.14 High-speed computer packaging

1 GHz are larger than the interconnection thicknesses. This is not the case for interconnection-resistance calculations for ceramic hybrid and printed wiring board; as a result, their effective H_{INT} is assumed to be equal to skin depth at 1 GHz. This takes into account enough harmonics in the Fourier expansion to ensure resistance values correct up to 250 MHz clock frequency. Skin depth is expressed as

$$\delta = \sqrt{\frac{\rho}{\pi f \mu}}, \tag{9.2}$$

where f is the frequency, and μ and ρ are the permeability and resistivity of the material.

9.7 Qualitative Description of the Model

This section qualitatively describes the SUSPENS model. Assuming that the total number of logic gates in a CPU is known (as small as 25,000 in a typical 16-bit microprocessor or as large as 1 million in a supercomputer), system performance is projected as a function of technology, design, and architecture parameters. The equations corresponding to the steps summarized below are listed in Table 9.8 for a CMOS-based system, and they will be derived in the following sections. The interactions among various parameters are also illustrated in Fig. 9.15. The calculation steps are as follows.

TABLE 9.8 SUSPENS Calculations for a CMOS CPU

Step	Calculation
1	$\overline{R} = \dfrac{2}{9}\left(7\dfrac{N_g{}^{p-0.5}-1}{4^{p-0.5}-1} - \dfrac{1-N_g{}^{p-1.5}}{1-4^{p-1.5}}\right)\dfrac{1-4^{p-1}}{1-N_g{}^{p-1}}$
2a	$d_g = \dfrac{f_g \overline{R} p_w}{e_w n_w}$
2b	$D_c = \sqrt{N_g}d_g$
2c	$l_{av} = \overline{R}d_g$
3a	$R_{gout} = f_g \dfrac{R_{tr}}{k}$
3b	$C_{gin} = 3kC_{tr}$
3c	$T_g = f_g R_{gout}l_{av}C_{int} + f_g R_{gout}C_{gin} + R_{int}C_{int}\dfrac{l_{av}^2}{2} + R_{int}l_{av}C_{gin}$
4a	$f_c = \left(f_{ld}T_g + R_{int}C_{int}\dfrac{D_c^2}{2} + \dfrac{D_c}{v_c}\right)^{-1}$
4b	$C_c = \dfrac{D_c^2 n_w e_w C_{int}}{p_w} + 3C_{tr}kN_g f_g$
4c	$P_c = \frac{1}{2}f_c f_d C_c V_{DD}^2$
5	$N_p = K_p N_g^\beta$
6a	$N_c = \dfrac{N_{gTOT}}{N_g}$
6b	$\overline{R}_M = \dfrac{2}{9}\left(7\dfrac{N_c{}^{\eta-0.5}-1}{4^{\eta-0.5}-1} - \dfrac{1-N_c{}^{\eta-0.75}}{1-4^{\eta-0.75}}\right)\dfrac{1-4^{\eta-1}}{1-N_c{}^{\eta-1}}$
6c	$F_P = max\left\{D_c,\ \dfrac{F_c}{F_c+1}\dfrac{\overline{R}_M N_p P_w}{E_w N_w},\ \sqrt{N_p}P_P,\ FP_{CC}\right\}$
6d	$D_M = \sqrt{N_c}F_P$
7a	$L_{INT} = D_M$
7b	$f_M = \left[15(N-1)R_{tr}C_{tr} + 2Z_0 C_{PAD} + R_{INT}C_{INT}L_{INT}^2 + \dfrac{L_{INT}}{v_M}\right]^{-1}$
7c	$C_M = \dfrac{F_c}{F_c+1}N_c N_p\left(3\dfrac{1-5^N}{1-5}C_{tr} + 2C_{PAD} + \overline{R}_M F_P C_{INT}\right)$
7d	$P_M = \frac{1}{2}F_D f_s C_M V_{DD}^2$
8	$f_S = min\{f_c, f_M\}$

1. Based on the number of logic gates on the chip N_g, the average inter-connection length in units of logic gate pitch $\overline{R}$ is calculated via Rent's rule.

2. Given the interconnection pitch p_w, number of interconnection layers n_w, utilization efficiency of interconnections e_w, and the fan-out of gates f_g, and using the average interconnection length established in step 1, one

can obtain the gate dimension d_g, chip size D_c, and average interconnection length in actual units l_{av}, by setting the supply and demand of interconnections equal.

3. Average gate delay T_g is then determined from the wiring load calculated in the previous step and from transistor and circuit parameters.

4. Maximum chip clock frequency f_c depends on logic depth f_{ld} and average gate delay T_g. The power consumption of the chip P_c is determined by f_c and total chip capacitance C_c.

5. The number of pins per chip N_p is derived as a function of gate count via Rent's rule.

6. Module-level calculations begin with this step. Chip count N_c is equal to the ratio of total number of logic gates on the module N_{gTOT} to number of logic gates per die N_g. Based on the pin counts (from step 5) and the number and pitch of module wiring levels (N_W, P_W), one can calculate the footprint size of the chips F_P and module size D_M by using Rent's rule in a way similar to steps 1 and 2.

7. The clock frequency that can be supported by the module f_M is determined by the maximum wire length on the module, the electrical properties of module interconnections, and the output resistance and input capacitance of I/O buffers. Module capacitance C_M (including those of I/O buffers) and power consumption P_M are determined from f_M.

8. The clock frequency of the system f_S is the lower of the two clock frequencies calculated in steps 4 and 7.

Figure 9.15 summarizes the interactions among the model parameters. The primary variables are on the left-hand side. Device technology, represented by minimum feature size F, expands into a set of nine parameters. Chip architecture is defined by logic depth f_{ld}, gate fan-out f_g, and Rent's constant p. The architecture parameters, combined with the integration level N_g, and device technology yield clock frequency f_c, die size D_c, and power dissipation P_c. System architecture is defined by Rent's constants for pin count (β, K_p) and module interconnections η. Total gate count N_{gTOT}, packaging parameters M, and Rent's constants obtain the size D_M and clock frequency f_M of the module. System performance f_S is determined by the combined effects of the chip and package limitations.

For a more detailed calculation, this hierarchy can be extended to the board level where a board contains the interconnections between modules. The module- and board-level calculations can be easily combined because no active devices exist between the module and the board.

As illustrated by Table 9.8, the equations of the SUSPENS model fit on a page, and the model is simple enough to be programmed or entered

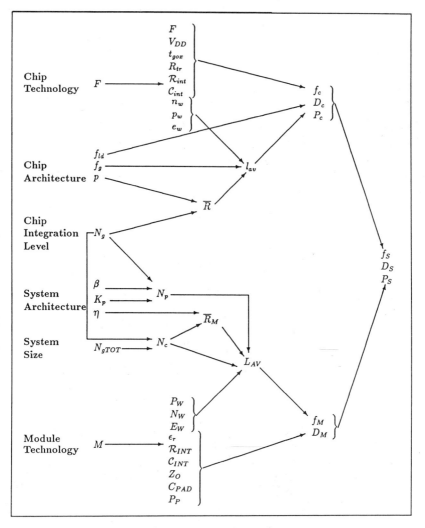

FIGURE 9.15   SUSPENS system model

in a spreadsheet program in a few hours. Using this spreadsheet program, effects of various technology and architecture changes on the overall system performance can be calculated, and "what-if" questions can be answered. The principal advantage of this model is that it includes all of the major technology, design, and packaging parameters, and the effect of each on system performance (clock frequency, power dissipation, system size) is expressed by analytical formulas. Using the model,

- The optimal integration level (number of logic gates per chip) can be determined for a given CPU size and for device and package technologies by varying the number of chips into which the CPU is partitioned and selecting the distribution with the best performance.

- The relative merits of circuit technologies (CMOS, nMOS, BiCMOS, cyrogenic CMOS, ECL, GaAs MESFET, HBT) can be compared by changing the electrical properties and circuit structures of the gates.

- The effects of scaling down the minimum feature size and interconnection pitch can be calculated.

- The effects of the number of chip wiring levels on overall system performance can be projected.

- Interconnection pitch and number of levels in the module can be optimized.

In addition, the model predicts such system parameters as clock frequency, functional throughput rate, computational capacity, packing density, power-dissipation density, total power dissipation, noise generation, overall size, yield measures, and cost indicators.

In all stages of this work, the agreement between the existing digital systems and the SUSPENS model was verified. At the chip level, more than 100 commercial and research microprocessors and gate arrays were studied, and the packing densities, clock frequencies, and pin counts were observed to agree with the model. At the system level, the IBM 3081, NEC SX, and Hitachi M-68X were examined. After observing the agreement between existing chips/systems and SUSPENS, the effects of device scaling were investigated, future CMOS microprocessor performances and requirements were predicted, and 1 μm CMOS, silicon bipolar, and GaAs heterojunction technologies were compared. The model is suitable for mainframes, minicomputers, workstations, and personal computers. It can describe the CPU of a very complicated mainframe, as well as a microprocessor.

9.8 Chip-Level Model

In this section a detailed description of chip-level calculations is presented. Parameters such as average interconnection length, gate packing density, and gate delay are calculated.

9.8.1 Average Interconnection Length

Average interconnection length can be determined by partitioning the logic design into hierarchical divisions (as illustrated in Fig. 9.16) and calculating the number of connections between the partitions via Rent's rule [9.27], [9.29].

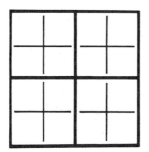

FIGURE 9.16 Calculation of average wire length using Rent's rule

The maximum interconnection length between two divisions is the size of the division that is one hierarchy level above them. The size of a block is the product of the square root of the number of logic gates it contains and the logic gate pitch. The number of connections between a logic block and the rest of the system is calculated by using Rent's rule and from the number of logic gates it contains. The gate count increases by a factor of four from one hierarchy level to the next. First, the gates are divided into groups of four and interconnected; then four of these are grouped together, and so on. Based on this approach, a series sum can be formed that gives the upper bound for the total interconnection length. Dividing this sum with the number of wires, an upper limit to average wire length in units of gate pitch is obtained:

$$\overline{R} = \frac{2}{9}\left(7\frac{N_g^{p-0.5} - 1}{4^{p-0.5} - 1} - \frac{1 - N_g^{p-1.5}}{1 - 4^{p-1.5}}\right)\frac{1 - 4^{p-1}}{1 - N_g^{p-1}} \qquad p \neq 0.5 \qquad (9.3)$$

$$\overline{R} = \frac{2}{9}\left(7\log_4 N_g - \frac{1 - N_g^{p-1.5}}{1 - 4^{p-1.5}}\right)\frac{1 - 4^{p-1}}{1 - N_g^{p-1}} \qquad p = 0.5, \qquad (9.4)$$

where p is the Rent's constant for on-chip wire length calculations. For $p > 0.5$ and $N_g \gg 1$,

$$\overline{R} = \frac{14}{9}\frac{1 - 4^{p-1}}{4^{p-0.5} - 1}N_g^{p-0.5};$$

when $p = 2/3$, this becomes

$$\overline{R} = 2.2N_g^{1/6}.$$

For $p < 0.5$ and $N_g \gg 1$,

$$\overline{R} = \frac{14}{9}\frac{1 - 4^{p-1}}{1 - 4^{p-0.5}},$$

and, when $p = 1/5$, this becomes

$$\overline{R} = 3.$$

The average interconnection length in actual units rather than in gate pitch is

$$l_{av} = \overline{R}d_g, \qquad (9.5)$$

where d_g is the logic gate dimension in, for example, microns.

Different cost-performance constraints are involved in the choices made at the chip, chip carrier, and board levels. Accordingly the SUSPENS model defines three separate Rent's constants: one for on-chip interconnection length calculations (p), one for pin count calculations (β), and one for module and/or board level wire length calculations (η).

9.8.2 Packing Density

Packing density is limited by transistors when all the circuit books can be placed right next to each other. Packing density is limited by wiring when the books cannot be placed right next to each other because it was not possible to wire them in that manner.

Transistor Packing Density Limited Chip

If the chip is transistor packing density limited, the gate dimension takes the form of

$$d_{gtrlim} = \sqrt{k_g}F, \qquad (9.6)$$

where F is the minimium feature size and k_g is a proportionality constant between gate area and F determined by transistor packing density and gate layout $(A_g = k_g F^2)$.

Interconnection Capacity Limited Chip

In logic-intensive VLSI chips, area is normally limited by wiring capacity [9.53]. Accordingly the effective logic gate size can be calculated by setting the supply and demand of interconnections equal.

Wiring length available per level per gate area is $e_w(d_g/p_w)d_g$ (see Fig. 9.17). Here p_w is wiring pitch, e_w is wiring efficiency, and d_g is the gate dimension. The efficiency factor accounts for the lines not utilized and the interconnections used for power and clock distribution,

$$\begin{array}{c} \text{interconnection available per gate} = e_w \dfrac{d_g}{p_w} d_g n_w \\ \text{interconnection required per gate} = f_g l_{av} \end{array} \qquad (9.7)$$

where n_w is the number of wiring levels and f_g is the fan-out of a typical gate. By setting the available interconnection length to the required amount and using Eq. 9.5,

$$\frac{e_w d_g{}^2 n_w}{p_w} = f_g \overline{R} d_g.$$

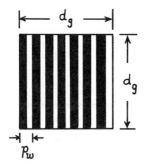

FIGURE 9.17 Calculation of the amount of available interconnections at a given level

As a result,

$$d_{gintlim} = \frac{f_g \overline{R} p_w}{e_w n_w}. \tag{9.8}$$

Logic gate pitch will be the larger of the two d_g in Eqs. 9.6 and 9.8.

9.8.3 Average Gate Delay and Optimization of Transistor Sizes

The circuit model in Fig. 9.18 is used to calculate the average gate delay. The time required for the output to reach 50 percent of its final value will be used as gate delay because, as described in Chapter 5, this is when the gates in the next logic state respond to the change in their inputs,

$$T_{50\%} = f_g R_{gout} C_{int} + f_g R_{gout} C_{gin} + \frac{R_{int} C_{int}}{2} + R_{int} C_{gin}. \tag{9.9}$$

Here, the first term is the delay due to wiring capacitance, the second term is the delay due to input capacitance of the gates at the next stage, the third term is the distributed-RC delay of the wires, and the fourth term is due to the resistance of the wires and the input capacitance of the gates. Individual terms can be calculated using the following equations:

$$\begin{aligned}
C_{int} &= l_{av} C_{int} \\
R_{int} &= l_{av} R_{int} \\
R_{gout} &= R_{tr}/k_o \\
C_{gin} &= k_i C_{tr}
\end{aligned} \tag{9.10}$$

At this point, a specific gate layout must be considered. Buffered domino logic yields a fast CMOS gate [9.54]. Another attractive option is the fully static buffered CMOS circuit illustrated in Fig. 9.19 along with other CMOS gate configurations. The following calculations are developed for domino gates, but they also apply to static logic. The buffered domino gate model of SUSPENS is shown in Fig. 9.20.

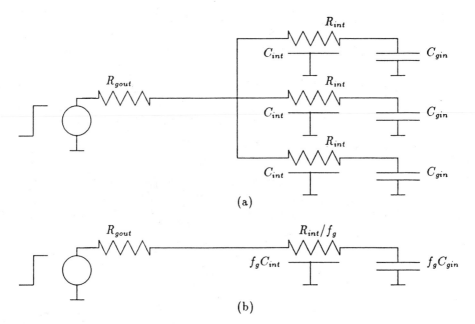

FIGURE 9.18   RC circuit model of an average gate: (a) actual driver with a fan-out of 3, (b) equivalent circuit with fan-out branches lumped together

The sizes of the logic and driver stages will be optimized separately. The factors k_i and k_o are the proportionality constants that represent the sizes of the input (logic) and output (driver) stages, respectively. For example, when k_i is increased by a factor of two, the width-to-length ratios of all transistors in the input stage are doubled. The constants k_l and k_d are the number of minimum-feature-size squares in the gate area with $k_i = 1$ and $k_o = 1$. This yields

$$
\begin{aligned}
A_l &= k_i k_l F^2 \\
A_d &= k_o k_d F^2 \\
A_g &= A_l + A_d
\end{aligned}
\tag{9.11}
$$

where A_l and A_d are the areas of the logic and driver circuitry in Fig. 9.20. An example layout for the domino gate yields $k_i = 2$, $k_l = 135$, $k_o = 11$, and $k_d = 14$.

In the following section, gate delay is calculated and optimized for both the transistor packing density and interconnection capacity limited chips.

Transistor Packing-Density-Limited Chip

The gate pitch for the transistor packing-density-limited chip is

$$
d_{gtrlim} = \sqrt{k_i k_l + k_o k_d} F.
\tag{9.12}
$$

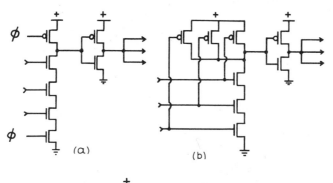

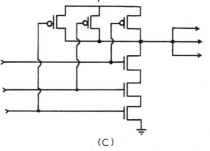

FIGURE 9.19 CMOS gate configurations: (a) domino logic, (b) buffered static CMOS, (c) low-speed static CMOS

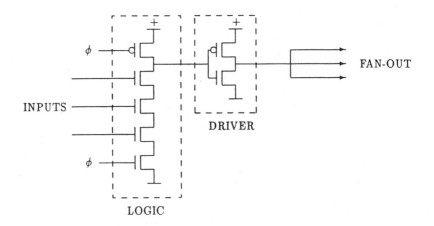

FIGURE 9.20 Domino logic gate

Input stage delay is

$$T_i = f_g \frac{R_{tr}}{k_i} 3k_o C_{tr} \tag{9.13}$$

whose components are accounted for as:

- f_g: there can be f_g nMOS transistors in series
- R_{tr}: on-resistance of a minimum-size nMOS transistor
- k_i: width-to-length (W/L) ratio of input transistors
- 3: one for the input gate capacitance of the nMOS transistor in the output buffer plus two for the gate of the pMOS transistor (1+2=3) for a buffer with balanced pull-up and pull-down delays
- k_o: size of the output buffer
- C_{tr}: gate input capacitance of a minimum-size nMOS transistor.

Output stage delay is

$$T_o = f_g R_{gout} C_{int} + f_g R_{gout} C_{gin} + \frac{R_{int} C_{int}}{2} + R_{int} C_{gin}$$

$$= \frac{f_g R_{tr}}{k_o}(l_{av} C_{int} + k_i C_{tr}) + l_{av} R_{int}\left(\frac{l_{av} C_{int}}{2} + k_i C_{tr}\right). \tag{9.14}$$

Substituting the expression for average interconnection length,

$$l_{av} = \overline{R}d_g = \overline{R}\sqrt{k_i k_l + k_o k_d}F,$$

the output stage delay becomes

$$T_o = \frac{f_g R_{tr}}{k_o}\left(\overline{R}FC_{int}\sqrt{k_i k_l + k_o k_d} + k_i C_{tr}\right)$$

$$+ \overline{R}FR_{int}\sqrt{k_i k_l + k_o k_d}\left(\frac{\overline{R}FC_{int}}{2}\sqrt{k_i k_l + k_o k_d} + k_i C_{tr}\right). \tag{9.15}$$

Total gate delay is then expressed as

$$T_g = T_i + T_o$$

$$= \frac{3f_g k_o R_{tr} C_{tr}}{k_i} + \frac{f_g R_{tr}}{k_o}\left(\overline{R}FC_{int}\sqrt{k_i k_l + k_o k_d} + k_i C_{tr}\right) \tag{9.16}$$

$$+ \overline{R}FR_{int}\sqrt{k_i k_l + k_o k_d}\left(\frac{\overline{R}FC_{int}}{2}\sqrt{k_i k_l + k_o k_d} + k_i C_{tr}\right).$$

The gate parameters can be optimized by obtaining the partial derivatives of gate delay with respect to k_i and k_o and setting them to zero. It is sufficient to choose a reasonable value for k_i (such as $k_i = 3$) and then optimize the more critical parameter k_o.

Interconnection Capacity-Limited Chip

When the chip is interconnection capacity limited, the average wire length is independent of output buffer size k_o,

$$d_{gintlim} = \frac{f_g \overline{R} p_w}{e_w n_w}$$
$$l_{av} = \overline{R} d_g = \frac{f_g \overline{R}^2 p_w}{e_w n_w}. \tag{9.17}$$

The only reason k_o may not be made as large as possible is the delay from the input to output stage. The maximum value of k_o that fills all the area available can be obtained by setting d_{gtrlim} to $d_{gintlim}$:

$$k_{omax} = \frac{1}{k_d} \left[\left(\frac{f_g \overline{R} p_w}{e_w n_w F} \right) - k_i k_l \right].$$

Total gate delay is then expressed as

$$\begin{aligned}
T_g &= T_i + T_o \\
&= \frac{3 f_g k_o R_{tr} C_{tr}}{k_i} + \frac{f_g R_{tr}}{k_o} \left(\frac{f_g \overline{R}^2 p_w}{e_w n_w} C_{int} + k_i C_{tr} \right) \\
&+ \frac{f_g \overline{R}^2 p_w}{e_w n_w} R_{int} \left(\frac{f_g \overline{R}^2 p_w}{e_w n_w} \frac{C_{int}}{2} + k_i C_{tr} \right).
\end{aligned} \tag{9.18}$$

Again, the gate parameters can be optimized by setting the partial derivatives of the gate delay with respect to the parameters to zero.

9.8.4 Chip Properties

In this section, chip parameters, such as chip size, clock frequency, power dissipation, and various performance indicators are derived.

Chip Size

The chip size is calculated from the number of gates and the effective gate area:

$$D_c = d_g \sqrt{N_g}.$$

Clock Frequency

As described in Chapter 8, the minimum cycle time in a flip-flop-based system is given by

$$T_c = t_{latch} + t_{logic} + t_{setup} + t_{skew}.$$

This can also be expressed as

$$T_c = f_{ld}T_g + \mathcal{R}_{int}\mathcal{C}_{int}\frac{D_c{}^2}{2} + \frac{D_c}{v_c}. \tag{9.19}$$

The second equation applies to single-phase latch, flip-flop, or two-phase single- or double-latch machines described in Chapter 8.

Let us examine the terms of this cycle time equation. The first term, $f_{ld}T_g$, takes into account delay through the logic gates. The latch delay t_{latch}, combinational logic delay t_{logic}, and setup time t_{setup} are all included in this term. Here, f_{ld} is the logic depth, and T_g is the average gate delay. The logic depth is defined as the number of logic stages between two clocked latches, including the latch delay and setup time. Average gate delay is used because, for critical paths that involve large fan-outs or long interconnections with large capacitive loads, delay can be reduced by increasing the driver size appropriately.

The second term is the distributed-RC delay of an interconnection that crosses the chip halfway diagonally. This represents the portion of the global wire delay that cannot be reduced by increasing the driver size. On-chip clock skew t_{skew} is usually proportional to this term. It also accounts for the delay of a global signal wire that crosses the chip. In submicron CMOS chips with large die sizes, there should not be more than one (and preferably none) such wire in a latch-to-latch path. Here, the worst-case delay instead of a multiple of the average delay is considered because it is not possible to reduce distributed-RC delays by changing the size of the driver or interconnection. A larger driver has no effect on the RC delay of the wire. When a wire is widened, its resistance goes down; however, its capacitance rises approximately by the same factor (may be slightly reduced because of fringing field effects), and the RC constant remains approximately the same. As a result, distributed-RC delays can be reduced mainly by increasing the vertical dimensions (interconnection and insulator thicknesses), which are technology parameters and cannot be changed at the layout level.

The last term accounts for the speed-of-light limit. Unless the chip is very large or the cycle time is less than 1 nsec (higher than 1 GHz clock frequency), this term should be negligible but it is included for the sake of completeness. The distributed-RC and time-of-flight delays combined approximate the lossy transmission line delay described in Chapter 6.

Logic depth f_{ld} is difficult to determine. Figure 9.21 illustrates the logic depth concept used in determining the maximum clock frequency, and Table 9.9 lists a sampling of chips and systems and their approximate logic depths. As illustrated in Table 9.9, logic depth may vary significantly among different types of chips and systems. CMOS microprocessors have a logic depth of 15 to 30, TTL parts have a logic depth of 15 to 20, ECL bipolar mainframes keep the logic depth 8 to 12, and in supercomputers it can be as low as 6. The logic depth may not be exactly the actual number of gates in the critical path. This is true especially for CMOS chips. Even if the actual gate count in

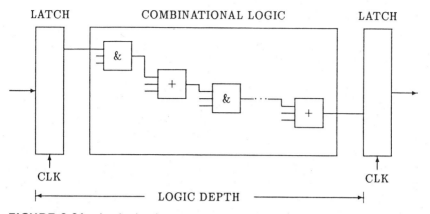

FIGURE 9.21 Logic depth concept

the critical path is not very large, on-chip buffers that drive high-capacitance lines may not be truly optimized because of area limitations. This results in a large "effective" f_{ld}. In other words, a logic depth of 30 does not always mean that there are 30 stages of logic in the critical path; it may consist of a long

TABLE 9.9 Logic Depths of Various Chips and Systems

Chip or System Type	Gate Delay T_g (nsec)	Clock Frequency f_c (MHz)	Logic Depth $f_{ld} \approx 1/T_g f_c$
Microprocessors			
Intel 8008	70	0.5–0.8	18–29
Intel 8080	20	2–3	17–25
Intel 8085	12	3–5	17–28
Intel 8086	8	5–8	16–25
Intel 80386	2.5	12–16	25–33
Motorola 68000	4	8–12.5	20–31
Motorola 68020	2.5	17	24
Fairchild Clipper	2	33 (16.5)	30
Stanford MIPSX	2.5	20	20
Supercomputers			
NEC SX	0.5	167	12
Cray-1	1	80	12.5
Cray-2	0.6	250	6.7
Cray-3		500	6
Cray-4		1000	

or nonoptimized stage with a delay of 20 average gates in addition to a few average stages. On-chip RAM accesses are also included in the logic depth.

Pipelining is one way of minimizing the logic depth and improving the execution rate. To reduce the number of gates in a path, the path can be divided into smaller subsections that operate concurrently, as described in Section 9.3. As the system becomes more and more pipelined, the overhead portion of the delay, such as latch or flip-flop delay and clock skew, becomes more important because it remains constant and only the combinational logic delay is reduced. Pipelining also requires additional hardware and interconnections to control the pipeline dependencies and to forward information when necessary. In order to yield a net gain, the cycle time penalty due to additional complexity introduced by adding another stage of pipeline should be significantly less than the savings. Also to be considered when adding a pipeline stage is the CPI penalty due to dependencies and the discontinuities in the instruction stream. If the average cycles-per-instruction ratio degrades more than the cycle time improvement, one can end up with a net performance loss rather than a gain.

Parallelism is also very effective in reducing the logic depth and increasing the execution rate. This was also described in in section 9.3. Figure 9.10 illustrates how parallelism can reduce the logic depth at the cost of increased hardware. Parallel designs are faster, and they require more hardware than less parallel implementations that execute in a more serial fashion.

The techniques that increase concurrency (pipelining and parallelism) had limited usage in early microprocessor designs because of cost and die area limitations. To conserve area and to utilize the limited hardware efficiently, data buses were shared for many purposes, and register files had limited capabilities. Microprocessors also did not implement very heavily pipelined designs with complicated bypassing schemes, again because of area and interconnection limitations. Performance issues similar to those described for data buses and register files apply to the instruction-decoding units of microprocessors and to floating-point and memory-management units. As a result microprocessors were implemented with much larger logic depths than higher-performance CPUs. All of these are currently changing, and advanced implementation concepts are showing up in microprocessors as advances in silicon technology allow more function to be integrated on the same chip.

Reduced-instruction-set (RISC) microprocessors have smaller logic depths than their complex-instruction-set (CISC) counterparts because of their simpler instruction-decoding and execution units and because of the additional pipelining and parallelism made possible by the freed-up hardware.

Because gate arrays are used for many applications (from inexpensive consumer products to high-end supercomputers), they exhibit a spectrum of logic depths. Supercomputers are finely tuned designs that require a great amount of hardware and heavy pipelining. As a result they have relatively small logic depths.

Power Calculations

External capacitance per logic gate is expressed as

$$C_{ext} = f_g l_{av} C_{int} + f_g C_{gin}$$
$$= f_g \overline{R} d_g C_{int} + f_g k_i C_{tr}, \tag{9.20}$$

and internal capacitance per logic gate is

$$C_{inter} = 3 k_o C_{tr} + 5 C_{tr} \tag{9.21}$$

Here, $3k_o$ accounts for the output stage with size k_o, and the factor 3 is the result of the balanced output buffer with symmetric rising and falling delays (1 for nMOS + 2 for pMOS = 3). The last term (factor 5) represents the gate capacitance of clocked transistors and parasitic drain capacitances. The total capacitance per logic gate is

$$C_g = f_g \overline{R} d_g C_{int} + (f_g k_i + 3 k_o + 5) C_{tr}. \tag{9.22}$$

The time average of dynamic power dissipation per gate takes the form of

$$P_g = \frac{1}{2} f_c f_d C_g V_{DD}^2, \tag{9.23}$$

where f_d is the portion of the gates that switches during a clock period (duty factor). The power dissipation of the chip is

$$P_c = N_g P_g, \tag{9.24}$$

and power-dissipation density is

$$Q_c = \frac{P_g}{d_g^2}. \tag{9.25}$$

The relation between the power dissipation and clock frequency is derived as follows. The total charge and energy stored in the capacitor shown in Fig. 9.22 is given by the following equations:

$$Q = C V_{DD}$$
$$E = \frac{1}{2} C V_{DD}^2. \tag{9.26}$$

It is important to note that when a capacitor is being charged, an energy exactly equal to the energy being stored in the capacitor is dissipated at the resistive element. The half that is stored at the capacitor is later dissipated when the capacitance is discharged, again at the resistive element. As a result, exactly half the energy is dissipated when the capacitor is charged, and exactly half is dissipated when the capacitance is discharged. Two logic state changes are required to dissipate a total energy of CV_{DD}^2. If the node switched every

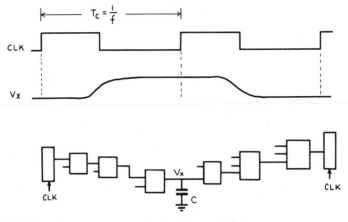

FIGURE 9.22   Dynamic power-dissipation calculations

cycle $(1010 \ldots 1010)$, the power dissipation would be

$$\Delta t = \frac{1}{f_c}$$

$$P = \frac{E}{\Delta t} = \frac{1}{2} f_c C V_{DD}^2 . \tag{9.27}$$

Since the node does not switch every cycle, a duty factor can be used to adjust the power dissipation:

$$P = \frac{1}{2} f_c f_d C V_{DD}{}^2 . \tag{9.28}$$

Duty factor is hard to estimate but 25 to 30 percent is a good approximation. In power calculations, one needs to take into account *dynamic* rather than *static* duty factors. In other words, a node "logically" switches only once during a cycle. Electrically, however, a node may switch multiple times before it settles in a machine cycle depending on the arrival times of the inputs. Every time the node switches, energy will be dissipated.

Performance Indicators

Figure-of-merit factors and performance metrics can be defined to measure and compare the functionality of chips. The *functional throughput rate* of a chip FTR_c is a measure of the computation power per unit area of silicon,

$$FTR_c = \frac{f_c N_g}{A_c}$$

$$= \frac{f_c}{d_g{}^2} , \tag{9.29}$$

and *computational capacity* CC_c is a measure of the computation power of the entire chip:

$$CC_c = f_c N_g. \tag{9.30}$$

The *power efficiency* of a chip PE_c is the reciprocal of the system-level version of the power-delay product:

$$PE_c = \frac{CC_c}{P_c}$$
$$= \frac{1}{P_g T_c}. \tag{9.31}$$

9.9   Package-Level Model

In this section, the calculations are extended to the package level.

9.9.1   Number of Pins per Chip

Estimated number of pins per chip [9.26] is determined by Rent's rule:

$$N_p = K_p N_g{}^\beta. \tag{9.32}$$

Here, β is Rent's constant for the number of pins, and K_p is a multiplicative constant. The Rent's constant for pins is differentiated from the Rent's constant for average on-chip wire length calculations. Depending on the way the system is partitioned into chips and the use of bidirectional, multiplexed, or serial data transfer between chips, Rent's constant for pin count β may differ from Rent's constant for on-chip interconnection lengths earlier referred to as p. Figure 9.11 and Table 9.5 present pin-count data obtained from a wide variety of products.

9.9.2   Average Interconnection Length at the Module Level

The approach applied to calculate the average interconnection length at the chip level can also be used at the module level. The number of gates is replaced by the number of chips, and gate pitch is replaced by chip footprint size. Rent's constant for package-level interconnections η may be larger than p because it is easier to lay out a chip than a board to minimize wire lengths and the granularity of circuit building blocks on the chips is smaller than the chip carriers. The average inteconnection length at the module level in units of

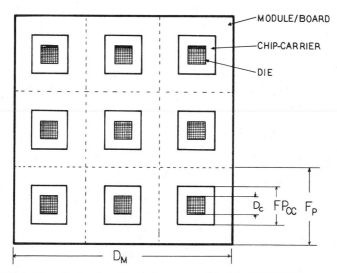

FIGURE 9.23   Packaging hierarchy as modeled in SUSPENS

chip pitches is

$$\overline{R}_M = \frac{2}{9}\left(7\frac{N_c^{\eta-0.5}-1}{4^{\eta-0.5}-1} - \frac{1-N_c^{\eta-0.75}}{1-4^{\eta-0.75}}\right)\frac{1-4^{\eta-1}}{1-N_c^{\eta-1}}. \qquad (9.33)$$

The average interconnection length in real units is

$$L_{AV} = \overline{R}_M F_P, \qquad (9.34)$$

where F_P is the footprint size.

9.9.3   Chip Footprint Size

If there is only a single layer of chip carriers on the module or board, footprint size cannot be smaller than chip carrier size. It is more likely that the footprint will be limited by the interconnection capacity of the board and will be larger than chip carrier size. One possible packaging hierarchy is illustrated in Fig. 9.23. Chip carriers can be eliminated by flip-mounting or tab bonding the dice directly on the module or the board.

The number of interconnections can be obtained by assuming a fan-out of F_c for the chip pins:

$$\text{number of interconnections} = \frac{F_c}{F_c+1}N_cN_p.$$

Based on Eq. 9.34,

$$\text{total interconnection length} = \frac{F_c}{F_c + 1} N_c N_p \overline{R}_M F_P,$$

and

$$\text{total interconnection available} = N_c \frac{F_P{}^2}{P_W} N_W E_W.$$

By setting the supply and demand of interconnections equal, wiring-capacity-limited F_P is calculated as

$$F_P = \frac{F_c}{F_c + 1} \frac{\overline{R}_M N_p P_W}{E_W N_W}. \tag{9.35}$$

Footprint size also can be limited by the pad contact pitch P_P or a chip carrier-related minimium value FP_{CC},

$$F_P = max \left\{ D_c, \quad \frac{F_c}{F_c + 1} \frac{\overline{R}_M N_p P_W}{E_W N_W}, \quad \sqrt{N_p} P_P, \quad FP_{CC} \right\}. \tag{9.36}$$

The terms in the braces represent die size-limited, module/board wiring capacity-limited, pad pitch-limited, and chip carrier-size-limited cases, respectively. In an optimally designed system, the last two should not play a major role, and the first two must be approximately equal (footprint should be slightly larger than die size to provide some spacing between neighboring dice for engineering changes). In most of the current midrange and low-end systems, however, footprint size is determined by the last two terms. High-end mainframe computers are better optimized, and the first two terms become dominant.

9.9.4   Chip-to-Chip Transmission Delay

Chip-to-chip delay in a high-end CMOS system is derived from the circuit model in Fig. 9.24. The output pad is driven by a cascaded driver, and the on-resistance of the final stage is equal to the characteristic impedance of the transmission line Z_0. The ratio of two consecutive drivers is 5, which produces a near minimal delay with small area. The optimal ratio is e, the base of a natural logarithm; however, total delay increases slowly as a function of this ratio. Consequently, with a ratio larger than e, much area can be saved with little increase in delay [9.57]–[9.60].

The number of stages N can be calculated by setting the on-resistance of the output stage to Z_0 and pessimistically assuming that the first inverter of

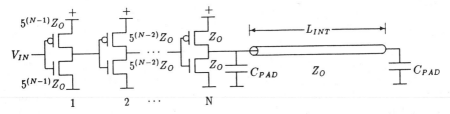

FIGURE 9.24 Circuit model for chip-to-chip delay

the cascaded chain is a minimum-sized device:

$$R_{tr} = 5^{(N-1)} Z_0$$

$$N = 1 + \frac{\ln\left(\frac{R_{tr}}{Z_0}\right)}{\ln(5)}. \tag{9.37}$$

If the effective on-resistance of a minimum-sized pMOS transistor is twice that of an nMOS transistor, total delay in the inverter chain is expressed as

$$T_{chain} = 15(N - 1)R_{tr}C_{tr},$$

where R_{tr} and C_{tr} are the on-resistance and input capacitance of a minimum-sized nMOS transistor. The factor 15 is the result of the ratio of consecutive inverters ($=5$) multiplied by the sum of the W/L ratios of nMOS and pMOS transistors ($1 + 2 = 3$). Total transmission delay then becomes:

$$T_M = 15(N - 1)R_{tr}C_{tr} + 2Z_0 C_{PAD} + R_{INT}C_{INT}L_{INT}^2 + \frac{L_{INT}}{v_M}. \tag{9.38}$$

The first term is the delay of the inverter chain, the second is the delay caused by capacitive loading of the I/O pins and contact pads, the third is the distributed-RC delay of module or board interconnections, and the last is the transmission line time-of-flight delay. The time-of-flight delay assumes a point-to-point net with first incident switching at the receiver. The delay equation can be modified for lower-performance systems where the driver is not as strong and the receivers do not switch at the first incident of the waveform. Then multiple reflections may be required, and the line may act as a lumped capacitor.

In current CMOS systems, the delay of the inverter chain, capacitive loading of the pins, and the board wiring are the dominant delay factors. In a future more-optimized design using submicron device technology and some form of a direct-chip-attach packaging, parasitic pad capacitances and module-level distributed-RC delays will be minimal, and the delay of the inverter chain and time-of-flight delay will be the dominant delay terms. In thin film packages, distributed-RC delays of the module wires can be significant.

9.9.5 Module Clock Frequency

To achieve the highest possible speed, the "logic depth" of off-chip communication must be equal to 1. In a given path, there must be only one chip-to-chip information transfer during a clock cycle. In addition, any path that contains a chip-to-chip connection should not have too much on-chip logic. To some extent, chip-to-chip communication must be pipelined with on-chip information processing to optimize the performance.

With 2 μm CMOS output buffers, the delay of the inverter chain can be as long as 15 nsec. In 1 μm CMOS, the off-chip driver delay goes down to 5–10 nsec. When chip-to-chip delay becomes comparable or greater than the cycle time, more than one cycle can be taken to transmit information from one chip to another. This can be acceptable to access main memory chips and maybe even to cache chips, but it should be avoided between the CPU chips if at all possible. The introduction of many stages of pipelining will hurt the overall performance because performance is degraded due to dependencies and when the pipeline is broken. Additional cycles should be avoided in critical operations such as during a conditional branch instruction or in a cache access that fetches an instruction or executes a load that will bring in data needed by a subsequent operation. All these will degrade the CPI of the machine. Taking these into consideration, worst-case chip-to-chip information transfer is assumed to take exactly one cycle in the SUSPENS model, and the worst-case chip-to-chip delay path is assumed to be an output buffer and an interconnection that extends across the module,

$$L_{INTMAX} = D_M$$
$$D_M = \sqrt{N_c} F_P. \tag{9.39}$$

The module clock frequency f_M is determined by the worst-case chip-to-chip delay determined by combining Eqs. 9.38 and 9.39:

$$f_M = \left. \frac{1}{T_M} \right|_{L_{INT}=L_{INTMAX}}. \tag{9.40}$$

It should be noted that these module clock frequency calculations differ from the chip calculations by taking a path of *maximum* length rather than a multiple of an *average* path delay because only one chip-to-chip information transfer occurs in a given clock period but several gates exist in an on-chip path. The difference in granularity requires a maximum delay calculation for chip-to-chip paths rather than a multiple of an average delay used for on-chip paths.

Delay equations similar to the ones in the SUSPENS model are used by CPU design engineers. For example, as reported in reference [9.62], in the initial stages of designing a software-compatible multichip version of Intel 80386/80486, Nexgen engineers started with a thorough analysis of CMOS processing technology and typical processor architectures to obtain an accurate estimate of the machine cycle time and the expected system performance. Nexgen engineers used two separate equations to calculate the on-chip and chip-to-chip machine cycle times. Their on-chip delay equation is

$$T_{mc1} = N_g(T_{pd} + L_c T_{ic}) + (T_{su} + T_{cko}) + T_{ck-skew} \tag{9.41}$$

where T_{mc1} is the on-chip machine cycle time, N_g is the number of logic levels on the chip, T_{pd} is the fixed propagation delay per internal gate level, L_c is the average on-chip interconnection length, and T_{ic} is the on-chip gate delay adder per unit length of interconnection, T_{su} is the latch set-up time, T_{cko} is the latch clock-to-output delay, and $T_{ck-skew}$ is the on-chip clock skew. The chip-to-chip delay equation is

$$T_{mc2} = (T_{obf} + L_b T_{bic}) + T_{ibf} + T_{bck-skew} + (T_{su} + T_{cko}) + N_{g2}(T_{pd} + L_c T_{ic}) \tag{9.42}$$

where T_{mc2} is the chip-to-chip machine cycle time, T_{obf} is the output buffer delay, L_b is the board interconnection length, T_{bic} is the interchip delay adder per unit length of board wire, T_{ibf} is the input buffer delay, $T_{bck-skew}$ is board-level chip-to-chip clock skew including on-chip clock skew and N_{g2} is the number of on-chip logic levels in the chip-to-chip path. Again as reported in reference [9.62], for a 1 μm CMOS technology, by assuming $N_g = 10$, $T_{pd} = 1$ nsec, $L_c = 1000$ μm, $T_{ic} = 0.7$ psec/μm, $T_{su} = 2$ nsec, $T_{cko} = 2$ nsec, and $T_{ck-skew} = 1$ ns, the on-chip cycle time is obtained as 22 nsec. Assuming no logic in the chip-to-chip path ($N_{g2} = 0$), and assuming $T_{obf} = 2.5$ nsec, $L_b = 25$ cm, $T_{bic} = 100$ psec/cm, $T_{ibf} = 2$ nsec, and $T_{bck-skew} = 2$ nsec, the minimum chip-to-chip machine cycle time is obtained as 13 nsec. If there are two chip crossings in a path, this yields a cycle time of approximately 26 nsec [9.62].

9.9.6   Module Power Dissipation

Termination at the driving end assumed in the above I/O buffer calculations is the most efficient transmission line driving scheme, and its power dissipation is expressed as

$$P_M = \frac{1}{2}F_D f_S C_M V_{DD}^2, \tag{9.43}$$

where F_D is the percentage of the lines that switches, f_S is the system clock frequency, and C_M is the total capacitance related to chip-to-chip lines, including both the interconnection and gate capacitances of the output buffers and the pad capacitances. Total capacitance related to a single line is

$$\begin{aligned}
C &= 3(1 + 5 + 5^2 + \cdots + 5^{N-1})C_{tr} + 2C_{PAD} + L_{INT}C_{INT} \\
&= 3\frac{1 - 5^N}{1 - 5}C_{tr} + 2C_{PAD} + L_{INT}C_{INT}.
\end{aligned} \tag{9.44}$$

To determine the total module capacitance, L_{INT} in Eq. 9.44 is replaced by

L_{AV},

$$L_{AV} = \overline{R}_M F_P,$$

and C is multiplied by

$$\text{number of interconnections} = \frac{F_c}{F_c + 1} N_c N_p,$$

which results in

$$C_M = \frac{F_c}{F_c + 1} N_c N_p \left(3 \frac{1 - 5^N}{1 - 5} C_{tr} + 2C_{PAD} + \overline{R}_M F_P C_{INT} \right).$$

The power dissipation related to the package is then given by Eq. 9.43.

9.10   System Model

The clock frequency of the system is determined by the lower of the chip and module clock frequencies,

$$f_S = min \ \{ \ f_c, \ f_M \ \}. \tag{9.45}$$

Total power dissipation is the sum of chip- and module-level power dissipation in Eqs. 9.23 and 9.43, respectively. Both power equations must be evaluated at the system clock frequency in Eq. 9.45,

$$P_S = P_g N_{gTOT} + P_M.$$

The overall functional throughput rate, which indicates the computational capacity per unit area, is

$$FTP_S = \frac{f_S N_{gTOT}}{D_M{}^2}$$

$$= \frac{f_S N_g}{F_P{}^2}, \tag{9.46}$$

and the total computational capacity and power efficiency of the system are

$$CC_S = f_S N_{gTOT}$$

$$PE_S = \frac{CC_S}{P_S}. \tag{9.47}$$

9.11 Comparison of the Model to Actual Microprocessors

In this section, microprocessors are analyzed using the SUSPENS model. This section has two purposes: (1) to illustrate the application of the SUSPENS model with a step-by-step examples and (2) to analyze an important class of electronic components—microprocessors—in order to predict some of the trends.

Microprocessors are densely packed low-cost single-chip CPUs. Their die sizes are limited by yield considerations, and their power dissipations and pin counts are constrained by packaging costs. As a result, some of the above calculations must be modified before they are applied to microprocessors because microprocessors are not optimized for ultimate speed with no cost constraints. To represent the circuit design philosophy of microprocessors accurately, the gates in Fig. 9.25 are used to determine the gate delay instead of the buffered gates in Fig. 9.20. The calculations for microprocessors are derived for both nMOS and CMOS circuits because nearly all early microprocessors were designed with nMOS technology, and most current and next-generation chips are in CMOS.

The average interconnection length in units of gate pitch is first obtained as a function of gate count and Rent's constant (Eq. 9.3):

$$\overline{R} = f(N_g, p).$$

Assuming an interconnection-limited chip, gate and die dimensions are then

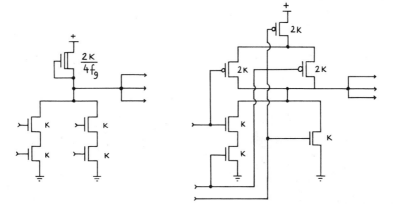

FIGURE 9.25 nMOS (left) and CMOS (right) gate circuits in microprocessors. The factors next to the transistors indicate their width-to-length (W/L) ratios—inversely proportional to their on-resistances.

derived by setting the supply and demand for interconnections equal (Eq. 9.8):

$$d_g = \frac{f_g \overline{R} p_w}{e_w n_w}$$

$$D_c = \sqrt{N_g} d_g,$$

which yields the average interconnection length (Eq. 9.5),

$$l_{av} = \overline{R} d_g.$$

In an nMOS gate, the average number of pull-down transistors is f_g, which is the average fan-in/fan-out count. In nMOS there is one depletion mode pull-up per gate. In a CMOS gate, there is a pair of nMOS/pMOS transistors per fan-in. As a result, the relationship between the number of transistors and the number of gates is

$$\begin{aligned} N_{tr} &= (f_g + 1)N_g \\ N_{tr} &= 2f_g N_g \end{aligned} \tag{9.48}$$

for the static nMOS and CMOS chips, respectively.

The 50 percent gate delay in nMOS and CMOS gates is defined as

$$T_{50\%} = f_g R_{gout} l_{av} C_{int} + f_g R_{gout} C_{gin} + \frac{R_{int} C_{int}}{2} + R_{int} C_{gin}.$$

Gate output resistance R_{gout} and input capacitance C_{gin} are different in nMOS and CMOS gates. In a static nMOS gate,

$$R_{gout} = \frac{1}{2}\left(4\frac{f_g}{2}\frac{R_{dep}}{k} + \frac{f_g}{2}\frac{R_{enh}}{k}\right)$$

$$C_{gin} = kC_{tr}. \tag{9.49}$$

The factor of 4 is the ratio of the pull-down to pull-up W/L ratios in E/D nMOS gates, which ensures that the "0" value of the gate output is low enough to satisfy the noise margin requirements, and the factor $f_g/2$ accounts for an average of $f_g/2$ nMOS pull-down transistors in series. The first term in the parentheses is the pull-up resistance, and the second term is the pull-down resistance. The pull-up and pull-down delays are averaged.

In static CMOS gates,

$$R_{gout} = \frac{2f_g R_{trNMOS}}{2k}$$

$$C_{gin} = 3kC_{tr}. \tag{9.50}$$

The factor of 2 in the numerator of the R_{gout} equation accounts for pMOS W/L ratio twice that of the nMOS in order to balance the pull-up and pull-down delays. The factor $f_g/2$ accounts for an average of $f_g/2$ pull-up transistors in series. The factor of 3 in C_{gin} also accounts for pMOS W/L ratio being twice that of the nMOS ($3 = 2 + 1$).

The clock frequency is determined using Eq. 9.19,

$$f_c = \left(f_{ld}T_g + R_{int}C_{int}\frac{D_c^2}{2} + \frac{D_c}{v_c} \right)^{-1}.$$

Assuming that half of the pull-up transistors on the chip would be conducting at a given time, power in an nMOS chip is

$$P_c = \frac{N_g}{2}V_{DD}\frac{V_{DD}}{4\frac{f_g}{2}\frac{R_{dep}}{k}} + \frac{1}{3}\frac{1}{2}N_p f_c f_d C_{OUT}V_{DD}^2,$$

where the second term accounts for the dynamic power consumption at the I/O buffers, and C_{OUT} is the total capacitance at an output pin.

The power consumption of a CMOS microprocessor is determined by the dynamic power dissipation at the total on- and off-chip capacitances associated with the microprocessor. Total capacitance of the chip is

$$C_{TOT} = \frac{D_c^2 n_w e_w C_{int}}{p_w} + 3kC_{tr}N_g f_g,$$

and CMOS power dissipation is

$$P_c = \frac{1}{2}f_c f_d C_{TOT}V_{DD}^2 + \frac{1}{3}\frac{1}{2}N_p f_c f_d C_{OUT}V_{DD}^2.$$

Tables 9.10 and 9.11 present step-by-step calculations for nMOS and CMOS microprocessors.

Figures 9.26, 9.27, and 9.28 plot the model predictions for chip size, clock frequency, and power dissipation as a function of gate count. The solid lines are the results obtained from MOS technology and microprocessor parameters in Tables 9.6 and 9.12. The data points represent a variety of commercial and research microprocessors. The dotted line is the trend curve across the technologies as minimum feature size is scaled down.

For a given gate count on the x-axis, the year in which a state-of-the-art microprocessor with that many gates was first announced is indicated at the top. Because the "number of gates" axis is scaled logarithmically, the equal spacings between the five-year periods on the time axis are a restatement of Moore's well-known law stating that the number of devices on a specific class of chips (microprocessor, DRAM, SRAM) increases exponentially with time [9.61].

The dotted lines plot chip size, clock frequency, and power consumption for the corresponding gate count and year. These trend lines do not follow any of the solid lines because a trend line encompasses all the technologies from 7 μm nMOS to 0.7 μm CMOS as minimum feature size and technology changes with time. Consequently a trend line is made up of pieces of different solid lines. For example, early in time, the 7 μm nMOS determines the trend line, later 3 μm nMOS takes over, and so forth.

TABLE 9.10   SUSPENS Calculations for nMOS Microprocessors

Step	Calculation
1	$\overline{R} = \dfrac{2}{9}\left(7\dfrac{N_g{}^{p-0.5}-1}{4^{p-0.5}-1} - \dfrac{1-N_g{}^{p-1.5}}{1-4^{p-1.5}}\right)\dfrac{1-4^{p-1}}{1-N_g{}^{p-1}}$
2	$d_g = \dfrac{f_g\overline{R}p_w}{e_w n_w}$
3	$D_c = \sqrt{N_g}d_g$
4	$l_{av} = \overline{R}d_g$
5	$R_{gout} = \dfrac{1}{2}\left(4\dfrac{f_g}{2}\dfrac{R_{dep}}{k} + \dfrac{f_g}{2}\dfrac{R_{enh}}{k}\right)$
6	$C_{gin} = kC_{tr}$
7	$T_g = f_g R_{gout}l_{av}C_{int} + f_g R_{gout}C_{gin} + R_{int}C_{int}\dfrac{l_{av}^2}{2} + R_{int}l_{av}C_{gin}$
8	$f_c = \left(f_{ld}T_g + R_{int}C_{int}\dfrac{D_c^2}{2} + \dfrac{D_c}{v_c}\right)^{-1}$
9	$N_p = K_p N_g^{\beta}$
10	$P_c = \dfrac{N_g}{2}V_{DD}\dfrac{V_{DD}}{4\frac{f_g}{2}\frac{R_{dep}}{k}} + \dfrac{1}{3}\dfrac{1}{2}N_p f_c f_d C_{OUT}V_{DD}^2$

The data points for chip size in Fig. 9.26 agree closely with the model predictions (solid lines). For a given gate count, die size decreases with advancing technology because of reduced minimum feature dimension and more levels of interconnections.

The data and model predictions for clock frequency in Fig. 9.27 are in good agreement. For a given gate count, clock frequency improves with advancing technology (from one solid line to the next) because of faster transistors, reduced gate input capacitances, and shorter interconnection lengths resulting from smaller die sizes. For a given technology (represented by a single solid line in this plot), clock frequency degrades with increasing gate count because of longer wires. The dotted trend line is extrapolated from the curves for chip size in Fig. 9.26. The gate count and technology for a specific year are first obtained from the trend curve in Fig. 9.26 (for example, for the year 1995, the gate count is 6 million, technology is 0.7 μm CMOS, and die size is 1.5 × 1.5 cm). The same gate count and technology is then used to extrapolate the trend curve for clock frequency by forcing the dotted line to intersect that technology at the given gate count in Fig. 9.27 (in this example, $f_c = 60$ MHz).

The data points and model predictions for power dissipation in Fig. 9.28

TABLE 9.11 SUSPENS Calculations for CMOS Microprocessors

Step	Calculation
1	$\overline{R} = \dfrac{2}{9}\left(7\dfrac{N_g{}^{p-0.5}-1}{4^{p-0.5}-1} - \dfrac{1-N_g{}^{p-1.5}}{1-4^{p-1.5}}\right)\dfrac{1-4^{p-1}}{1-N_g{}^{p-1}}$
2	$d_g = \dfrac{f_g\overline{R}p_w}{e_w n_w}$
3	$D_c = \sqrt{N_g}\,d_g$
4	$l_{av} = \overline{R}d_g$
5	$R_{gout} = f_g\dfrac{R_{tr}}{k}$
6	$C_{gin} = 3kC_{tr}$
7	$T_g = f_g R_{gout} l_{av} C_{int} + f_g R_{gout} C_{gin} + R_{int} C_{int}\dfrac{l_{av}^2}{2} + R_{int} l_{av} C_{gin}$
8	$f_c = \left(f_{ld}T_g + R_{int}C_{int}\dfrac{D_c^2}{2} + \dfrac{D_c}{v_c}\right)^{-1}$
9	$C_{TOT} = \dfrac{D_c^2 n_w e_w C_{int}}{p_w} + 3C_{tr}kN_g f_g$
10	$N_p = K_p N_g^{\beta}$
11	$P_c = \dfrac{1}{2}f_c f_d C_{TOT} V_{DD}^2 + \dfrac{1}{3}\dfrac{1}{2}N_p f_c f_d C_{OUT} V_{DD}^2$

follow each other closely; however, the agreement is not as good as chip size and clock frequency. This is not the result of any fundamental flaws in the model; it reflects the fact that the designers were forced to invent "clever circuit tricks" to "violate" the assumptions of the SUSPENS model so as to constrain power dissipation. As can be seen, early nMOS data points at the left-hand side of Fig. 9.28 follow the SUSPENS predictions; however, the

TABLE 9.12 Architecture- and Implementation-Dependent Parameters

System or Chip Type	Logic Depth f_{ld}	On-chip Wire Length Rent's Constant p	Pin Count Rent's Constant β	Pin Count Multiplicative Constant K_p	Module Wire Length Rent's Constant η
Microprocessor	15-30	0.4	0.45	0.82	—
Gate array	15-40	0.5	0.50	1.9	—
High-speed computer	8-12	0.6	0.63	1.4	0.65

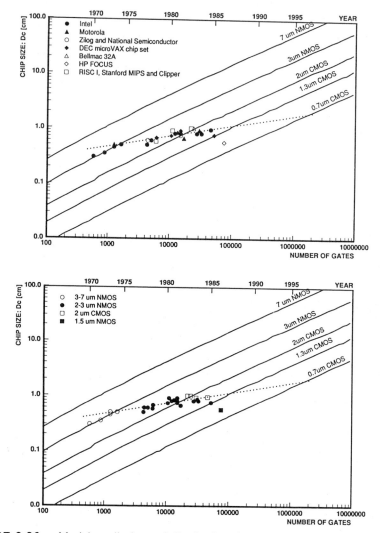

FIGURE 9.26 Model predictions of die size in microprocessors. Solid lines are the results of the model, and the dotted line is the trend curve.

power consumption of the *n*MOS designs with more gates is suppressed by clocking techniques that turn off the sections of the chip that are not active or other similar power-saving methods. The zigzag pattern of the trend curve indicates the power savings achieved by changing the technology from *n*MOS to CMOS between 1980 and 1985. The data points for CMOS microprocessors follow the model predictions more closely than do the *n*MOS points. The trend curve for power dissipation is extrapolated from Fig. 9.26 in a way similar to the extrapolation of the clock frequency curve.

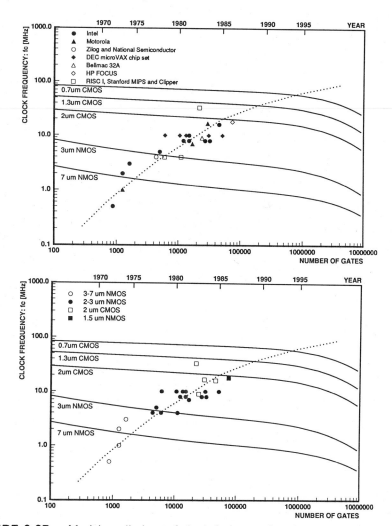

FIGURE 9.27   Model predictions of clock frequency in microprocessors. Solid lines are the results of the model for different technologies. The dotted line is the trend curve, which goes across technologies with time.

The results of the predictions for a state-of-the-art commercial microprocessor that will be available in large quantities in 1995 are summarized in Table. 9.13. These predictions were made in 1985. The instruction execution rate is derived from clock frequency based on the assumption that 0.5 to 1.0 instruction will be executed during each clock period as a result of improved RISC techniques. With the announcements of Intel 80486 (10.5 mm × 15.7 mm, 25–33 MHz, 1.2 million transistors, 15–19 MIPS), Intel 80860 (10 mm

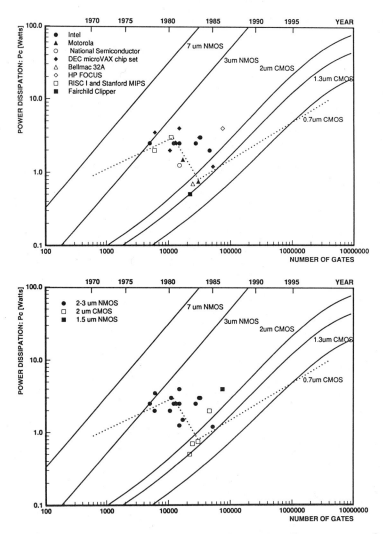

FIGURE 9.28 Model predictions of power dissipation in microprocessors. Solid lines are the results of the model, and the dotted line is the trend curve.

× 13 mm, 33 MHz, 1 million transistors), Motorola 68040 (12.7 mm × 12.7 mm, 1.2 million transistors) four years after the initial SUSPENS predictions, the die size, clock frequency, and power dissipation curves seem to be fairly accurate. The clock frequency curves predict 35 MHz microprocessors in 1989-90, and a number of vendors already announced 33 MHz microprocessors to be available in 1989-90. Die size predictions are for 1990 is 12 mm, and there are microprocessors announced with die sizes of 10–13 mm. Power dissipation of 2 W also seems to be on target.

TABLE 9.13 Commercial Microprocessors in 1995

(Generated in 1985 using the SUSPENS model)

Year	1995
Technology	CMOS
Minimum feature size	0.7 μm (drawn)
	0.5 μm (L-effective)
Number of metal layers	4
Interconnection pitch	2 μm
Number of transistors	6,000,000
Chip size	1.5 × 1.5 cm
Power dissipation	4 W
Clock frequency	60 MHz
Instruction execution rate	30–60 MIPS
Number of pins	410

Table 9.14 compares step by step the SUSPENS predictions with the design data from Intel 80386 [9.20]–[9.21] and Fairchild Clipper [9.19] microprocessors. Chip size, clock frequency,[3] power dissipation, and pin count predictions are very close to the actual values. The accuracy can be improved by taking the data from the designs performed by a team or division and extrapolating to predict the nature of their future designs based on this data. One can investigate the benefits of new technologies, as well as the direction of the current design methology. This is one way the SUSPENS model can be used by a specific company or division.

Figures 9.29 and 9.30 plot die size and clock frequency as a function of gate count for microprocessors in 1.3 μm and 0.7 μm CMOS technologies. The number of metal layers is varied from one to four, and one layer of gate material (polysilicon or silicide) is assumed. The impact of multilevels of interconnections on the chip parameters is significant. Despite their smaller feature sizes, 0.7 μm CMOS chips with one level of metal are larger than 1.3 μm chips with four levels of metal. Because of the larger distributed-RC delays of narrower lines, the difference between the clock frequencies is more dramatic. At integration densities above a few 100,000 gates, 1.3 μm CMOS microprocessors with four layers of metal are faster than 0.7 μm chips with one or two layers. The knee point in the clock frequency plot is a result of the capacitive and resistive delays associated with long interconnections on large dice. The

---

[3]Clipper receives a clock signal with a 33 MHz frequency and divides it by two. As a result, its operation is more like a 16.5 MHz microprocessor.

TABLE 9.14 SUSPENS Calculations for two Commercial Microprocessors

$F = 2 \ \mu m$, $V_{DD} = 5 \ V$, $t_{gox} = 400 \ A$,
$R_{NMOS} = 10 \ k\Omega$, $R_{PMOS} = 20 \ k\Omega$, $C_{tr} = 4 \ fF$,
$C_{int} = 2 \ pF/cm$, $p_w = 6 \ \mu m$, $n_w = 1 + 2$,
$e_w = 0.4$, $p = 0.4$, $\beta = 0.45$, $K_p = 0.82$,
$f_g = 3$, $f_d = 0.3$, $k = 9$, $C_{OUT} = 50 \ pF$

Parameter	Intel 80386 SUSPENS Prediction	Actual Design	Percent Difference
N_{tr}		180,000	
N_g	30,000		
$\overline{R}$	4.2		
$d_g \ (\mu m)$	63		
$D_c \ (cm)$	1.1	0.96	13%
$R_{gout} \ (\Omega)$	5000		
$C_{gin} \ (fF)$	108		
$C_{int} \ (fF)$	53		
$T_g \ (nsec)$	2.4		
f_{ld}	25		
$f_c \ (MHz)$	16.6	12–16	4%
$\sum C_{tr} \ (pF)$	9,720		
$\sum C_{int} \ (pF)$	4,800		
$C_c \ (pF)$	14,520		
N_P	85–106	120	12%
$P_c \ (W)$	1	1–2	30%

Parameter	Fairchild Clipper SUSPENS Prediction	Actual Design	Percent Difference
N_{tr}		132,000	
N_g	22,000		
$\overline{R}$	4.1		
$d_g \ (\mu m)$	62		
$D_c \ (cm)$	0.92	1.0	8%
$R_{gout} \ (\Omega)$	5000		
$C_{gin} \ (fF)$	108		
$C_{int} \ (fF)$	52		
$T_g \ (nsec)$	2.4	0.5–3.0	25%
f_{ld}	25		
$f_c \ (MHz)$	16.6	16.5	1%
$\sum C_{tr} \ (pF)$	7,128		
$\sum C_{int} \ (pF)$	3,358		
$C_c \ (pF)$	10,486		
N_P	73–93	132	29%
$P_c \ (W)$	0.77	0.5	35%

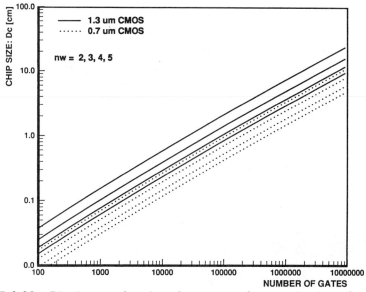

FIGURE 9.29   Die size as a function of gate count for several values of n_w

SUSPENS model can be used to evaluate the impact of various technology parameters on the chip and system variables in a similar manner.

9.12   Comparison of CMOS and Bipolar Technologies

The performances of 1 μm CMOS, silicon bipolar (Si BJT), and GaAs heterojunction bipolar transistor (GaAs HBT)-based VLSI chips are compared in Fig. 9.31. These three circuits use the Rent's constants from the high-speed computer line of Table 9.12 and assume the 1 μm CMOS device parameters in Table 9.6 and silicon and GaAs bipolar parameters in Table 9.15. The device cross-sections of the assumed silicon and GaAs bipolar transistors are illustrated in Figs. 9.32 and 9.33. The silicon BJT is oxide isolated and utilizes self-aligned polysilicon base and emitter contacts. The GaAs device has proton isolation and consists of several layers deposited by molecular beam epitaxy. Its base is 5 percent graded GaAlAs to enhance carrier transport with a built-in electric field. Both devices are assumed to have 1000 A base width. All chips have the same architectures (logic depth and Rent's constants) and number of interconnection levels; only the circuit and device technologies are different. The CMOS chip follows the calculation steps in

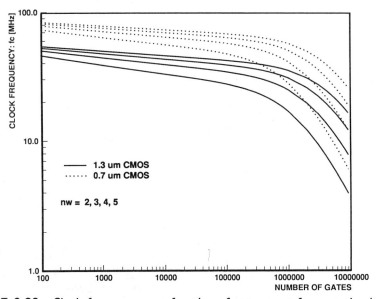

FIGURE 9.30   Clock frequency as a function of gate count for several values of n_w

Table 9.11 but with high-speed computer instead of microprocessor parameters. Silicon bipolar and GaAs heterojunction bipolar chips assume a current-mode logic (CML) gate illustrated in Fig. 9.34. The calculations for these CML cir-

TABLE 9.15   Silicon and GaAs Bipolar
Technology Parameters
(emitter area: $1 \times 2 \ \mu m$; base area: $4 \times 4 \ \mu m$)

Parameter		Silicon Bipolar	GaAs Heterojunction Bipolar [9.65]
τ_B	$(psec)$	3	0.6
R_B	(Ω)	500	400
C_t	(fF)	20	6
C_{CB}	(fF)	5	5
C_{CS}	(fF)	6	0
p_w	(μm)	4	4
n_w		4	4
$\mathcal{R}_{int}$	(Ω/cm)	375	375
$\mathcal{C}_{int}$	(pF/cm)	2.0	2.0

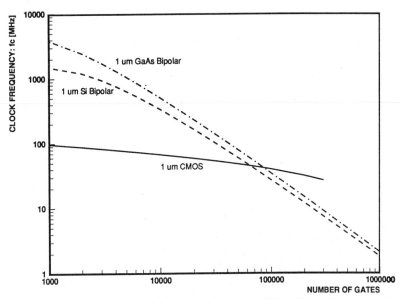

FIGURE 9.31 Comparison of clock frequencies in CMOS, silicon bipolar, and GaAs heterojunction bipolar chips as a function of gate count

cuits are summarized in Table 9.16, and the delay expressions are described in reference [9.63]. Here, I_{DC} is the value of the gate current source, I_{IO} is the total current of an output buffer, V_S is the logic swing (typically 0.6 V), R_B is the base resistance, V_{EE} is the power supply (typically 5.2 V), τ_B is the base delay ($\tau_B = R_B C_E$). The parameter C_t includes the base-emitter tran-

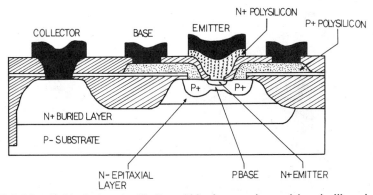

FIGURE 9.32 Oxide-isolated self-aligned bipolar transistor with polysilicon base and emitter contacts

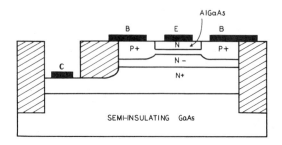

FIGURE 9.33   GaAs heterojunction bipolar transistor

sition capacitance, as well as the base-emitter stray capacitance, C_{CB} is the collector-to-base capacitance, C_{CS} is the collector-to-substrate capacitance, and C_L is the load capacitance. T_i is the delay of an inverting output, and T_n is the delay of a noninverting output.

Actually, an ECL gate is more suitable for high-performance design because it isolates the logic function from the driver by placing an emitter follower in front of a CML gate (see Chapter 2). In this way the logic current source and the driver current source can be independently optimized. This further complicates the calculations, and a CML gate is sufficient for the purposes of this study. The model can be extended to include ECL gates by simply replacing the delay equation and introducing an optimization scheme that selects the gate and emitter follower currents.

The power dissipation of the silicon and GaAs bipolar chips is held constant at 10 W, and the CMOS curve stops when its power consumption reaches 10 W (it begins at 0.75 W at 1000 gates and reaches 10 W at 300,000 gates). The CMOS circuit surpasses the performance of its bipolar counterparts at

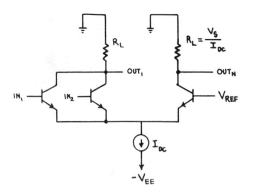

FIGURE 9.34   Current-mode logic gate

TABLE 9.16   SUSPENS Calculations for CML Circuits

Step	Calculation

1. $\overline{R} = \dfrac{2}{9}\left(7\dfrac{N_g^{p-0.5}-1}{4^{p-0.5}-1} - \dfrac{1-N_g^{p-1.5}}{1-4^{p-1.5}}\right)\dfrac{1-4^{p-1}}{1-N_g^{p-1}}$

2. $d_g = \dfrac{f_g \overline{R} p_w}{e_w n_w}$

3. $D_c = \sqrt{N_g} d_g$

4. $l_{av} = \overline{R} d_g$

5. $C_L = l_{av} C_{int}$

6. $I_{DC} = \dfrac{P_{c,internal}}{N_g V_{EE}}$

7. $B^* = \dfrac{2R_B I_{DC}}{V_S}$

8. $T_i = (f_g + B^*)\left(\tau_B + \dfrac{f_g}{f_g+1}\dfrac{C_t V_S}{I_{DC}} + 1.7\dfrac{C_{CB}V_S}{I_{DC}}\right)$
   $\quad + 0.7\dfrac{(f_g C_{CB} + f_g C_{CS} + C_L)V_S}{I_{DC}}$

9. $T_n = (f_g + B^*)\left(\tau_B + \dfrac{f_g}{f_g+1}\dfrac{C_t V_S}{I_{DC}} + 0.7\dfrac{C_{CB}V_S}{I_{DC}}\right)$
   $\quad + 0.7\dfrac{(2C_{CB} + C_{CS} + C_L)V_S}{I_{DC}}$

10. $T_g = \dfrac{T_i + T_n}{2}$

11. $f_c = \left(f_{ld}T_g + R_{int}C_{int}\dfrac{D_c^2}{2} + \dfrac{D_c}{v_c}\right)^{-1}$

12. $N_p = K_p N_g^\beta$

13. $P_c = N_g I_{DC} V_{EE} + \frac{1}{3}\frac{1}{2} N_p I_{IO} V_{EE}$

an integration level of 50,000 gates because the gate current of bipolar chips drops with increasing circuit count to satisfy the power-dissipation limit. On the other hand, a CMOS gate conducts current during only 10 percent of the clock period ($f_{ld} = 10$) and consumes only a fraction of the power dissipated by bipolar gates at the same clock frequency. Of course, the cross-over point will move to larger gate counts as the power-dissipation limit is increased.

Figure 9.31 can be used to compare the performance of CMOS, Si BJT, and GaAs HBT chips not only for a single-chip design but also for multichip implementations of the same CPU in different technologies. When a 100,000-gate CPU, for example, is implemented on a single chip, the clock frequency is

TABLE 9.17 Comparison of Single- and Multichip CPUs in CMOS, Si BJT, and GaAs HBT Technologies

Technology	N_g	D_c [cm]	T_g [psec]	f_c [MHz]	P [W]
CMOS	1,000	0.1	1,000	97	1
Si BJT	1,000	0.1	65	1500	10
GaAs HBT	1,000	0.1	25	3800	10
CMOS	10,000	0.5	1,500	67	2
Si BJT	10,000	0.5	290	330	10
GaAs HBT	10,000	0.5	190	500	10
CMOS	100,000	2.3	2,300	40	8
Si BJT	100,000	2.3	3,500	27	10
GaAs HBT	100,000	2.3	2,700	34	10

40, 28, and 35 MHz for CMOS, Si BJT, and GaAs HBT circuits, respectively. A 10-chip implementation of the same CPU (with every chip limited to a 10 W power consumption), on the other hand, can achieve a clock frequency of 70, 350, and 500 MHz in CMOS, Si BJT, and GaAs technologies. (This assumes that the packaging technology will improve with the devices and that system performance will not be limited by packaging and chip-to-chip delays.) As a result, the best technology for a single-chip design is CMOS (40 MHz, 10 W); however, Si BJT (330 MHz, 100 W), and GaAs HBT (500 MHz, 100 W) CPUs implemented as a 10-chip set are faster. Data summarizing the attributes of CPUs in these three technologies and in several integration densities are listed in Table 9.17. It should be pointed out that, currently, when a function that is implemented as multiple CMOS chips is integrated on a single CMOS chip, the performance improves because chip crossings are eliminated. In high-performance bipolar ECL, when multiple chips are combined on a single chip, the performance may degrade because of the power dissipation limit.

The different power-dissipation constraints are the reason for the unequal slopes of the CMOS and CML curves in Fig. 9.31. These slopes can be derived as follows. Gate delay is directly proportional to the capacitive load and voltage swing and inversely proportional to the current drive of the gate:

$$T \propto \frac{C \Delta V}{I}.$$

Clock frequency is inversely proportional to gate delay:

$$f \propto \frac{1}{T} \propto \frac{I}{C \Delta V}.$$

From Eq. 9.3, when $p > 0.5$ and $N_g \gg 1$, the average interconnection length is

$$\overline{R} \propto N_g^{p-0.5}.$$

If wiring capacitance dominates the loading of the gate, this implies that

$$C \propto l_{av} \propto N_g^{p-0.5}.$$

To meet the power-consumption limit, the current per gate in CML circuits is reduced with gate count

$$I_{CML} \propto \frac{1}{N_g},$$

which implies that

$$f_{CML} \propto \frac{I}{C\Delta V} \propto N_g^{-p-0.5}.$$

On the other hand, CMOS circuits have no power-dissipation constraints,

$$I_{CMOS} \propto 1,$$

which indicates that

$$f_{CMOS} \propto \frac{I}{C\Delta V} \propto N_g^{-p+0.5}.$$

For the log-log plots in Fig. 9.31, the slopes of the lines when $p = 0.6$ should be:

$$\begin{aligned}
slope_{CML} &= -p - 0.5 = -1.1 \\
slope_{CMOS} &= -p + 0.5 = -0.1.
\end{aligned} \tag{9.51}$$

The measured slopes are within 10 percent of these simplified calculations. As can be seen in Fig. 9.31, CMOS circuits are the fastest at high integration levels, and GaAs HBT chips achieve the highest performance at low gate counts (3.8 GHz for 1000-circuit gate arrays).

9.13   Comparison of Packaging Technologies

The architecture and implementation parameters in Table 9.12 and the technology and packaging parameters in Tables 9.6 and 9.7 can be used to investigate the impact of device and technology developments on overall performance. Tables 9.18 and 9.19 list the results of the SUSPENS simulations for 2, 1.3, 0.7, and 0.35 μm CMOS single- and multichip CPUs in several packaging options.

The following observations are obtained from Tables 9.18 and 9.19:

TABLE 9.18   Simulation Results of the System Model

2 μm CMOS CPU SIMULATION RESULTS
($p = 0.6$, $\beta = 0.6$, $K_p = 2.5$, $\pi = 0.6$, $f_{ld} = 10$, $f_g = 3$, $f_d = 0.5$)

Parameter	$N_{gTOT} = 10,000$				$N_{gTOT} = 100,000$				$N_{gTOT} = 1,000,000$			
	Single Chip	WSI	Thin Film Hybrid	Ceramic Hybrid	Single Chip	WSI	Thin Film Hybrid	Ceramic Hybrid	Single Chip	WSI	Thin Film Hybrid	Ceramic Hybrid
N_c	1	1000	1000	1000	1	1000	10	100	1	10	10	10
f [MHz]	93	250	250	250	58	180	150	140	15	39	46	54
D_M [cm]	1.5	1.9	4.8	6.0	6.8	7.7	19	19	30	22	45	57
P [W]	4.2	46	52	85	50	220	240	270	240	490	660	1200
Q [W/cm^2]	1.8	13	2.2	2.4	1.1	3.7	0.7	0.7	0.28	1.0	0.3	0.4
CC [$Kgates - GHz$]	0.93	2.5	2.5	2.5	5.8	18	15	14	15	39	46	54
PE [$gates - GHz/W$]	220	56	48	30	120	82	62	51	6.3	79	71	47
FTR[$gates - GHz/cm^2$]	400	680	110	70	120	300	40	36	18	83	23	17

1.3 μm CMOS CPU SIMULATION RESULTS
($p = 0.6$, $\beta = 0.6$, $K_p = 2.5$, $\pi = 0.6$, $f_{ld} = 10$, $f_g = 3$, $f_d = 0.5$)

Parameter	$N_{gTOT} = 10,000$				$N_{gTOT} = 100,000$				$N_{gTOT} = 1,000,000$			
	Single Chip	WSI	Thin Film Hybrid	Ceramic Hybrid	Single Chip	WSI	Thin Film Hybrid	Ceramic Hybrid	Single Chip	WSI	Thin Film Hybrid	Ceramic Hybrid
N_c	1	1000	1000	1000	1	100	100	10	1	10	10	10
f [MHz]	180	440	440	400	110	260	210	180	27	54	48	56
D_M [cm]	0.34	1.9	4.8	6.0	1.5	6.2	15	14	6.6	18	45	57
P [W]	1.7	23	31	50	19	68	81	85	85	170	200	410
Q [W/cm^2]	15	6.4	1.3	1.4	8.2	1.8	0.36	0.43	2	0.5	0.1	0.13
CC [$Kgates - GHz$]	1.8	4.4	4.4	4	11	26	21	18	27	54	48	56
PE [$gates - GHz/W$]	1100	190	140	79	600	380	260	210	320	320	250	140
FTR[$gates - GHz/cm^2$]	16000	12000	190	110	4800	690	89	89	630	160	23	18

- A single-chip design is the slowest for the majority of cases because of the distributed-RC delays of long on-chip global interconnections and the differences in the driving methods and "logic depths" between on- and off-chip circuitry. The single-chip design, however, surpasses the performance of multichip CPUs for large gate counts ($N_{gTOT} = 1,000,000$) at reduced minimum feature sizes ($F= 0.7$ and 0.35 μm) because of small die size and short interconnections.

- With its densely packed but low-loss interconnections, the thin film hybrid produces the highest-performance system for most cases. Because WSI may require redundancy, it may never achieve the performance of hybrid systems. For example, when triple redundancy is applied, the performance measures of WSI (CC, PE, FTR) will degrade by a factor of three. In addition, the possibility of populating both sides of a hybrid substrate can improve the computational capacity by a factor of two, which is probably not possible with WSI.

- The power efficiency (PE) and functional throughput rate (FTR) of the single chip are superior to those of other approaches because its small size results in smaller total capacitance and lower power consumption.

TABLE 9.19   Simulation Results of the System Model

0.7 μm CMOS CPU SIMULATION RESULTS
($p = 0.6$, $\beta = 0.6$, $K_p = 2.5$, $\pi = 0.6$, $f_{ld} = 10$, $f_g = 3$, $f_d = 0.5$)

Parameter	$N_{gTOT} = 10,000$				$N_{gTOT} = 100,000$				$N_{gTOT} = 1,000,000$			
	Single Chip	WSI	Thin Film Hybrid	Ceramic Hybrid	Single Chip	WSI	Thin Film Hybrid	Ceramic Hybrid	Single Chip	WSI	Thin Film Hybrid	Ceramic Hybrid
N_c	1	100	100	100	1	10	10	10	1	10	10	10
f [MHz]	330	560	560	510	220	330	320	240	77	55	49	57
D_M [cm]	0.13	1.6	3.9	4.9	0.57	4.6	11	14	2.5	18	45	57
P [W]	0.47	4.5	6.8	12	5.6	17	24	36	35	43	54	130
Q [W/cm²]	29	1.8	0.45	0.5	17	0.8	0.2	0.2	5.7	0.13	0.03	0.04
CC [Kgates − GHz]	3.3	5.6	5.6	5.1	22	33	32	24	77	55	49	57
PE [gates − GHz/W]	7100	1200	820	430	4000	1900	1300	660	2200	1300	900	440
FTR[gates − GHz/cm²]	210000	2300	370	220	69000	1600	240	120	13000	170	24	18

0.35 μm CMOS CPU SIMULATION RESULTS
($p = 0.6$, $\beta = 0.6$, $K_p = 2.5$, $\pi = 0.6$, $f_{ld} = 10$, $f_g = 3$, $f_d = 0.5$)

Parameter	$N_{gTOT} = 10,000$				$N_{gTOT} = 100,000$				$N_{gTOT} = 1,000,000$			
	Single Chip	WSI	Thin Film Hybrid	Ceramic Hybrid	Single Chip	WSI	Thin Film Hybrid	Ceramic Hybrid	Single Chip	WSI	Thin Film Hybrid	Ceramic Hybrid
N_c	1	100	10	10	1	10	10	10	1	10	10	10
f [MHz]	740	1100	940	780	480	510	360	270	140	57	50	59
D_M [cm]	0.051	1.6	2.9	3.6	0.23	4.6	11	14	1.0	18	45	57
P [W]	0.11	1.6	1.2	2.1	1.2	5.1	5.8	9.4	6.3	8.7	12	31
Q [W/cm²]	43	0.63	0.14	0.16	24	0.24	0.05	0.05	6.3	0.03	0.01	0.01
CC [Kgates − GHz]	7.4	11	9.4	7.8	48	51	36	27	140	57	50	59
PE [gates − GHz/W]	67000	6800	7500	3700	38000	10000	6200	2900	22000	6500	4200	1900
FTR[gates − GHz/cm²]	2900000	4600	1100	610	920000	2400	280	130	140000	170	24	18

- The optimal integration level, represented by the number of chips into which the system is partitioned (N_c), is determined by simulating the system at various partition levels. When the system is partitioned into many chips, the module becomes very large and performance is limited by the package; when it is partitioned into only a few chips, performance is limited by on-chip interconnections. The number of chips decreases as the total number of gates in the system becomes larger to avoid an excessively large module dimension, which will limit performance. The multichip CPUs in these tables were simulated for $N_c = 10$, 100, 1000, and 10,000. The dependence of the clock frequency to N_c was generally not very strong.

- Power dissipation (P) never becomes a major problem in CMOS systems. For the majority of the cases presented in Table 9.18, the total power dissipation is less than 100 W and never rises above 1200 W. Although this is very high compared to current CMOS processors, packages that can dissipate up to 1000 W have been successfully manufactured for bipolar systems [9.47].

- Computational capacity (CC) increases more slowly than total gate count because of clock-frequency degradation.

9.14   Extensions to the SUSPENS Model

The SUSPENS model can be easily extended to include additional system parameters such as yield, cost, and reliability.

Yield and cost can be calculated by using the established relationships between area and yield. For example, assuming random defects with Poisson distribution, the following yield formula is obtained,

$$Y = e^{-D_0 A},$$

where Y is the yield, A is the chip area, and D_0 is the defect density per unit area. The product $D_0 A$ is the average number of defects per chip. According to this formula, to achieve a yield better than 10 percent, the defect density should be less than 2.3 defects per die. As an alternative, Murphy's yield law [9.66]

$$Y = \left(\frac{1 - e^{-D_0 A}}{D_0 A} \right)^2,$$

or Price yield law [9.67]

$$Y = \frac{1}{1 + D_0 A},$$

or other more recent formulas can be used.

As the integration level is increased, the chips will become larger, their yield will go down, and the silicon cost will go up. At the same time, packaging cost will go down because fewer chip carriers and less board wiring will be required, and mounting and assembly costs will be reduced. At a certain integration level, the sum of silicon and package cost will attain a minimum.

System reliability can be assessed by assigning relative failure rates for on-chip interconnections, bonding wires, chip carriers, connections between the chip carrier and the board, and board wires and by calculating the overall reliability as the chip integration level varies. The reliability figures can be also fed into the cost calculations in order to calculate not only the manufacturing cost but also the lifetime cost of the system, including service costs.

9.15   Summary

This chapter introduced a system-level circuit model that encompasses material, device, circuit, logic, packaging, and architecture parameters. The model agrees well with the existing systems and can accurately predict the overall performance of microprocessors, gate arrays, and mini- and mainframe computers for competing CMOS, bipolar, and GaAs technologies.

The key issues addressed were how to determine the impact of a proposed device or packaging technology on overall system performance, how best to partition a CPU to exploit a new technology, and how to optimize the number of transistors on a chip for maximum system throughput. Answering these questions requires a unified model that includes devices, circuits, logic, packaging, and architecture. The greatest challenge in creating such a model was linking the architecture parameters to the technology parameters. The SUSPENS model utilizes Rent's rule to provide this link. Rent's rule yields the number of connections to a logic block as a function of gate count, and average wire lengths can be calculated from this rule.

In summary,

- A system-level circuit model encompassing device, circuit, logic, packaging, and architecture parameters was developed that can predict clock frequency, power dissipation, and chip and module sizes.

- The model agrees well with existing microprocessors and predicts that, in ten years, a six-million-transistor microprocessor will be able to execute 30–60 MIPS.

- The performances of 1 μm CMOS, bipolar, and GaAs technologies were compared, and it was observed that the GaAs chips are the fastest at low integration levels (3.8 GHz for 1000-circuit gate arrays) and that CMOS circuits are superior when more than 50,000 gates are integrated on a chip.

- The SUSPENS model was also used to determine the impact of new device and packaging technologies on system performance and to optimize the number of transistors on a chip for maximum clock frequency.

A

USEFUL CONSTANTS AND CONVERSIONS

A.1   Physical Constants

Speed of light in free space	c_0	3.0×10^{10} cm/s
Permittivity of free space	ϵ_0	8.854×10^{-14} F/cm
Permeability of free space	μ_0	1.257×10^{-8} H/cm $= 4\pi \times 10^{-9}$ H/cm
Planck's constant	h	6.625×10^{-34} J·s $= 4.135 \times 10^{-15}$ eV·s
Electronic charge	q	1.6×10^{-19} C
Free electron mass	m_0	9.1×10^{-31} kg
Avagadro's number	N_A	6.023×10^{23} molecules/mol
Boltzmann's constant	k_B	1.38×10^{-23} J/K $= 8.62 \times 10^{-5}$ eV/K
Thermal energy	$k_B T$	0.02586 eV at 300°K
Thermal voltage	$\frac{k_B T}{q}$	25.86 mV at 300°K

A.2   Properties of Silicon

Bandgap of Si (E_{gSi})	1.12 eV
Temperature dependence of E_{gSi}	$E_{gSi} = \alpha - \beta T$
	$\alpha = 1.17$ eV,
	$\beta = -3 \times 10^{-4} \frac{eV}{K}$
Bandgap of SiO$_2$ (E_{gSiO_2})	8 eV
Bandgap of Si$_3$N$_4$ ($E_{gSi_3N_4}$)	5 eV
Intrinsic carrier concentration of Si (n_i)	1.5×10^{10} cm^{-3}
Temperature dependence of n_i	$3.9 \times 10^{16} \times T^{3/2} e^{-\frac{0.605eV}{k_B T}}$ cm^{-3}
Bulk mobility of electrons in Si (μ_n)	1350 cm^2/V·s
Bulk mobility of holes in Si (μ_p)	480 cm^2/V·s
Surface mobility of electrons in Si (μ_n)	600 cm^2/V·s
Surface mobility of holes in Si (μ_p)	200 cm^2/V·s
Breakdown field of Si ($\mathcal{E}_{Breakdown,Si}$)	3×10^5 V/cm
Breakdown field of SiO$_2$ ($\mathcal{E}_{Breakdown,SiO_2}$)	6×10^6 V/cm

A.3   Relative Dielectric Constants (ϵ_r)

Silicon carbide (SiC)	42.0
Alumina (Al$_2$O$_3$)	9.5
Beryllia (BeO)	6.7
Epoxy glass (PC board)	5.0
Polyimide	2.5–3.5
Silicon dioxide (SiO$_2$)	3.9
Silicon nitride (Si$_3$N$_4$)	7.5
Gallium arsenide (GaAs)	10.9
Silicon (Si)	11.7
Germanium (Ge)	16.0

A.4 Bulk Electrical Resistivities of Pure Metals at 22°C (ρ)

Silver (Ag)	1.6 $\mu\Omega\cdot$cm
Copper (Cu)	1.7 $\mu\Omega\cdot$cm
Gold (Au)	2.2 $\mu\Omega\cdot$cm
Aluminum (Al)	2.8 $\mu\Omega\cdot$cm
Tungsten (W)	5.3 $\mu\Omega\cdot$cm
Molybdenum (Mo)	5.3 $\mu\Omega\cdot$cm
Titanium (Ti)	43.0 $\mu\Omega\cdot$cm

A.5 Thermal Conductivities at 0°C (κ)

Silver (Ag)	4.3 W/cm$\cdot$K
Copper (Cu)	4.0 W/cm$\cdot$K
Gold (Au)	3.2 W/cm$\cdot$K
Aluminum (Al)	2.4 W/cm$\cdot$K
Tungsten (W)	1.7 W/cm$\cdot$K
Molybdenum (Mo)	1.4 W/cm$\cdot$K
Titanium (Ti)	0.22 W/cm$\cdot$K
Silicon (Si)	1.5 W/cm$\cdot$K
Germanium (Ge)	0.6 W/cm$\cdot$K
Gallium arsenide (GaAs)	0.5 W/cm$\cdot$K
Silicon carbide (SiC)	2.2 W/cm$\cdot$K
Beryllia (BeO)	2.0 W/cm$\cdot$K
Alumina (Al_2O_3)	0.3 W/cm$\cdot$K
Silicon dioxide (SiO_2)	0.01 W/cm$\cdot$K
Polyimide	0.004 W/cm$\cdot$K
Epoxy glass (PC board)	0.003 W/cm$\cdot$K

A.6   Coefficients of Linear Thermal Expansion at 20°C (α)

Silver (Ag)	19.0×10^{-6} K^{-1}
Copper (Cu)	17.0×10^{-6} K^{-1}
Gold (Au)	14.2×10^{-6} K^{-1}
Aluminum (Al)	23.0×10^{-6} K^{-1}
Tungsten (W)	4.6×10^{-6} K^{-1}
Molybdenum (Mo)	5.0×10^{-6} K^{-1}
Titanium (Ti)	8.6×10^{-6} K^{-1}
Silicon (Si)	2.5×10^{-6} K^{-1}
Germanium (Ge)	5.7×10^{-6} K^{-1}
Gallium arsenide (GaAs)	5.8×10^{-6} K^{-1}
Silicon dioxide (SiO$_2$)	0.5×10^{-6} K^{-1}
Silicon carbide (SiC)	3.7×10^{-6} K^{-1}
Beryllia (BeO)	6.0×10^{-6} K^{-1}
Alumina (Al$_2$O$_3$)	6.0×10^{-6} K^{-1}
Invar	0.7×10^{-6} K^{-1}
Epoxy glass (PC board)	15.0×10^{-6} K^{-1}

A.7   Useful Conversions

$1 \ \mu m = 1 \times 10^{-4}$ cm $= 1 \times 10^{-6}$ m

$1 \ \text{Å} = 1 \times 10^{-8}$ cm $= 1 \times 10^{-4} \ \mu m = 10$ nm

1 mil $= 0.001$ inch $= 25.4 \ \mu m$

1 eV $= 1.602 \times 10^{-19}$ J

$273°$K $= 0°$C

A.8   Mathematical Constants

e (the root of natural logarithm) $= 2.718$

$\pi = 3.142$

B

SYMBOLS

A_c	chip area (cm^2)
A_d	area of the output buffer (driver) of a gate (cm^2)
A_g	gate area (cm^2)
A_l	area of the input (logic) section of a gate (cm^2)
c_0	speed of light in free space (cm/sec)
C_g	total capacitance per gate (F)
C_{gin}	input capacitance of a gate (F)
C_{int}	capacitance of a chip interconnection (F)
C_{INT}	capacitance of a module interconnection (F)
C_L	load capacitance (F)
C_M	total capacitance of the module (F)
C_o	input capacitance of a minimum-sized repeater (F)
C_{PAD}	I/O pad capacitance (F)
C_{tr}	input capacitance of a minimum-sized transistor (F)
CC_c	computational capacity of the chip $(gates\text{-}Hz)$
CC_S	computational capacity of the system $(gates\text{-}Hz)$
c_{int}	chip interconnection capacitance per unit length (F/cm)
c_{INT}	module interconnection capacitance per unit length (F/cm)
d_g	gate pitch (cm)
D_c	chip size (cm)

D_M	module size (cm)
e	root of the natural logarithm
e_w	utilization efficiency of chip interconnections
E_f	efficiency factor for driver and receiver circuits
E_W	utilization efficiency of module interconnections
f	ratio of consecutive gates in cascaded drivers
f_c	maximum chip clock frequency (Hz)
f_d	portion of the on-chip gates that switch during a clock period
f_g	fan-out of the gates
f_{ld}	logic depth of the on-chip gates
f_M	maximum module clock frequency (Hz)
f_S	maximum system clock frequency (Hz)
F	minimum feature size (cm)
F_c	fan-out of the chip output drivers
F_D	portion of the output buffers that switch during a clock period
F_P	footprint size of the chips (cm)
FP_{CC}	chip carrier-size-limited footprint (cm)
FTR_c	functional throughput rate of the chip $(gates\text{-}Hz/cm^2)$
FTR_S	functional throughput rate of the system $(gates\text{-}Hz/cm^2)$
h	size of the repeaters
H_{int}	thickness of the on-chip interconnections (cm)
H_{INT}	thickness of the module interconnections (cm)
I	electric current (A)
I_{av}	average current drive (A)
I_{DD}	current from the positive power supply (A)
I_{SS}	current to the negative power supply (usually ground) (A)
ΔI_{max}	maximum current surge (A)
k	number of repeaters
k_d	minimum feature-sized squares in the output stage with $k_o = 1$
k_g	number of minimum feature-sized squares in the gate
k_i	ratio of the input stage
k_l	minimum feature-sized squares in the input stage with $k_i = 1$
k_o	ratio of the output stage
K_p	multiplicative constant for the pin count
l_{av}	average chip interconnection length (cm)
l_{int}	global interconnection length (cm)
l_{loc}	local interconnection length (μm)

L	drawn transistor length (μm)
L_{AV}	average module interconnection length (cm)
L_{diff}	lateral diffusion length (μm)
L_{eff}	effective transistor length (μm)
$\mathcal{L}_{int}$	interconnection inductance per unit length (H/cm)
n_w	number of interconnection levels on the chip
N_c	number of chips in the system
N_g	number of gates per chip
N_{gTOT}	total number of gates in the system
N_p	number of I/O pins per chip
N_{SUB}	substrate doping concentration ($1/cm^3$)
N_W	number of interconnection levels on the module
p	Rent's constant for chip interconnections
p_w	wiring pitch of chip interconnections (cm)
P_c	chip power dissipation (W)
P_g	power dissipation per gate (W)
P_P	contact pad pitch (cm)
P_S	system power dissipation (W)
P_W	wiring pitch of module interconnections (cm)
PE_c	power efficiency of the chip ($gates$-Hz/W)
PE_S	power efficiency of the system ($gates$-Hz/W)
PT_c	minimum chip clock period (sec)
Q_c	power-dissipation density of the chip (W/cm^2)
Q_S	power-dissipation density of the system (W/cm^2)
R_{gout}	output resistance of a gate (Ω)
R_{int}	resistance of a chip interconnection (Ω)
R_{INT}	resistance of a module interconnection (Ω)
R_o	output resistance of a minimum-sized repeater (Ω)
R_S	source resistance of a transmission line driver (Ω)
R_{tr}	on-resistance of a minimum-sized nMOS transistor (Ω)
$\overline{R}$	average chip interconnection length in units of gate pitch
$\overline{R}_M$	average module line length in units of chip footprint size
$\mathcal{R}_{int}$	chip interconnection resistance per unit length (Ω/cm)
$\mathcal{R}_{INT}$	module interconnection resistance per unit length (Ω/cm)
S	scaling factor for device dimensions ($S > 1$)
S_C	scaling factor for chip dimensions ($S_C > 1$)
t_f	time-of-flight delay (sec)

t_{gox}	gate oxide thickness (A)
t_{ox}	interconnection insulator layer thickness (μm)
T_g	average gate delay (sec)
T_i	input-stage delay of a gate (sec)
T_M	chip-to-chip transmission delay on the module (sec)
T_o	output-stage delay of a gate (sec)
v	propagation speed of electromagnetic waves (cm/sec)
v_c	propagation speed of electromagnetic waves on the chip (cm/sec)
v_M	propagation speed of waves on the module (cm/sec)
V_{DD}	power-supply voltage (V)
V_{pp}	peak-to-peak potential swing (V)
V_T	threshold potential of transistors (V)
ΔV	potential swing for a logic state change (V)
W	transistor width (μm)
W_{int}	width of chip interconnections (cm)
W_{INT}	width of module interconnections (cm)
W_{sp}	spacing of chip interconnections (cm)
W_{SP}	spacing of module interconnections (cm)
X_j	junction depth (μm)
Z_0	characteristic impedance of interconnections (Ω)
α	coefficient of linear thermal expansion (K^{-1})
β	Rent's constant for the chip pin count
β	current gain of a bipolar transistor
γ	reflection coefficient for transmission line termination
δ	skin depth (cm)
ϵ	electric permittivity (dielectric constant) (F/cm)
ϵ_{ox}	dielectric constant of oxide (F/cm)
ϵ_r	relative permittivity
η	Rent's constant for module interconnections
κ	thermal conductivity $(W/cm \cdot K)$
μ	magnetic permeability $(Henry/cm)$
μ_n, μ_p	mobility of the carriers $(cm^2/V\text{-}s)$
μ_r	relative permeability
ϕ	magnetic flux per unit length (Wb/cm)
ρ	resistivity $(\Omega\text{-}cm)$
ρ_{int}	interconnection resistivity $(\Omega\text{-}cm)$
τ_r	rise time (sec)

REFERENCES

Chapter 1

[1.1] G. Moore, "VLSI: some fundamental challenges," *IEEE Spectrum,* vol. 16, p. 30, 1979.

[1.2] J. Meindl, "Theoretical, practical, and analogical limits in ULSI," *IEEE International Electron Devices Meeting (IEDM'83),* pp. 8–13, Dec. 1983.

[1.3] G.H. Heilmeier, "Microelectronics: end of the beginning or beginning of the end," *IEEE International Electron Devices Meeting (IEDM'84),* pp. 2–5, Dec. 1984.

[1.4] R.H. Dennard, F.H. Gaensslen, H.N. Yu, V.L. Rideout, E. Bassous, and A.R. LeBlanc, "Design of ion implanted MOSFET's with very small physical dimensions," *IEEE Journal of Solid-State Circuits,* vol. SC-9, pp. 256–268, Oct. 1974.

[1.5] D. Pramatik and A.N. Saxena, "VLSI metallization using aluminum and its alloys, Part I," *Solid-State Technology,* pp. 127–133, Jan. 1983.

[1.6] D. Pramatik and A.N. Saxena, "VLSI metallization using aluminum and its alloys, Part II," *Solid-State Technology,* pp. 131–138, March 1983.

[1.7] J.W. Goodman, F.I. Leonberger, S.Y. Kung, and R.A. Athale, "Optical interconnections for VLSI systems," *Proceedings of the IEEE,* vol. 72, no. 7, pp. 850–866, July 1984.

[1.8] H.B. Bakoglu and J.D. Meindl, "CMOS driver and receiver circuits for reduced interconnection delays," in *1985 International Symposium on VLSI Technology, Systems and Applications Proc. Tech Papers,* pp. 171–175, Taipei, Taiwan, May 1985.

[1.9] H.B. Bakoglu and J.D. Meindl, "New CMOS driver and receiver circuits reduce interconnection propagation delays," in *1985 Symposium on VLSI Techology Dig. of Tech. Papers,* pp. 54–55, Kobe, Japan, May 1985.

[1.10] K.C. Saraswat and F. Mohammadi, "Effect of scaling of interconnections on the time delay of VLSI circuits," *IEEE Journal of Solid-State Circuits,* vol. SC-17, pp. 275–280, April 1982.

[1.11] H.B. Bakoglu and J.D. Meindl, "Optimal interconnect circuits for VLSI," *IEEE International Solid State Circuits Conference (ISSCC'84) Dig. Tech. Papers,* pp. 164-165, San Francisco, Feb. 1984.

[1.12] H.B. Bakoglu and J.D. Meindl, "Optimal interconnection circuits for VLSI," *IEEE Transactions on Electron Devices,* vol. ED-32, pp. 903–909, May 1985.

[1.13] H.B. Bakoglu, "Packaging for high-speed systems," *IEEE International Solid State Circuits Conference (ISSCC'88) Dig. Tech. Papers,* pp. 100–101, San Francisco, Feb. 1988.

[1.14] "Bipolar packaging lags technology," *Electronic Engineering Times,* p. 75 October 3, 1988.

[1.15] C.J. Bartlett, J.M. Segelken, and N.A. Teneketges, "Multichip packaging design for VLSI-based systems," *IEEE Transactions on Components, Hybrids, and Manufacturing Technology,* pp. 647–653, vol. CHMT-12, no. 4, December 1987.

[1.16] I. Catt, D. Walton and M. Davidson, *Digital Hardware Design,* Macmillan, London, 1979.

[1.17] A.J. Blodgett, "Microlectronic packaging," *Scientific American,* pp. 86–96, July 1983.

[1.18] C.W. Ho, D.A. Chance, C.H. Bajorek, and R.E. Acosta, "The thin film module as a high-performance semiconductor package," *IBM Journal of Research and Development*, vol. 26, pp. 286–296, May 1982.

[1.19] H. Hasegawa, M. Furukawa, and H. Yanai, "Properties of microstrip line on Si-SiO_2 system," *IEEE Transactions on Microwave Theory and Techniques*, vol. MTT-19, pp. 869–881, Nov. 1971.

[1.20] C. Mead and L. Conway, *Introduction to VLSI Systems*, Reading, Mass., Addison-Wesley Publishing Co., Chapter 7, 1980.

[1.21] C.L. Seitz, "Self timed VLSI systems," *Caltech Conference on VLSI*, pp. 345–355, Jan. 1979.

[1.22] F. Anceau, "A synchronous approach for clocking VLSI systems," *IEEE Journal of Solid-State Circuits*, vol. SC-17, pp. 51–56, Feb. 1982.

[1.23] B. Randell and P.C. Treleaven (eds.), *VLSI Architecture*, Chapter 15 by J.A. Marques and A. Cunha, "Clocking of VLSI circuits," Prentice-Hall, Englewood Cliffs, NJ, 1983.

[1.24] H.B. Bakoglu, J.T. Walker, and J.D. Meindl, "A symmetric clock-distribution tree and optimized high-speed interconnections for reduced clock skew in ULSI and WSI circuits," *IEEE International Conference on Computer Design: VLSI in Computers and Processors (ICCD'86)*, pp. 118–122, Rye Brook, NY, Oct. 1986.

[1.25] K.D. Wagner, "Clock system design," *IEEE Design and Test of Computers*, pp. 9–27, October 1988.

[1.26] L.W. Schaper and D.I. Amey, "Improved electrical performance required for future MOS packaging," *IEEE Transactions on Components, Hybrids and Manufacturing Technology*, vol. CHMT-6, pp. 282–289, Sept. 1983.

[1.27] T.C. May and M.H. Woods, "Alpha-particle-induced soft errors in dynamic memories," *IEEE Transactions on Electron Devices*, ED-26, no. 1, Jan. 1979.

[1.28] T.C. May, "Soft errors in VLSI—Present and future," *IEEE Transactions on Components, Hybrids, and Manufacturing Technology*, pp. 377–387, Dec. 1979.

[1.29] R.W. Keyes, "The wire-limited logic chip," *IEEE Journal of Solid-State Circuits*, vol. SC-17, pp. 1232–1233, Dec. 1982.

[1.30] H.B. Bakoglu and J.D. Meindl, "A system level circuit model for multi- and single-chip CPUs," *IEEE International Solid State Circuits Conference (ISSCC'87) Dig. Tech. Papers,* New York, Feb. 1987.

[1.31] H.B. Bakoglu, "Circuit and system performance limits on ULSI: Interconnections and packaging," Ph.D. dissertation, Stanford University, 1987.

Chapter 2

[2.1] A.S. Grove, *Physics and Technology of Semiconductor Devices,* Wiley, New York, NY, 1967.

[2.2] R.S. Muller and T.I. Kamins, *Device Electronics for Integrated Circuits,* Wiley, New York, NY, 1977.

[2.3] E.S. Yang, *Fundamentals of Semiconductor Devices,* McGraw-Hill, New York, 1978.

[2.4] B.G. Streetman, *Solid State Electronic Devices,* Prentice-Hall, Englewood Cliffs, NJ, 1980.

[2.5] T.E. Dillinger, *VLSI Engineering,* Prentice-Hall, Englewood Cliffs, NJ, 1988.

[2.6] S.M. Sze, *Physics of Semiconductor Devices,* 2d ed., Wiley, New York, 1981.

[2.7] R.H. Krambeck, C.M. Lee, and H.S. Law, "High-speed compact circuits with CMOS," *IEEE Journal of Solid-State Circuits,* vol. SC-17, pp. 614–619, June 1982.

[2.8] R.H. Dennard, F.H. Gaensslen, H.N. Yu, V.L. Rideout, E. Bassous, and A.R. LeBlanc, "Design of ion implanted MOSFET's with very small physical dimensions," *IEEE Journal of Solid-State Circuits,* vol. SC-9, pp. 256–268, Oct. 1974.

[2.9] P.A. Hart, T.V. Hof, and F.M. Klaassen, "Device down scaling and expected circuit performance," *IEEE Transactions on Electron Devices,* vol. ED-26, no. 4, pp. 421–429, April 1979.

[2.10] H. Yu, A. Reisman, C.M. Osburn, and D. Critchlow, "1 μm MOSFET VLSI technology: Part 1—an overview," *IEEE Transactions on Electron Devices,* vol. ED-26, no. 4, pp. 318–324, April 1979.

[2.11] Texas Instruments VLSI Laboratory, "Technology and design challenges of MOS VLSI," *IEEE Journal of Solid-State Circuits,* vol. SC-17, no. 3, pp. 442–448, June 1982.

[2.12] A. Reisman, "Device, circuit, and technology scaling to micron and submicron dimensions," *Proceedings of the IEEE,* vol. 71, no. 5, pp. 550–565, May 1983.

[2.13] G. Baccarani, M.R. Wordeman, and R.H. Dennard, "Generalized scaling theory and its application to a 1/4 micrometer MOSFET design," *IEEE Transactions on Electron Devices,* vol. ED-31, no. 4, pp. 452–462, April 1984.

[2.14] B. Hoeneisen and C.A. Mead, "Fundamental limitations in microelectronics-I. MOS technology," *Solid-State Electronics,* vol. 15, pp. 819–829, 1972.

[2.15] R.W. Keyes, "Physical limits in digital electronics," *Proceedings of the IEEE,* vol. 63, no. 5, pp. 740–767, May 1975.

[2.16] P.M. Solomon, "A comparison of semiconductor devices for high-speed logic," *Proceedings of the IEEE,* vol. 70, no. 5, pp. 489–509, May 1982.

[2.17] F.F. Fang and A.B. Fowler, "Hot electron effects and saturation velocities in silicon inversion layers," *Journal of Applied Physics,* vol. 41, pp. 1825–1831, March 1970.

[2.18] R.W. Coen and R.S. Muller, "Velocity of surface carriers in inversion layers in silicon," *Solid-State Electronics,* vol. 23, pp. 35–40, 1980.

[2.19] J.A. Cooper and D.F. Nelson, "Measurement of the high-field drift velocity of electrons in inversion layers in silicon," *IEEE Electron Device Letters,* vol. EDL-2, July 1981.

[2.20] F.F. Fang and A.B. Fowler, "Transport properties of electrons in inverted silicon surfaces," *Physics Reviews,* vol. 169, pp. 619–631, May 1968.

[2.21] A.G. Sabnis and J.T. Clemens, "Characterization of the electron mobility in the inverted (100) Si surface," *IEEE International Electron Devices Meeting (IEDM'79) Technical Digest,* pp. 18–21, Dec. 1979.

[2.22] S.C. Sun and J.D. Plummer, "Electron mobility in inversion and accumulation layers on thermally oxidized silicon surfaces," *IEEE Transactions on Electron Devices,* vol. ED-27, pp. 1497–1508, Aug. 1980.

[2.23] H.C. Pao and C.T. Shah, "Effects of diffusion current on characteristics of metal-oxide (insulator) semiconductor transistors," *Solid State Electronics,* vol. 9, pp. 927–937, 1966.

[2.24] Y. Hayashi, "Static characteristics of extremely thin gate oxide MOS transistors," *Electronics Letters,* vol. 11, pp. 618–620, 1975.

[2.25] Y. El-Mansy, "MOS device and technology constraints in VLSI," *IEEE Journal of Solid-State Circuits,* vol. SC-17, no. 12, pp. 197–203, April 1982.

[2.26] P.I. Suciu and R.L. Johnson, "Experimental derivation of the source and drain resistances of MOS transistors," *IEEE Transactions on Electron Devices,* vol ED-27, no. 9, pp. 1846–1848, Sept. 1980.

[2.27] P.K. Chatterjee, W.R. Hunter, T.C. Holloway, and Y.T. Lin, "The impact of scaling laws on the choice of n-channel and p-channel for MOS VLSI," *IEEE Electron Device Letters,* vol. EDL-1, no. 10, pp. 220–223, Oct. 1980.

[2.28] D.B. Scott, W.R. Hunter, and H. Shichijo, "A transmission-line model for silicided diffusions: impact on the performance of VLSI circuits," *IEEE Transactions on Electron Devices,* vol. ED-29, pp. 651–660, 1982.

[2.29] S. Ogura, P.J. Tsang, W.W. Walker, D.L. Critchlow, and J.F. Shepard, "Design and characteristics of the lightly doped drain source (LDD) insulated gate field effect transistors," *IEEE Transactions on Electron Devices,* ED-27, pp. 1359–1367, Aug. 1980.

[2.30] F.J. Lai, J.Y. Sun, and S.H. Dhong, "Design and characteristics of a lightly doped drain (LDD) device fabricated with self-aligned titanium disilicide," *IEEE Transactions on Electron Devices,* vol. ED-33, no. 3, pp. 345–353, March 1986.

[2.31] R.M. Swanson and J.D. Meindl, "Ion implanted complementary MOS transistors in low voltage circuits," *IEEE Journal of Solid-State Circuits,* SC-7, pp. 146–153, 1972.

[2.32] G.W. Taylor, "Subthreshold conduction in MOSFETs," *IEEE Transactions on Electron Devices,* ED-25, no.3, pp. 337–350, March 1978.

[2.33] J.R. Brews, "Subthreshold behavior of uniformly and nonuniformly doped long-channel MOSFET," *IEEE Transactions on Electron Devices,* vol. ED-26, no. 9, pp. 1282–1291, 1979.

[2.34] J.R. Brews, "Physics of the MOS transistor," *Applied Solid State Science, Supplement 2A,* Academic Press, London, pp. 2–120, 1981.

[2.35] J. Nishizawa, T. Ohmi, and H. Chen, "A limitation of channel length in dynamic memories," *IEEE Journal of Solid-State Circuits,* vol. SC-15, no. 4, pp. 705–714, Aug. 1980.

[2.36] R.R. Troutman, "VLSI limitations from drain-induced barrier lowering," *IEEE Journal of Solid-State Circuits,* SC-14, pp. 383–390, 1979.

[2.37] T.H. Ning, P.W. Cook, R.H. Dennard, C.M. Osburn, S.E. Shuster, and H.N. Yu, "1-μm MOSFET VLSI technology: Part IV: Hot-electron design constraints," *IEEE Transactions on Electron Devices,* vol. ED-26, pp. 346–353, April 1979.

[2.38] E. Takeda et al., "Role of hot-hole injection in hot carrier effects and small degraded channel region in MOSFETs," *IEEE Electron Device Letters,* vol. EDL-4, 1983.

[2.39] E. Takeda, et al., "An empirical model for device degradation due to hot-carrier injection," *IEEE Electron Device Letters,* vol. EDL-4, 1983.

[2.40] B. Hoefflinger, H. Sibbert, and G. Zimmer, "Model and performance of hot-electron MOS transistors for VLSI," *IEEE Transactions on Electron Devices,* vol. ED–26, pp. 513–520, April 1979.

[2.41] R.K. Pancholy, "The effects of VLSI scaling on EOS/ESD failure threshold," *1981 Electrical Overstress/Electrostatic Discharge (EOS/ESD) Symposium Proceedings,* pp. 85–89, 1981.

[2.42] T.V. Mulett, "On chip protection of high density NMOS devices," *1981 Electrical Overstress/Electrostatic Discharge (EOS/ESD) Symposium Proceedings,* pp. 90–96, 1981.

[2.43] C. Duvvury, "A summary of most effective electrostatic discharge protection circuits for MOS memories and their observed failure modes," *1983 Electrical Overstress/Electrostatic Discharge (EOS/ESD) Symposium Proceedings,* pp. 181–184, 1983.

[2.44] C.M. Lin, L. Richardson, K. Chi, and R. Simcoe, "A CMOS VLSI ESD input protection device, DIFIDW," *1984 Electrical Overstress/Electrostatic Discharge (EOS/ESD) Symposium Proceedings,* pp. 202–209, 1984.

[2.45] D.B. Estreich, "The physics and modeling of latch-up in CMOS integrated circuits and systems," Ph.D. dissertation, Stanford University, 1980.

[2.46] D.B. Estreich and R.W. Dutton, "Modeling latch-up in CMOS integrated circuits and systems," *IEEE Transactions on Computer-Aided Design of Integrated Circuits,* vol. CAD-1, no. 4, pp. 157–163, Oct. 1982.

[2.47] R.R. Troutman, "Recent developments and future trends in latch-up prevention in scaled CMOS," *IEEE Transactions on Electron Devices,* vol. ED-30, p. 1564, 1983.

[2.48] R.R. Troutman, "Recent developments in CMOS latch-up," *IEEE International Electron Devices Meeting (IEDM'84),* p. 264, 1984.

[2.49] J.Y. Chen, "CMOS—The emerging VLSI technology," *IEEE Circuits and Devices Magazine,* pp. 16–31, March 1986.

[2.50] P.B. Ghate, "Metalization for very-large-scale integrated circuits," *Thin Solid Films,* 93, Elsevier Sequoia Publishers, The Hague, pp. 359–383, 1982.

[2.51] A.K. Sinha, "Interconnect materials technology for VLSI," *Proceedings of the First International Symposium on VLSI Science and Technology/1982,* Electrochemical Society, pp. 173–193, 1982.

[2.52] D. Pramatik and A.N. Saxena, "VLSI metallization using aluminum and its alloys, Part I," *Solid-State Technology,* pp. 127–133, Jan 1983.

[2.53] D. Pramatik and A.N. Saxena, "VLSI metallization using aluminum and its alloys, Part II," *Solid-State Technology,* pp. 131–138, March 1983.

[2.54] P.S. Ho, "VLSI interconnect metalization," *Semiconductor International,* pp. 128–133, Aug. 1985.

[2.55] J.R. Black, "Electromigration—A brief survey on some recent results," *IEEE Transactions on Electron Devices,* vol. ED-16, p. 338, 1969.

[2.56] P.P. Merchant, "Electromigration: an overview (VLSI metalization)," *Hewlett-Packard Journal,* vol. 33, no. 8, p. 28, Aug. 1982.

[2.57] P.S. Ho, "Basic problems for electromigration in VLSI," *Proceedings of 20th Annual International Reliability Physics Symposium,* p. 288, April 1982.

[2.58] M.H. Woods, "The implications of scaling on VLSI reliability," *Proceedings of 22th Annual International Reliability Physics Symposium,* 1984.

[2.59] S. Vaidya, D.B. Frazer, and A.K. Sinha, "Electromigration resistance of fine-line Al for VLSI applications," *Proceedings of 18th Annual International Reliability Physics Symposium*, pp. 165–167, 1980.

[2.60] S.I. Subramanian and C.Y. Ting, "Electromigration study of the Al-Cu/Ti/Al-Cu system," *1984 International Reliability Physics Symposium*, April 1984.

[2.61] R.E. Jones and L.D. Smith, "Contact spiking and electromigration passivation cracking observed for titanium layered aluminum metallization," *IEEE VLSI Multilevel Interconnection Conference*, pp. 194–200, Santa Clara, CA, June 1985.

[2.62] D.S. Gardner, T.L. Michalka, K.C. Saraswat, T.W. Barbee, J.P. McVittie, and J.D. Meindl, "Layered and homogenous films of aluminum and aluminum/silicon with titanium and tungsten for multilevel interconnects," *IEEE Journal of Solid-State Circuits*, vol. SC-20, no. 1, pp. 94–103, Feb. 1985.

[2.63] S. Vaidya and A.K. Sinha, "Electromigration induced leakage at shallow junction contacts metallized with aluminum/poly-silicon," *Proceedings of 20th Annual International Reliability Physics Symposium*, p. 50, 1982.

[2.64] P.A. Gargini, C. Tseng, and M.H. Woods, "Elimination of silicon electromigration in contacts by the use of an interposed barrier metal," *Proceedings of 20th Annual International Reliability Physics Symposium*, p. 66, 1982.

[2.65] R.A. Levy and M.L. Green, "Characterization of LPCVD aluminum for VLSI processing," *Proceedings of 1984 Symposium on VLSI Technology*, pp. 32–33, Sept. 1984.

[2.66] F. Walcyzk and J. Rubinstein, "A merged CMOS/bipolar VLSI process," *IEEE International Electron Devices Meeting (IEDM'83)*, pp. 59–62, Dec. 1983.

[2.67] J. Miyamoto, "A 1.0 μm n-well CMOS/bipolar technology for VLSI circuits," *IEEE International Electron Devices Meeting (IEDM'83)*, pp. 63–66, Dec. 1983.

[2.68] A.R. Alvarez, P. Meller, and B. Tien, "2 micron merged bipolar-CMOS technology," *IEEE International Electron Devices Meeting (IEDM'84)*, pp. 761–764, Dec. 1984.

[2.69] H. Momose, H. Shibata, S. Saitoh, J. Miyamoto, K. Kanzaki, and S. Kohyama, "1.0-μm n-well CMOS/bipolar technology," *IEEE Journal of Solid-State Circuits,* vol. SC-20, no. 1, pp. 137–143, Feb. 1985.

[2.70] Y. Okada, K. Kaneko, S. Kudo, K. Yamazaki, and T. Okabe, "An advanced bipolar-MOS-I^2L technology with a thin epitaxial layer for anolog-digital VLSI," *IEEE Journal of Solid-State Circuits,* vol. SC-20, no. 1, pp. 152–156, Feb. 1985.

[2.71] A.R. Alvarez, "Technology considerations in Bi-CMOS integrated circuits," *IEEE International Conference on Computer Design (ICCD'85),* pp. 159–163, Oct. 1985.

[2.72] R.W. Keyes, E.P. Harris, and K.L. Konnerth, "The role of low temperatures in the operation of logic circuitry," *Proceedings of IEEE,* vol. 58, pp. 1914–1932, 1970.

[2.73] F.H. Gaensslen, V.L. Rideout, E.J. Walker, and J.J. Walker, "Very-small MOSFETs for low temperature operation," *IEEE Transactions on Electron Devices,* vol. ED-24, pp. 218–229, March 1977.

[2.74] J.A. Bracchita, T.L. Honan, and R.L. Anderson, "Hot-electron-induced degradation in MOSFETs at 77 K," *IEEE Transactions on Electron Devices,* vol. ED-32, no. 9, pp. 1850–1857, Sept. 1985.

[2.75] J.C.S. Wu and J.D. Plummer, "Short-channel effects in MOSFETs at liquid-nitrogen temperature," *IEEE Transactions on Electron Devices,* vol. ED-33, no. 7, pp. 1012–1019, July 1986.

[2.76] M. Aoki, S. Hanamura, T. Masuhara, and K. Yano, "Performance and hot-carrier effects of small CRYO-CMOS devices," *IEEE Transactions on Electron Devices,* vol. ED-34, no. 1, pp. 8–19, Jan. 1987.

[2.77] J.Y.C. Sun, Y. Taur, R.H. Dennard, and S.P. Klepner, "Submicrometer-channel CMOS for low temperature operation," *IEEE Transactions on Electron Devices,* vol. ED-34, no. 1, pp. 19–28, Jan. 1987.

[2.78] L. Krusin-Elbaum, J.Y.C. Sun, and C.Y. Ting, "On the resistivity of $TiSi_2$: The implication for low temperature applications," *IEEE Transactions on Electron Devices,* vol. ED-34, no. 1, pp. 58–64, Jan. 1987.

[2.79] Special issue on Low-Temperature Semiconductor Electronics, *IEEE Transactions on Electron Devices,* vol. ED-34, no. 1, Jan. 1987.

[2.80] R.C. Longsworth and W.A. Steyert, "Technology for liquid-nitrogen-cooled computers," *IEEE Transactions on Electron Devices,* vol. ED–34, no. 1, pp. 4–7, Jan. 1987.

[2.81] T.H. Ning, D.D. Tang, and P.M. Solomon, "Scaling properties of of bipolar devices," *IEEE International Electron Devices Meeting (IEDM '80),* pp. 61–64, Dec. 1980.

[2.82] D.A. Hodges and H.G. Jackson, *Analysis and Design of Digital Integrated Circuits,* McGraw-Hill, New York, 1983.

[2.83] S. Konaka, Y. Yamamoto, and T. Sakai, "A 30 ps Si bipolar IC using super self-aligned process technology," *Extended Abstracts of the 16th (1984 International) Conference on Solid State Devices and Materials,* pp. 209–212, Kobe, Japan, 1984.

[2.84] T. Tashiro, H. Takemura, T. Kamiya, F. Tokuyoshi, S. Ohi, H. Shiraki, M. Nakamae, and T. Nakamura, "An 80 ps ECL circuit with high current density transistor," *IEEE International Electron Devices Meeting (IEDM'84),* pp. 686–689, Dec. 1984.

[2.85] M. Vora, Y.L. Ho, S. Bhamre, F. Chien, G. Bakker, H. Hingarh, and C. Smitz, "A sub-100 picosecond bipolar ECL technology," *IEEE International Electron Devices Meeting (IEDM'85),* pp. 34–37, Dec. 1985.

[2.86] D.D. Tang, G.P. Li, C.T. Chuang, D.A. Danner, M.B. Ketchen, J.L. Mauer, M.J. Smyth, M.P. Manny, J.D. Cressler, B.J. Ginsberg, E.J. Petrillio, T.H. Ning, C.C. Hu, and H.S. Park, "73 ps bipolar ECL circuits," *IEEE International Solid-State Circuits Conference (ISSCC'86),* pp. 104–105, Feb. 1986.

[2.87] S.P. Gaur, "Performance limitations of silicon bipolar transistors," *IEEE Transactions on Electron Devices,* vol. ED-26, no. 4, pp. 415–421, April 1979.

[2.88] D.D. Tang, "Heavy doping effects in NPN bipolar transistors," *IEEE Transactions on Electron Devices,* vol. ED-27, no. 5, pp. 563–570, March 1980.

[2.89] K. Nakazato, T. Nakamura, T. Okabe, and M. Nagata, "Characteristics and scaling properties of n-p-n transistors with a sidewall base contact structure," *IEEE Journal of Solid-State Circuits,* vol. SC-20, no. 1, pp. 248–252, Feb. 1985.

[2.90] T. Nakamura, K. Nakazato, T. Miyazaki, T. Okabe, and M. Nagata, "High-speed IIL circuits using a sidewall base contact structure," *IEEE Journal of Solid-State Circuits*, vol. SC-20, no. 1, pp. 168–172, Feb. 1985.

[2.91] J. Graul, A. Glasl, and H. Murrmann, "Ion implanted bipolar high-performance transistors with polysil emitter," *IEEE International Electron Devices Meeting (IEDM'75)*, pp. 450–454, Dec. 1975.

[2.92] J. Graul, A. Glasl, and H. Murrmann, "High-performance transistors with arsenic-implanted polysil emitters," *IEEE Journal of Solid-State Circuits*, vol. SC-10, pp. 491–495, Aug. 1976.

[2.93] T.H. Ning and R. Issac, "Effect of emitter contact on current gain of silicon bipolar devices," *IEEE Transactions on Electron Devices*, vol. ED-27, no. 11, pp. 2051–2055, Nov. 1980.

[2.94] G.L. Patton, "Physics, technology, and modeling of polysilicon emitter contacts for VLSI bipolar transistors," Ph.D. dissertation, Stanford University, 1986.

[2.95] Special Issue on Power, Logic, and Special Devices in III-V Compounds, *IEEE Transaction on Electron Devices*, vol. ED-27, no. 6, June 1980.

[2.96] Special Issue on Heterojunction Field-Effect Transistors, *IEEE Transaction on Electron Devices*, vol. ED-33, no. 5, May 1986.

[2.97] C.A. Liechti, "GaAs IC technology— Impact on the semiconductor industry," *IEEE 1984 International Electron Devices Meeting (IEDM'84)*, pp. 13–18, 1984.

[2.98] Special issue on GaAs Microprocessor Technology, *IEEE Computer*, vol. 19, no. 10, Oct. 1986.

[2.99] L.E. Larson, J.F. Jensen, and P.T. Greiling, "GaAs high-speed digital IC technology: An overview," *IEEE Computer*, vol. 19, no. 10, pp. 21–27, Oct. 1986.

[2.100] Y. Ishii et al., "Processing technologies for GaAs memory LSIs," *GaAs IC Symposium Digest of Technical Papers*, Boston, Oct. 1984.

[2.101] H. Markoc and P.M. Solomon, "The HEMT, a superfast transistor," *IEEE Spectrum*, pp. 28–35, Feb. 1984.

[2.102] H. Kroemer, "Heterostructure bipolar transistors and integrated circuits," *Proceedings of IEEE*, vol. 70, no. 1, pp. 13–25, 1982.

[2.103] P.M. Pitner, "Molecular Beam Epitaxy of Ultra-High Speed Hetero-junction Bipolar Transistors," Ph.D. dissertation, Stanford University, 1987.

Chapter 3

[3.1] "Bipolar packaging lags technology," *Electronic Engineering Times*, p. 75, Oct. 3, 1988.

[3.2] H.B. Bakoglu, "Packaging for high-speed systems," *IEEE International Solid-State Circuits Conference (ISSCC'88)*, pp. 100-101, San Francisco, Feb. 1988.

[3.3] I. Catt, D. Walton, and M. Davidson, *Digital Hardware Design*, Macmillan, London, 1979.

[3.4] J. Lyman, "Surface mounting alters the PC-board scene," *Electronics*, pp. 113–116, Feb. 9, 1984.

[3.5] A.H. Mones and R.K. Spielberger, "Interconnecting and packaging VLSI chips," *Solid State Technology*, pp. 119–122, Jan. 1984.

[3.6] E.C. Blackburn, "VLSI packaging reliability," *Solid State Technology*, pp. 113–116, Jan. 1984.

[3.7] J. Lyman, "VLSI packages are presenting diversified mix," *Electronics*, pp. 67–73, Sept. 17, 1984.

[3.8] R. Bowlby, "The DIP may take its final bows," *IEEE Spectrum*, vol. 22, no.3, pp. 37–42, June 1985.

[3.9] C.L. Cohen, "Japan's packaging goes world class," *Electronics*, pp. 26–31, Nov. 11, 1985.

[3.10] W. Andrews, "High-density gate arrays tax utility, packaging, and testing," *Computer Design*, pp. 43–47, Aug. 1, 1988.

[3.11] A. DeSena, "Innovative packages emerge to carry faster, denser chips," *Computer Design*, pp. 35–40, Oct. 1, 1988.

[3.12] A.J. Blodgett and D.R. Barbour, "Thermal conduction module: a high performance multilayer ceramic package," *IBM Journal of Research and Development*, vol. 26, no. 1, pp. 30-36, Jan. 1982.

[3.13] A.J. Blodgett, "Microelectronic packaging," *Scientific American*, pp. 86–96, July 1983.

[3.14] S. Oktay and H.C. Kammerer, "A conduction cooled module for high-performance LSI devices," *IBM Journal of Research and Development,* vol. 26, no. 1, pp. 55–66, Jan. 1982.

[3.15] R.C. Chu, U.P. Hwang, and R.E. Simons, "Conduction cooling for an LSI package: A one-dimensional approach," *IBM Journal of Research and Development,* vol. 26, no. 1, pp. 45–54, Jan. 1982.

[3.16] R.F. Bonner, J.A. Asselta, and F.W. Haining, "Advanced printed-circuit board design for high-performance computer applications," *IBM Journal of Research and Development,* vol. 26, no. 3, pp. 297–305, May 1982.

[3.17] D.P. Seraphim, "A new set of printed-circuit technologies for the IBM 3081 processor unit," *IBM Journal of Research and Development,* vol. 26, no. 1, pp. 37–44, Jan. 1982.

[3.18] T. Watari and H. Murano, "Packaging technology for the NEC SX supercomputer," *IEEE Transactions on Components, Hybrids, and Manufacturing Technology,* vol. CHMT-8, no. 4, pp. 462–467, Dec. 1985.

[3.19] F. Kobayashi, K. Ogiue, G. Toda, and M. Wajima, "Packaging technology for the supercomputer Hitachi S-810 array-processor," *Proceedings of the Electronics Components Conference,* pp. 379–382, May 1984.

[3.20] C.W. Ho, D.A. Chance, C.H. Bajorek, and R.E. Acosta, "The thin film module as a high-performance semiconductor package," *IBM Journal of Research and Development,* vol. 26, pp. 286–296, May 1982.

[3.21] C.J. Bartlett, "Advanced packaging for VLSI," *Solid State Technology,* pp. 119-123, June 1986.

[3.22] C.J. Bartlett, J.M. Segelken, and N.A. Teneketges, "Multichip packaging design for VLSI-based systems," *IEEE Transactions on Components, Hybrids, and Manufacturing Technology,* pp. 647–653, vol. CHMT-12, no. 4, Dec. 1987.

[3.23] J. Lyman, "Multichip modules aim at next-generation VLSI," *Electronic Design,* pp. 33–34, March 9, 1989.

[3.24] R.K. Spielberger, C.D. Huang, W.H. Nunne, A.H. Mones, D.L. Fett, and F.L. Hampton, "Silicon-on-silicon packaging," *IEEE Transactions on Components, Hybrids, and Manufacturing Technology,* vol. CHMT-7, no. 2, pp. 193–196, June 1984.

[3.25] G.F. Taylor, B.J. Donlan, J.F. McDonald, A.S. Bergendahl, and R.H. Steinvorth, "The wafer transmission module—Wafer scale integration packaging," *IEEE 1985 Custom Integrated Circuits Conference (CICC '85)*, pp. 55–58, May 1985.

[3.26] A.S. Bergendahl, B.J. Donlan, J.F. McDonald, R.H. Steinvorth, and G.F. Taylor, "A thick film lift-off technique for high frequency interconnection in wafer scale integration," *IEEE 1985 VLSI Multilevel Interconnection Conference*, pp. 154–162, June 25–26, 1985.

[3.27] F.C. Chong, C.W. Ho, K. Liu, and S. Westbrook, "A high-density multichip memory module," *WESCON/85 Professional Program Session Record*, session 7, Nov. 1985.

[3.28] J. Lyman, "Silicon-on-silicon hybrids are coming into their own," *Electronics*, vol. 60, no. 11, pp. 47–48, May 28, 1987.

[3.29] R.W. Johnson, J.L. Davidson, R.C. Jaeger, and D.V. Kerns, "Hybrid silicon wafer-scale packaging technology," *IEEE International Solid-State Circuits Conference (ISSCC'86)*, pp. 166–167, Feb. 1986.

[3.30] B.A. Wooley, M.A. Horowitz, R.F. Pease, and T.S. Yang, "Active substrate system integration," *IEEE International Conference on Computer Design (ICCD'87)*, pp. 468–471, Rye Brook, NY, Oct. 1987.

[3.31] R.C. Aubusson and I. Catt, "Wafer-scale integration—A fault-tolerant procedure," *IEEE Journal of Solid-State Circuits*, vol. SC-13, pp. 339–344, June 1978.

[3.32] Y. Egawa, T. Wada, Y. Ohmori, N. Tsuda, and K. Masuda, "A 1-Mbit full-wafer MOS RAM," *IEEE Journal of Solid-State Circuits*, vol. SC-15, pp. 677–686, Aug. 1980.

[3.33] Y. Kitano, S. Kohda, H. Kikuchi, and S. Sakai, "A 4-Mbit full-wafer ROM," *IEEE Journal of Solid-State Circuits*, vol. SC-15, pp. 686, Aug. 1980.

[3.34] I. Catt, "Wafer-scale integration," *Wireless World* (Great Britain), vol. 87, no. 1546, pp. 57–59, July 1981.

[3.35] L. Snyder, "Introduction to the configurable, highly parallel computer," *IEEE Computer*, pp. 47–56, Jan. 1982.

[3.36] D.L. Peltzer, "Wafer-scale integration: the limits of VLSI?," *VLSI Design*, pp. 43–47, Sept. 1983.

[3.37] J.F. McDonald, E.H. Rodgers, K. Rose, and A.J. Steckl, "The trials of wafer scale integration," *IEEE Spectrum,* pp. 32–39, Oct. 1984.

[3.38] R.R. Johnson, "The significance of wafer-scale integration in computer design," *IEEE 1984 International Conference on Computer Design (ICCD '84),* pp. 101–105, Oct. 1984.

[3.39] H. Stopper, "A wafer with electrically programmable interconnections," *IEEE International Solid-State Circuits Conference (ISSCC'85),* pp. 268–269, Feb. 1985.

[3.40] J.I. Raffel, A.H. Anderson, G.H. Chapman, K.H. Konkle, B. Mathur, A.M. Soares, and P.W. Wyatt, "A wafer-scale integrator using restructurable VLSI," *IEEE Journal of Solid-State Circuits,* vol. SC-20, no. 1, pp. 399–406, Feb. 1985.

[3.41] T. Leighton and C.E. Leiserson, "Wafer-scale integration of systolic arrays," *IEEE Transactions on Computers,* vol. C-34, no. 5, pp. 448–461, May 1985.

[3.42] *American Institute of Physics Handbook, 3d Edition,* McGraw-Hill, New York, 1972.

[3.43] M. Mahalingam, "Thermal management in semiconductor device packaging," *Proceedings of the IEEE,* vol. 73, no. 9, pp. 1396–1404, Sept. 1985.

[3.44] G. Fehr, J. Long, and A. Tippetts, "New generation of high pin count packages," *Proceedings of IEEE 1985 Custom Integrated Circuits Conference (CICC'85),* pp. 46–49, 1985.

[3.45] C. Mitchell and H. Berg, "Thermal studies of a plastic dual-in-line package," *IEEE Transactions on Components, Hybrids, and Manufacturing,* vol. CHMT-2, no. 4, pp. 500–511, Dec. 1979.

[3.46] J. Andrews, M. Mahalingam, and H. Berg, "Thermal characteristics of 16- and 40-pin plastic DIPs," *IEEE Transactions on Components, Hybrids, and Manufacturing,* vol. CHMT-4, no. 4, pp. 455–461, Dec. 1981.

[3.47] M. Mahalingam, J. Andrews, and J. Drye, "Thermal studies on pin grid array packages for high density LSI and VLSI logic circuits," *IEEE Transactions on Components, Hybrids, and Manufacturing,* vol. CHMT-6, no. 3, pp. 246–256, Sept. 1983.

[3.48] K. Smith, "An inexpensive high frequency, high power, VLSI chip carrier," *Proceedings of IEEE 1985 Custom Integrated Circuits Conference,* pp. 42–45, 1985.

[3.49] D.B. Tuckerman and F. Pease, "High performance heat sinking for VLSI," *IEEE Electron Device Letters,* vol. EDL-2, no. 5, pp. 126–129, May 1981.

[3.50] D.B. Tuckerman, "Heat-transfer microstructures for integrated circuits," Ph.D. dissertation, Stanford University, 1984.

[3.51] R.F. Pease and O.K. Kwon, "Physical limits to the useful packaging density of electronic systems," *IBM Journal of Research and Development,* pp. 636–646, vol. 32, no. 5, Sept. 1988.

[3.52] E.A. Wilson, "True liquid cooling of computers," *Proceedings of the 1977 National Computer Conference,* pp. 341–348, AFIPS Press, Montvale, NJ, 1977.

[3.53] R.J. Petschauer, "Evolution of high performance computer packaging," *Professional Program Session Record of WESCON/85,* session 7, Nov. 1985.

[3.54] N.W. Ashcroft and N.D. Mermin, *Solid State Physics,* Chap. 34, Holt, Rinehart, and Winston, Saunders College, Philadelphia, 1976.

[3.55] J.W. Goodman, F.I. Leonberger, S.Y. Kung, and R.A. Athale, "Optical interconnections for VLSI systems," *Proceedings of the IEEE,* vol. 72, no. 7, pp. 850–866, July 1984.

[3.56] J.W. Goodman, R.K. Kostuk, and B. Clymer, "Optical interconnects: an overview," *IEEE VLSI Multilevel Interconnection Conference (V-MIC'85),* pp. 219–224, Santa Clara, CA, June 1985.

[3.57] T. Bell, "Optical computing: a field in flux," *IEEE Spectrum,* Aug. 1986.

[3.58] L.D. Hutcheson, P. Haugen, and A. Husain, "Optical interconnects replace hardwire," *IEEE Spectrum,* pp. 30–35, March 1987.

[3.59] Special issue on optical interconnections, *Optical Engineering,* Oct. 1986.

Chapter 4

[4.1] H. Sachs and W. Holligsworth, "A high performance 846,000 transistor UNIX engine—the Fairchild CLIPPER," *IEEE International Conference on Computer Design: VLSI in Computers (ICCD'85),* Oct. 1985.

[4.2] L.W. Schaper and D.I. Amey, "Improved electrical performance required for future MOS packaging," *IEEE Transactions on Components, Hybrids, and Manufacturing Technology,* vol. CHMT-6, pp. 282–289, Sept. 1983.

[4.3] L.A. Glasser and D.W. Dobberpuhl, *The Design and Analysis of VLSI Circuits,* Addison-Wesley Publishing Co., Reading, Mass., 1985.

[4.4] C.P. Yuan and T.N. Trick, "A simple formula for the estimation of the capacitance of two-dimensional interconnects in VLSI circuits," *IEEE Electron Device Letters,* vol. EDL-3, pp. 391–393, Dec. 1982.

[4.5] T. Sakurai and K. Tamaru, "Simple formulas for two- and three-dimensional capacitances," *IEEE Transactions on Electron Devices,* vol. ED-30, pp. 183–185, Feb. 1983.

[4.6] A.E. Ruehli, "Survey of computer-aided electrical analysis of integrated circuit interconnections," *IBM Journal of Research and Development,* vol. 23, pp. 626–639, Nov. 1979.

[4.7] A.E. Ruehli and P.A. Brennan, "Accurate metallization capacitances for integrated circuits and packages," *IEEE Journal of Solid-State Circuits,* vol. SC-8, pp. 289–290, Aug. 1973.

[4.8] A.E. Ruehli and P.A. Brennan, "Capacitance models for integrated circuit metallization wires," *IEEE Journal of Solid-State Circuits,* vol. SC-10, Dec. 1975.

[4.9] R.L.M. Dang and N. Shigyo, "Coupling capacitances for two-dimensional wires," *IEEE Electron Device Letters,* vol. EDL-2, pp. 196–197, Aug. 1981.

[4.10] W.H. Chang, "Analytical IC metal-line capacitance formulas," *IEEE Transactions on Microwave Theory Technology,* vol. MTT-24, pp. 608–611, Sept. 1976.

[4.11] M. Inoue et al., "A 16 Mb DRAM with an open bit-line architecture," *IEEE International Solid-State Circuits Conference (ISSCC'88),* pp. 246–247, Feb. 1988.

[4.12] S. Fujii et al., "A 45ns 16Mb DRAM with triple-well structure," *IEEE International Solid-State Circuits Conference (ISSCC'89),* pp. 248–249, Feb. 1989.

[4.13] M. Aoki et al., "An experimental 16 Mb DRAM with transposed data-line structure," *IEEE International Solid-State Circuits Conference (ISSCC'88),* pp. 250–251, Feb. 1988.

[4.14] O. Minato, T. Masuhara, T. Sasaki, Y. Sakai, T. Hayashida, K. Naga-sawa, K. Nishimura, and T. Yasui, "A Hi-CMOSII 8K×8b static RAM," *IEEE International Solid-State Circuits Conference (ISSCC'82),* pp. 256–257, Feb. 1982.

[4.15] O. Minato, T. Masuhara, T. Sasaki, Y. Sakai, and T. Hayashida, "A 20ns 64K CMOS SRAM," *IEEE International Solid-State Circuits Conference (ISSCC'84),* pp. 222–223, Feb. 1984.

[4.16] H. Shinohara, K. Anami, K. Ichinose, T. Wada, Y. Kohno, Y. Kawai, Y. Akasaka, and S. Kayano, "A 45 ns 256K CMOS SRAM with tri-level word line," *IEEE International Solid-State Circuits Conference (ISSCC'85),* pp. 62–63, Feb. 1985.

[4.17] H. Shinohara, K. Anami, K. Ichinose, T. Wada, Y. Kohno, Y. Kawai, Y. Akasaka, and S. Kayano, "A 45 ns 256K CMOS SRAM with a tri-level word line," *IEEE Journal of Solid-State Circuits,* vol. SC-20, no. 5, pp. 929–934, Oct. 1985.

[4.18] T. Komatsu, H. Taniguchi, N. Okazaki, T. Nishihara, S. Kayama, N. Hoshi, J. Aoyama, and T. Shimada, "A 35-ns 128K×8 CMOS SRAM," *IEEE Journal of Solid-State Circuits,* vol. SC-22, no. 5, pp. 721–726, Oct. 1987.

[4.19] F. Miyaji et al., "A 25ns 4Mb CMOS SRAM with dynamic bit line loads," *IEEE International Solid-State Circuits Conference (ISSCC'89),* pp. 250–251, Feb. 1989.

[4.20] T. Wada, T. Hirose, H. Shinohara, Y. Kawai, K. Yuzuriha, Y. Kohno, and S. Kayano, "A 34-ns 1-Mbit CMOS SRAM using triple polysilicon," *IEEE Journal of Solid-State Circuits,* vol. SC-22, no. 5, pp. 727–732, Oct. 1987.

[4.21] M. Horowitz, P. Chow, D. Stark, R.T. Simoni, A. Salz, S. Przybylski, J. Hennessy, G. Gulak, A. Agarwal, and J.M. Acken, "MIPS-X: A 20-MIPs peak, 32-bit microprocessor with on-chip cache," *IEEE Journal of Solid-State Circuits,* vol. SC-22, no. 5, pp. 790–799, Oct. 1987.

[4.22] S. Gumm and C.T. Dreher, "Designer's guide to dynamic RAMs—Part 1: Unraveling the intricacies of dynamic RAMs," *EDN,* vol. 34, no. 7, pp. 155–166, March 30, 1989.

[4.23] R. Wawrzynek, "Designer's guide to dynamic RAMs—Part 2: Tailor memory-system architecture for your chosen DRAM," *EDN*, vol. 34, no. 8, pp. 157–164, April 13, 1989.

[4.24] R. Adams and G. Scavone, "Designer's guide to dynamic RAMs—Part 3: Design a DRAM controller from the top down," *EDN*, vol. 34, no. 9, pp. 183–188, April 27, 1989.

[4.25] F. Tabaian and M. Toomer, "Designer's guide to dynamic RAMs—Part 4: Attention to layout details facilitates DRAM board design," *EDN*, vol. 34, no. 10, pp. 179–186, May 11, 1989.

[4.26] R. Foss, "The design of MOS dynamic memories," *IEEE International Solid-State Circuits Conference (ISSCC'79)*, pp. 140–141, Feb. 1979.

[4.27] R. Foss, "The evolution of dynamic RAM," *1985 International Symposium on VLSI Technology, Systems and Applications*, pp. 9–13, Taipei, May 1985.

[4.28] T. Mano, J. Yamada, J. Inoue, and S. Nakajima, "Circuit techniques for a VLSI memory," *IEEE Journal of Solid-State Circuits*, vol. SC-18, no. 5, pp. 463–469, Oct. 1983.

[4.29] S. Saito, S. Fujii, Y. Okada, S. Sawada, S. Shinozaki, K. Natori, and O. Osawa, "A 1-Mbit CMOS DRAM with fast page mode and static column mode," *IEEE Journal of Solid-State Circuits*, vol. SC-20, no. 5, pp. 903–908, Oct. 1985.

[4.30] T. Mano, T. Matsumura, J. Yamada, J. Inoue, S. Nakajima, K. Minegishi, K. Miura, T. Matsuda, C. Hashimoto, and H. Namatsu, "Circuit technologies for 16Mb DRAMs," *IEEE International Solid-State Circuits Conference (ISSCC'87)*, pp. 22–23, Feb. 1987.

[4.31] K.U. Stein, A. Sihling, and E.Doering, "Storage array and sense/refresh circuits for single-transistor memory cells," *IEEE Journal of Solid-State Circuits*, vol. SC-7, pp. 336–340, Oct. 1972.

[4.32] K.U. Stein and H. Friedrich, "A 1 mil^2 single transistor memory cell in silicon-gate technology," *IEEE Journal of Solid-State Circuits*, vol. SC-8, pp. 319–323, Oct. 1973.

[4.33] S. Suzuki, M. Nakao, T. Takeshima, and M. Yoshida, "A 128K word × 8b DRAM," *IEEE International Solid-State Circuits Conference (ISSCC '84)*, pp. 106–107, Feb. 1984.

[4.34] N.C. Lu and H.H. Chao, "Half-V_{DD} bit-line sensing scheme in CMOS DRAMs," *IEEE Journal of Solid-State Circuits,* vol. SC-19, no. 4, pp. 451–454, Aug. 1984.

[4.35] M. Takada, T. Takeshima, M. Sakamoto, T. Shimizu, H. Abiko, T. Katoh, M. Kikuchi, S. Takahashi, Y. Sato, and Y. Inoue, "A 4-Mbit DRAM with half-internal-voltage bit-line precharge," *IEEE Journal of Solid-State Circuits,* vol. SC-21, no. 5, pp. 612–617, Oct. 1986.

[4.36] S. Fujii, S. Saito, Y. Okada, M. Sato, S. Sawada, S. Shinozaki, K. Natori, and O. Ozawa, "A 50 μA standby 1 MW $\times$ 1b/256KW $\times$ 4b CMOS DRAM," *IEEE International Solid-State Circuits Conference (ISSCC'86),* pp. 266-267, Feb. 1986.

[4.37] T. Fruyama, T. Ohsawa, Y. Watanabe, H. Ishiuchi, T. Watanabe, T. Tanaka, K. Natori, and O. Ozawa, "An experimental 4-Mbit CMOS DRAM," *IEEE Journal of Solid-State Circuits,* vol. SC-21, no. 5, pp. 605–611, Oct. 1986.

[4.38] F. Masuoka, S. Ariizumi, T. Iwase, M. Ono, and N. Endo, "An 80 ns 1Mb ROM," *IEEE International Solid-State Circuits Conference (ISSCC'84),* pp. 146–147, San Francisco, 1984.

[4.39] E.M. Blaser and D.A. Conrad, "FET logic configuration," *IEEE International Solid-State Circuits Conference (ISSCC'78),* pp. 14–15, 1978.

[4.40] P.W. Cook, S.E. Schuster, J.T. Parrish, V. DiLonardo and D.R. Freedman, "1 μm MOSFET VLSI technology: part III- logic circuit design methodology and applications," *IEEE Transactions on Electron Devices,* vol. ED-26, pp. 333–345, April 1979.

[4.41] L.C. Pfennings, W.G. Mol, J.J. Bastiaens, and J.M. vanDijk, "Differential split-level CMOS logic for sub-nanosecond speeds," *IEEE International Solid-State Circuits Conference (ISSCC'85),* pp. 212–213, Feb. 1985.

[4.42] L.G. Heller, W.R. Griffin, J.W. Davis, and N.G. Thoma, "Cascode voltage switch logic: a differential CMOS logic family," *IEEE International Solid-State Circuits Conference (ISSCC'84),* pp. 16–17, Feb. 1984.

[4.43] R.H. Krambeck, C.M. Lee, and H.S. Law, "High-speed compact circuits with CMOS," *IEEE Journal of Solid-State Circuits,* vol. SC-17, pp. 614-619, June 1982.

[4.44] N.F. Goncalves and H.J. deMan, "NORA: a racefree dynamic CMOS technique for pipelined logic structures," *IEEE Journal of Solid-State Circuits*, vol. SC-18, no. 3, pp. 261-266, June 1983.

[4.45] C.L. Chen and R.H.J.M. Otten, "Considerations for implementing CMOS processors," *IEEE International Conference on Computer Design (ICCD'84)*, pp. 48–53, Oct. 1984.

[4.46] V.G. Oklobdzija and R.K. Montoye, "Design-performance trade-offs in CMOS domino logic," *IEEE Custom Integrated Circuits Conference (CICC'85)*, pp. 334–337, 1985.

[4.47] C.M. Lee and E.W. Szeto, "Zipper CMOS," *IEEE Circuits and Devices Magazine*, pp. 10–17, May 1986.

[4.48] J.A. Pretorius, A.S. Shubart, and C.A.T. Salama, "Latched domino CMOS logic," *IEEE Journal of Solid-State Circuits*, vol. SC-21, no. 4, pp. 514–522, Aug. 1986.

[4.49] H.B. Bakoglu and J.D. Meindl, "CMOS driver and receiver circuits for reduced interconnection delays," *1985 International Symposium on VLSI Technology, Systems and Applications*, pp. 171–175, Taipei, Taiwan, May 1985.

[4.50] H.B. Bakoglu and J.D. Meindl, "New CMOS driver and receiver circuits reduce interconnection propagation delays," *1985 Symposium on VLSI Technology*, pp. 54–55, Kobe, Japan, May 1985.

[4.51] H.C. Lin and L.W. Linholm, "An optimized output stage for MOS Integrated Circuits," *IEEE Journal of Solid-State Circuits*, vol. SC-10, no. 2, pp. 106–109, April 1975.

[4.52] R.C. Jeager, "Comments on 'An optimized output stage for MOS integrated circuits'," *IEEE Journal of Solid-State Circuits*, vol. SC-10, pp. 185–186, June 1975.

[4.53] C. Mead and M. Rem, "Minimum propagation delays in VLSI," *IEEE Journal of Solid-State Circuits*, vol. SC-17, pp. 773–775, Aug. 1982.

[4.54] E.T. Lewis, "Optimization of device area and overall delay for CMOS VLSI design," *IEEE Proceedings*, vol. 72, pp. 670–689, June 1984.

[4.55] R. Poujours et al., "Low level MOS transistor amplifier using storage techniques," *IEEE International Solid-State Circuits Conference (ISSCC'73)*, pp. 152–153, Feb. 1973.

[4.56] J.L. McCreary and P.R. Gray, "All-MOS charge redistribution analog-to-digital conversion techniques—part I," *IEEE Journal of Solid-State Circuits,* vol. SC-10, pp. 371–379, Dec. 1975.

[4.57] Y.S. Yee, L.M. Terman, and L.G. Heller, "A 1 mV MOS comparator," *IEEE Journal of Solid-State Circuits,* vol. SC-13, pp. 294–297, June 1978.

[4.58] L.W. Nagel, "SPICE 2: A computer program to simulate semiconductor circuits," Electronics Research Laboratory, University of California, Berkeley, memo. ERL-M510, May 1975.

[4.59] J.R. Pfiester, "Performance Limits of CMOS Very Large Scale Integration," Technical Report No. G541-1, Stanford Electronics Laboratories, Stanford University, Stanford, 1984.

[4.60] J. Miyamoto, S. Saitoh, H. Momose, H. Shibata, K. Kanzaki, and T. Iizuka, "A 28ns CMOS SRAM with bipolar sense amplifiers," *IEEE International Solid-State Circuits Conference (ISSCC'84),* pp. 224–225, Feb. 1984.

[4.61] K. Ogiue, M. Odaka, S. Miyaoka, I. Masuda, I. Ikeda, K. Tonomura, and T. Ohba, "A 13ns/500mW 64 Kb ECL RAM," *IEEE International Solid-State Circuits Conference (ISSCC'86),* pp. 212–213, Feb. 1986.

[4.62] S. Miyaoka, M. Odaka, K. Ogiue, T. Ikeda, M. Suzuki, H. Higuchi, and M. Hirao, "A 7-ns/350-mW 64-kbit ECL-compatible RAM," *IEEE Journal of Solid-State Circuits,* vol. SC-22, no. 5, pp. 847–849, Oct. 1987.

[4.63] M. Matsui, T. Ohtani, J.I. Tsujimoto, H. Iwai, A. Suzuki, K. Sato, M. Isobe, T. Matsuno, J.I. Matsunaga, and T. Iizuka, "A 25-ns 1-Mbit CMOS SRAM with loading-free bitlines," *IEEE Journal of Solid-State Circuits,* vol. SC-22, no. 5, pp. 733–740, Oct. 1987.

[4.64] T. Hotta, I. Masuda, and H. Maejima, "CMOS/bipolar circuits for 60 MHz digital processing," *IEEE International Solid-State Circuits Conference (ISSCC'86),* pp. 190–191, Feb. 1986.

[4.65] S.C. Lee, D.W. Schucker, and P.T. Hickman, "Bi-CMOS circuits for high performance VLSI," *1984 Symposium on VLSI Technology,* pp. 46–47, Sept. 1984.

[4.66] Y. Kowase, Y. Yanagawa, T. Inaba, J. Mameda, N. Horie, S. Ueda, and M. Nagata, "A Bi-CMOS analog/digital LSI with the programmable

280 bit SRAM," *IEEE Custom Integrated Circuits Conference (CICC'85)*, pp. 170–173, 1985.

[4.67] M.S. Adler, "A comparison between BIMOS device types," *IEEE Transactions on Electron Devices*, vol. ED-33, no. 2, pp. 286–293, Feb. 1986.

[4.68] H.J. DeLosSantos, and B. Hoefflinger, "Optimization and scaling of CMOS-bipolar drivers for VLSI interconnections," *IEEE Transactions on Electron Devices*, vol. ED-33, no. 11, pp. 1722–1730, , Nov. 1986.

[4.69] B.C. Cole, "Mixed-process chips are about to hit the big time," *Electronics*, pp. 27–31, March 3, 1986.

[4.70] "How Motorola moved BIMOS up to VLSI levels," *Electronics*, pp. 67–70, July 10, 1986.

[4.71] W.J. Kitchen and A.R. Alvarez, "Bi-CMOS for computer applications," *IEEE International Conference on Computer Design (ICCD'87)*, pp. 244–249, Oct. 1987.

[4.72] C.L. Cohen, "NEC's BiCMOS arrays shatter record," *Electronics*, pp. 82–83, Aug. 6, 1987.

[4.73] F. Walcyzk and J. Rubinstein, "A merged CMOS/bipolar VLSI process," *IEEE International Electron Devices Meeting (IEDM'83)*, pp. 59–62, Dec. 1983.

[4.74] J. Miyamoto, "A 1.0 μm n-well CMOS/bipolar technology for VLSI circuits," *IEEE International Electron Devices Meeting (IEDM'83)*, pp. 63–66, Dec. 1983.

[4.75] A.R. Alvarez, P. Meller, and B. Tien, "2 micron merged bipolar-CMOS technology," *IEEE International Electron Devices Meeting (IEDM'84)*, pp. 761–764, Dec. 1984.

[4.76] H. Momose, H. Shibata, S. Saitoh, J. Miyamoto, K. Kanzaki, and S. Kohyama, "1.0-μm n-well CMOS/bipolar technology," *IEEE Journal of Solid-State Circuits*, vol. SC-20, no. 1, pp. 137–143, Feb. 1985.

[4.77] Y. Okada, K. Kaneko, S. Kudo, K. Yamazaki, and T. Okabe, "An advanced bipolar-MOS-I^2L technology with a thin epitaxial layer for anolog-digital VLSI," *IEEE Journal of Solid-State Circuits*, vol. SC-20, no. 1, pp. 152–156, , Feb. 1985.

[4.78] A.R. Alvarez, "Technology considerations in Bi-CMOS integrated circuits," *IEEE International Conference on Computer Design (ICCD'85)*, pp. 159–163, Oct. 1985.

Chapter 5

[5.1] R.H. Dennard, F.H. Gaensslen, H.N. Yu, V.L. Rideout, E. Bassous, and A.R. LeBlanc, "Design of ion implanted MOSFET's with very small physical dimensions," *IEEE Journal of Solid-State Circuits,* vol. SC-9, pp. 256–268, Oct. 1974.

[5.2] M.I. Elmasry, "Interconnection delays in MOSFET VLSI," *IEEE Journal of Solid-State Circuits,* vol. SC-16, pp. 585–591, Oct. 1981.

[5.3] K.C. Saraswat and F. Mohammadi, "Effect of scaling of interconnections on the time delay of VLSI circuits," *IEEE Journal of Solid-State Circuits,* vol. SC-17, pp. 275–280, April 1982.

[5.4] H.B. Bakoglu and J.D. Meindl, "Optimal interconnect circuits for VLSI," *IEEE International Solid-State Circuits Conference (ISSCC'84),* pp. 164-165, San Francisco, Feb. 1984.

[5.5] H.B. Bakoglu and J.D. Meindl, "Optimal interconnection circuits for VLSI," *IEEE Transactions on Electron Devices,* vol. ED-32, pp. 903–909, May 1985.

[5.6] C.D. Phillips, W.C. Sellbach, J.A. Narud, and K.K. Lynn, "Complex monolithic arrays, some aspects of design and fabrication," *IEEE Journal of Solid-State Circuits,* vol. SC-2, pp. 156–172, Dec. 1967.

[5.7] J.T. Wallmark, "Noise spikes in digital VLSI circuits," *IEEE Transactions on Electron Devices,* vol. ED-29, no. 3, pp. 451–458, March 1982.

[5.8] P.M. Solomon, "A comparison of semiconductor devices for high-speed logic," *Proceedings of the IEEE,* vol. 70, pp. 489–509, May 1982.

[5.9] H. Yuan, Y. Lin, and S. Chiang, "Properties of interconnection on silicon, sapphire and semi-insulating gallium arsenide substrates," *IEEE Transactions on Electron Devices,* vol. ED-29, pp. 639–644, April 1982.

[5.10] T. Sakurai and T. Tamuru, "Simple formulas for two- and three-dimensional capacitances," *IEEE Transactions on Electron Devices,* vol. ED-30, pp. 183–185, Feb. 1983.

[5.11] C.P. Yuan and T.N. Trick, "A simple formula for the estimation of the capacitance of two-dimensional interconnects in VLSI circuits," *IEEE Electron Device Letters,* vol. EDL-3, pp. 391–393, Dec. 1982.

[5.12] A.E. Ruehli, "Survey of computer-aided electrical analysis of integrated circuit interconnections," *IBM Journal of Research and Development,* vol. 23, pp. 626–639, Nov. 1979.

[5.13] A.E. Ruehli and P.A. Brennan, "Accurate metallization capacitances for integrated circuits and packages," *IEEE Journal of Solid-State Circuits,* vol. SC-8, pp. 289–290, Aug. 1973.

[5.14] A.E. Ruehli and P.A. Brennan, "Capacitance models for integrated circuit metallization wires," *IEEE Journal of Solid-State Circuits,* vol. SC-10, Dec. 1975.

[5.15] R.L.M. Dang and N. Shigyo, "Coupling capacitances for two-dimensional wires," *IEEE Electron Device Letters,* vol. EDL-2, pp. 196–197, Aug. 1981.

[5.16] W.H. Chang, "Analytical IC metal-line capacitance formulas," *IEEE Transactions on Microwave Theory and Technology,* vol. MTT-24, pp. 608–611, Sept. 1976.

[5.17] D.B. Scott, W.R. Hunter, and H. Shichijo, "A transmission line model for silicided diffusions: impact on the performance of VLSI circuits," *IEEE Transactions on Electron Devices,* vol. ED-29, pp. 651–661, April 1982.

[5.18] A.K. Sinha, J.A. Cooper, and H.J. Levinstein, "Speed limitations due to interconnect time constants in VLSI integrated circuits," *IEEE Electron Device Letters,* vol. EDL-3, pp. 90–92, April 1982.

[5.19] A. Mayadas and M. Sahtzkes, "Electrical resistivity model for polycrystalline films: the case of arbitrary reflection at external surface," *Physical Review,* B-1, pp. 1382–1389, 1970.

[5.20] R.W. Keyes, E.P. Harris, and K.L. Konnerth, "The role of low temperatures in the operation of logic circuitry," *Proceedings of the IEEE,* vol. 58, no. 12, pp. 1914–1932, Dec. 1970.

[5.21] A. Wilnai, "Open-ended RC line model predicts MOSFET IC response," *EDN,* pp. 53–54, Dec. 1971.

[5.22] W.E. Engeler and D.M. Brown, "Performance of refractory metal multilevel interconnection system," *IEEE Transactions on Electron Devices,* vol. ED-19, no. 1, pp. 54–61, Jan. 1972.

[5.23] Y.V. Rajput, "Modelling distributed RC lines for the transient analysis of complex networks," *International Journal of Electronics,* vol. 36, no. 5, pp. 709–717, 1974.

[5.24] P. Penfield and J. Rubinstein, "Signal delay in RC tree networks," *IEEE 18th Design Automation Conference,* pp. 613–617, 1981.

[5.25] J. Rubinstein, P. Penfield, and M.A. Horowitz, "Signal delay in *RC* tree networks," *IEEE Interactions on Computer-Aided Design,* vol. CAD-2, no. 3, pp. 202–210, July 1983.

[5.26] R.J. Antinone and G.W. Brown, "The modeling of resistive interconnects for integrated circuits," *IEEE Journal of Solid-State Circuits,* vol. SC-18, no. 2, pp. 200–203, April 1983.

[5.27] G. De Mey, "A comment on 'The modeling of resistive interconnects for integrated circuits,' " *IEEE Journal of Solid-State Circuits,* vol. SC-19, no. 4, pp. 542–543, Aug. 1984.

[5.28] A.J. Walton, R.J. Holwill, and J.M. Robertson, "Numerical simulation of resistive interconnects for integrated circuits," *IEEE Journal of Solid-State Circuits,* vol. SC-20, no. 6, pp. 1252–1258, Dec. 1985.

[5.29] T. Sakurai, "Approximation of wiring delay in MOSFET LSI," *IEEE Journal of Solid-State Circuits,* vol. SC-18, pp. 418–426, Aug. 1983.

[5.30] M.A. Horowitz, "Timing models for MOS circuits," Ph.D. dissertation, Technical Report No. SEL83-003, Stanford University, 1983.

[5.31] "TRW's superchip passes first milestone," *Electronics,* pp. 49–54, July 10, 1986.

[5.32] R.W. Keyes, "The wire-limited logic chip," *IEEE Journal of Solid-State Circuits,* vol. SC-17, pp. 1232–1233, Dec. 1982.

[5.33] H.C. Lin and L.W. Linholm, "An optimized output stage for MOS Integrated Circuits," *IEEE Journal of Solid-State Circuits,* vol. SC-10, no. 2, pp. 106–109, April 1975.

[5.34] R.C. Jeager, "Comments on 'An optimized output stage for MOS integrated circuits,' " *IEEE Journal of Solid-State Circuits,* vol. SC-10, pp. 185–186, June 1975.

[5.35] C. Mead and M. Rem, "Minimum propagation delays in VLSI," *IEEE Journal of Solid-State Circuits,* vol. SC-17, pp. 773–775, Aug. 1982.

[5.36] E.T. Lewis, "Optimization of device area and overall delay for CMOS VLSI design," *Proceedings of the IEEE,* vol. 72, pp. 670–689, June 1984.

[5.37] D. Pramanik and A.N. Saxena, "VLSI metallization using aluminum and its alloys: Part I," *Solid-State Technology,* pp. 127–133, Jan. 1983.

[5.38] D. Pramanik and A.N. Saxena, "VLSI metallization using aluminum and its alloys: Part II," *Solid-State Technology,* pp. 131–138, March 1983.

[5.39] C.L. Seitz, "Self-timed VLSI systems," *Proceedings of Caltech Conference on VLSI,* C.L. Seitz, ed., pp. 345–355, Jan. 1980.

[5.40] C. Mead and L. Conway, *Introduction to VLSI Systems,* Addison-Wesley Publishing Co., Reading, Mass., 1980, Chap. 7.

[5.41] R.W. Keyes, "The evolution of digital electronics towards VLSI," *IEEE Transactions on Electron Devices,* vol. ED-26, pp. 271–278, 1979.

[5.42] W.S. Song and L.A. Glasser, "Power distribution techniques for VLSI circuits," *IEEE Journal of Solid-State Circuits,* vol. SC-21, no. 1, pp. 150–156, Feb. 1986.

[5.43] D. Johannsen, "Hierarchical power routing," Computer Science Department, memo 2069, California Institute of Technology, Pasadena, Oct. 1978.

Chapter 6

[6.1] C.W. Ho, D.A. Chance, C.H. Bajorek, and R.E. Acosta, "The thin film module as a high-performance semiconductor package," *IBM Journal of Research and Development,* vol. 26, pp. 286–296, May 1982.

[6.2] A. Deutsch and C.W. Ho, "Triplate structure design for thin film lossy unterminated transmission lines," *1981 International Symposium on Circuits and Systems,* pp. 1–5, Chicago, April 27-29, 1981.

[6.3] H. Hasegawa, M. Furukawa, and H. Yanai, "Properties of microstrip line on Si-SiO_2 system," *IEEE Transactions on Microwave Theory and Techniques,* vol. MTT-19, pp. 869–881, Nov. 1971.

[6.4] S. Seki and H. Hasegawa, "Pico-second pulse response of interconnections of high-speed GaAs and Si LSIs," in *Technical Digest of 1982 GaAs IC Symposium,* pp. 119–122, New Orleans, Nov. 9–11, 1982.

[6.5] H. Hasegawa and S. Seki, "On-chip pulse transmission in very high speed LSI/VLSI," *Technical Digest of the IEEE Microwave and Milli-meter-wave Monolitic Circuits Symposium,* pp. 29-33, San Francisco, 1984.

[6.6] S. Seki and H. Hasegawa, "Analysis of crosstalk in very high-speed LSI/VLSI's using a coupled multiconductor MIS microstrip line model," *IEEE Transactions on Electron Devices,* vol. ED-31, pp. 1948–1953, Dec. 1984.

[6.7] H. Hasegawa and S. Seki, "Analysis of interconnection delay on very high-speed LSI/VLSI chips using an MIS microstrip line model," *IEEE Transactions on Electron Devices,* vol. ED-31, pp. 1954–1960, Dec. 1984.

[6.8] H.T. Yuan, Y.T. Lin, and S.Y. Chiang, "Properties of interconnection on silicon, sapphire, and semi-insulating gallium arsenide substrates," *IEEE Transactions on Electron Devices,* vol. ED-29, pp. 639–644, April 1982.

[6.9] Y. Fukuoka, Q. Zhang, D.P. Neikirk, and T. Itoh, "Analysis of multi-layer interconnection lines for a high-speed digital integrated circuit," *IEEE Transactions on Microwave Theory and Techniques,* vol. MTT-33, pp. 527–532, June 1985.

[6.10] A.J. Blodgett, "Microlectronic packaging," *Scientific American,* pp. 86–96, July 1983.

[6.11] H.A. Wheeler, "Transmission-line properties of parallel strips separated by a dielectric sheet," *IEEE Transactions on Microwave Theory and Techniques,* pp. 172–185, March 1965.

[6.12] T.M. Hyltin, "Microstrip transmission on semiconductor dielectrics," *IEEE Transactions on Microwave Theory and Techniques,* pp. 777–781, Nov. 1965.

[6.13] H. Guckel, P.A. Brennan, and I. Palocz, "A parallel-plate waveguide approach to microminiaturized, planar transmission lines for integrated circuits," *IEEE Transactions on Microwave Theory and Techniques,* vol. MTT-15, pp. 468–476, Aug. 1967.

[6.14] M. Caulton and H. Sobol, "Microwave integrated-circuit technology— A survey," *IEEE Journal of Solid-State Circuits,* vol. SC-5, no. 6, pp. 292–303, Dec. 1970.

[6.15] M.V. Schneider, "Microstrip lines for microwave integrated circuits," *Bell System Technical Journal,* pp. 1421–1444, May–June 1969.

[6.16] A.J. Rainal, "Transmission properties of various styles of printed wiring boards," *Bell System Technical Journal,* vol. 58, no. 5, pp. 995–1025, May–June 1979.

[6.17] I.T. Ho and S.K. Mullick, "Analysis of transmission lines on integrated circuit chips," *IEEE Journal of Solid-State Circuits,* vol. SC-2, no. 4, pp. 201–208, Dec. 1967.

[6.18] R.A. Pucel, D.J. Masse and C.P. Hartwig, "Losses in microstrip," *IEEE Transactions on Microwave Theory and Techniques,* vol. MTT-16, pp. 342–350, June 1968.

[6.19] R.F. Harrington and C. Wei, "Losses on multiconductor transmission lines in multilayered dielectric media," *IEEE Transactions on Microwave Theory and Techniques,* vol. MTT-32, pp. 705–710, July 1984.

[6.20] M.A.R. Gunston, *Microwave Transmission-Line Impedance Data,* Van Nostrand Reinhold Company, London, 1972.

[6.21] T.C. Edwards, *Foundations for Microstrip Circuit Design,* Wiley, New York, 1984.

[6.22] K.C. Gupta, R. Garg, and I.J. Bahl, *Microstrip Lines and Slotlines,* Artech House, Dedham, Mass., 1979.

[6.23] K.C. Gupta, R. Garg, and R. Chadha, *Computer-Aided Design of Microwave Circuits,* Artech House, Dedham, Mass., 1981.

[6.24] L.V. Blake, *Transmission Lines and Waveguides,* Wiley, New York, 1969.

[6.25] S. Ramo, J.R. Whinnery, and T. Van Duzer, *Fields and Waves in Communication Electronics,* Wiley, New York, 1965.

[6.26] B.D. Popovic, *Introductory Engineering Electromagnetics,* Addison-Wesley Publishing Co., Reading, Mass., 1971.

[6.27] I. Catt, D. Walton, and M. Davidson, *Digital Hardware Design,* Macmillan, London, 1979.

[6.28] T.R. Blakeslee, *Digital Design with Standard MSI and LSI,* Wiley, New York, 1979.

[6.29] W.C. Johnson, *Transmission Lines and Networks,* Wiley, New York, 1950.

[6.30] S.R. Seshadri, *Fundamentals of Transmission Lines and Electromagnetic Fields,* Addison-Wesley Publishing Co., Reading, Mass., 1971.

[6.31] S.M. Sze, *VLSI Technology,* McGraw-Hill, New York, 1983.

[6.32] R.E. Matick, *Transmission Lines for Digital and Communications Networks,* McGraw-Hill, New York, 1969.

[6.33] B.M. Oliver, "Time domain reflectometry," *Hewlett-Packard Journal,* vol. 15, no. 6, Feb. 1964.

[6.34] A. Barna, *High Speed Pulse and Digital Techniques,* Wiley, New York, 1980.

[6.35] A. Feller, H.R. Kaupp, and J.J. Digiacomo, "Crosstalk and reflections in high-speed digital systems," *AFIPS Conference Proceedings, 1965 Fall Joint Computer Conference,* vol. 27, pt. 1, pp. 511–525, Spartan Books, Washington D.C., 1965.

[6.36] E.E. Davidson, "Electrical design of a high speed computer packaging system," *IEEE Transactions on Components, Hybrids, and Manufacturing Technology,* vol. CHMT-6, no. 3, pp. 272–282, Sept. 1983.

Chapter 7

[7.1] D.B. Jarvis, "The effects of interconnections on high-speed logic circuits," *IEEE Transactions on Electronic Computers,* pp. 476–487, Oct. 1963.

[7.2] A. Feller, H.R. Kaupp, and J.J. Digiacomo, "Crosstalk and reflections in high-speed digital systems," *1965 Fall Joint Computer Conference, AFIPS Proceedings,* vol. 27, pt. 1, pp. 511–525, Spartan Books, Washington D.C., Dec. 1965.

[7.3] I. Catt, "Crosstalk (noise) in digital systems," *IEEE Transactions on Electronic Computers,* vol. EC-16, no. 6, pp. 743–763, Dec. 1967.

[7.4] A.J. Rainal, "Transmission properties of various styles of printed wiring boards," *Bell System Technical Journal,* vol. 58, no. 5, pp. 995–1025, May–June 1979.

[7.5] K.C. Gupta, R. Garg, and I.J. Bahl, *Microstrip Lines and Slotlines,* Artech House, Dedham, Mass., 1979.

[7.6] T.C. Edwards, *Foundations of Microstrip Circuit Design*, Wiley, New York, 1981.

[7.7] E.T. Lewis, "An analysis of interconnect line capacitance and coupling for VLSI circuits," *Solid State Electronics*, vol. 27, nos. 8/9, pp. 741–749, 1984.

[7.8] R. Garg and I.J. Bahl, "Characteristics of coupled microstriplines," *IEEE Transactions on Microwave Theory and Techniques*, vol. MTT-27, no. 7, pp. 700–705, July 1979.

[7.9] W.R. Smith and D.E. Snyder, "Circuit loading and crosstalk signals from capacitance in SOS and bulk-silicon interconnect channels," *IEEE VLSI Multilevel Interconnection Conference (V-MIC'84)*, pp. 218–227, June 21–22, 1984.

[7.10] S. Seki and H. Hasegawa, "Analysis of crosstalk in very high-speed LSI/VLSI's using a coupled multiconductor MIS microstrip line model," *IEEE Transactions on Electron Devices*, vol. ED-31, no. 12, pp. 1948–1953, Dec. 1984.

[7.11] W.S. Song and L.A. Glasser, "Power distribution techniques for VLSI circuits," *IEEE Journal of Solid-State Circuits*, vol. SC-21, no. 1, pp. 150–156, Feb. 1986.

[7.12] E.E. Davidson, "Electrical design of a high speed computer packaging system," *IBM Journal of Research and Development*, vol. 26, no. 3, pp. 349–361, May 1982.

[7.13] A.J. Blodgett, "Microlectronic packaging," *Scientific American*, pp. 86–96, July 1983.

[7.14] C.W. Ho, D.A. Chance, C.H. Bajorek, and R.E. Acosta, "The thin film module as a high-performance semiconductor package," *IBM Journal of Research and Development*, vol. 26, pp. 286–296, May 1982.

[7.15] N. Raver, "FET off-chip drivers and package disturbs," *IEEE Custom Integrated Circuits Conference (CICC'84)*, pp. 574–579, May 1984.

[7.16] G.A. Katopis, "Delta-I noise specification for a high-performance computing machine," *Proceedings of the IEEE*, vol. 75, no. 9, pp. 1405–1415, Sept. 1985.

[7.17] I. Catt, D. Walton, and M. Davidson, *Digital Hardware Design*, Macmillan, London, 1979.

[7.18] J.H. Hackenberg, "Signal integrity in the VAX 8600 system," *Digital Technical Journal,* no. 1, pp. 61–65, Aug. 1985.

[7.19] D.B. Tuckerman and R.F.W. Pease, "High-performance heat sinking for VLSI," *IEEE Electron Device Letters,* vol. EDL-2, pp. 126–129, May 1981.

[7.20] L.W. Schaper and D.I. Amey, "Improved electrical performance required for future MOS packaging," *IEEE Transactions on Components, Hybrids and Manufacturing Technology,* vol. CHMT-6, no. 3, pp. 283–289, Sept. 1983.

[7.21] C.M. Val and J.E. Martin, "A new chip carrier for high performance applications: integrated decoupling capacitor chip carrier (IDCCC)," *IEEE Transactions on Components, Hybrids and Manufacturing Technology,* vol. CHMT-6, No. 3, pp. 290–297, Sept. 1983.

[7.22] T. Watari and H. Murado, "Packaging technology for the NEC SX supercomputer," *IEEE Transactions on Components, Hybrids, and Manufacturing Technology,* vol. CHMT-8, no. 4, pp. 462–467, Dec. 1985.

[7.23] *Memory Design Handbook,* Intel Corporation, Santa Clara, CA, 1981.

[7.24] T. Gabara and D. Thompson, "Ground bounce control in CMOS integrated circuits," *IEEE International Solid-State Circuits Conference (ISSCC'88),* Feb. 17–19, 1988.

[7.25] R.F. Sechler and B.L. Krauter, "Power distribution resonance in high-performance VLSI CMOS," forthcoming.

[7.26] B.C. Cole, "Alliance's 1-Mbit DRAM runs fast as a static RAM," *Electronics,* pp. 107–109, Jan. 7, 1988.

Chapter 8

[8.1] E.E. Davidson, "Electrical design of a high speed computer packaging system," *IEEE Transactions on Components, Hybrids, and Manufacturing Technology,* vol. CHMT-6, no. 3, pp. 272–282, Sept. 1983.

[8.2] R.H. Dennard, F.H. Gaensslen, H.N. Yu, V.L. Rideout, E. Bassous, and A.R. LeBlanc, "Design of ion implanted MOSFET's with very small physical dimensions," *IEEE Journal of Solid-State Circuits,* vol. SC-9, pp. 256–268, Oct. 1974.

[8.3] K.C. Saraswat and F. Mohammadi, "Effects of scaling of interconnections on the time delay of VLSI circuits," *IEEE Journal of Solid-State Circuits*, vol. SC-17, pp. 275–280, April 1982.

[8.4] H.B. Bakoglu and J.D. Meindl, "Optimal interconnect circuits for VLSI," in *IEEE International Solid-State Circuits Conference (ISSSC'84)*, pp. 164–165, San Francisco, Feb. 1984.

[8.5] H.B. Bakoglu and J.D. Meindl, "Optimal interconnection circuits for VLSI," *IEEE Transactions on Electron Devices*, vol. ED–32, no. 5, pp. 903–909, May 1985.

[8.6] M.G.H. Katevenis and M.G. Blatt, "Switch design for soft-configurable WSI systems," *Proceedings of the 1985 VLSI Conference*, University of North Carolina, Chapel Hill, May 1985.

[8.7] C. Mead and L. Conway, *Introduction to VLSI Systems*, Chapter 7 by C. Seitz, Addison-Wesley Publishing Company, Reading, MA, 1980.

[8.8] P.M. Kogge, *The Architecture of Pipelined Computers*, McGraw-Hill, New York, 1981.

[8.9] L.A. Glasser and D.W. Dobberpuhl, *The Design and Analysis of VLSI Circuits*, Addison-Wesley Publishing Company, Reading, MA, 1985.

[8.10] N. Weste and K. Eshraghian, *Principles of CMOS VLSI Design: A Systems Perspective*, Addison-Wesley Publishing Company, Reading, MA, 1985.

[8.11] A. Mukherjee, *Introduction to nMOS and CMOS VLSI Systems Design*, Prentice-Hall, Englewood Cliffs, NJ, 1986.

[8.12] B. Randell and P.C. Treleaven, eds., *VLSI Architecture*, Chapter 15 by J.A. Marques and A. Cunha, "Clocking of VLSI circuits," Prentice-Hall Inc., Englewood Cliffs, NJ, 1983.

[8.13] J. Peatman, *The Design of Digital Systems*, McGraw-Hill, New York, 1972.

[8.14] S. Waser and M.J. Flynn, *Introduction to Arithmetic for Digital Systems Designers*, CBS College Publishing, New York, 1982.

[8.15] G. Langdon, Jr., *Computer Design*, Computeach, San Jose, CA, 1982.

[8.16] S.H. Unger and C.J. Tan, "Optimal clocking schemes for high speed digital systems," *Proceedings of the IEEE International Conference on Computer Design: VLSI in Computers (ICCD'83)*, Oct.–Nov. 1983.

[8.17] S.H. Unger and C.J. Tan, "Clocking schemes for high-speed digital systems," *IEEE Transactions on Computers,* vol. C-35, no. 10, pp. 880–895, Oct. 1986.

[8.18] K.D. Wagner, "Clock system design," *IEEE Design and Test of Computers,* pp. 9–27, Oct. 1988.

[8.19] D.C. Noice, R. Mathews, and J. Newkirk, "A clocking discipline for two-phase digital systems," *IEEE International Conference on Circuits and Computers (ICCC'82),* Sept.–Oct. 1982.

[8.20] "Electrical design of the BELLMAC-32A microprocessor," *IEEE International Conference on Circuits and Computers (ICCC'82),* Sept.–Oct. 1982.

[8.21] D.C. Noice, "A clocking discipline for two-phase digital integrated circuits," Ph.D. dissertation, Stanford University, Jan. 1983.

[8.22] T.J. Chaney and C.E. Molnar, "Anomalous behavior of synchronizer and arbiter circuits," *IEEE Transactions on Computers,* vol. C-22, pp. 421–422, April 1973.

[8.23] G.R. Couranz and D.F. Wann, "Theoretical and experimental behavior of synchronizers operating in the metastable region," *IEEE Transactions on Computers,* vol. C-24, pp. 604–616, June 1975.

[8.24] D.J. Kinniment and J.V. Woods, "Synchronisation and arbitration circuits in digital systems," *Proceedings of Institute of Electrical Engineers,* (England), vol. 123, pp. 961–966, Oct. 1976.

[8.25] H.J. Veendrick, "The behavior of flip-flops used as synchronizers and prediction of their failure rate," *IEEE Journal of Solid-State Circuits,* vol. SC-15, pp. 169–176, April 1980.

[8.26] F. Rosenberger and T. Chaney, "Flip-flop resolving time test circuits," *IEEE Journal of Solid-State Circuits,* vol. SC-17, pp. 731–738, Aug. 1982.

[8.27] S.T. Flannagan, "Synchronization reliability in CMOS technology," *IEEE Journal of Solid-State Circuits,* vol. SC-20, pp. 880–882, Aug. 1985.

[8.28] P.A. Stoll, "How to avoid synchronization problems," *VLSI Design,* pp. 56–58, Nov.–Dec. 1982.

[8.29] C.L. Seitz, "Self timed VLSI systems," *Caltech Conference on VLSI*, pp. 345–355, Jan. 1979.

[8.30] F.U. Rosenberger, C.E. Molnar, T.J. Chaney, and T.P. Fang, "Q-modules: internally clocked delay-insensitive modules," *IEEE Transactions on Computers*, vol. 37, no. 9, pp. 1005–1019, Sept. 1988.

[8.31] F. Anceau, "A synchronous approach for clocking VLSI systems," *IEEE Journal of Solid-State Circuits*, vol. SC-17, pp. 51–56, Feb. 1982.

[8.32] J.A. Marques and A. Cunha, "Clocking of VLSI circuits," Chap. 15 of *VLSI Architecture*, B. Randell and P.C. Treleaven, eds., Prentice-Hall Inc., Englewood Cliffs, NJ, 1983.

[8.33] D.F. Wann and M.A. Franklin, "Asynchronous and clocked control structures for VLSI-based interconnection networks," *IEEE Transaction on Computers*, vol. C-32, pp. 284–293, March 1983.

[8.34] C. Barney, "Logic designers toss out the clock," *Electronics*, pp. 42–45, Dec. 9, 1985.

[8.35] R. Woudsma and J.M. Noteboom, "The modular design of clock-generator circuits in a CMOS building block system," *IEEE Journal of Solid-State Circuits*, vol. SC-20, no. 3, pp. 770–774, June 1985.

[8.36] M. Bazes, "A novel precision MOS synchronous delay line," *IEEE Journal of Solid-State Circuits*, vol. SC-20, no. 6, pp. 1265–1272, Dec. 1985.

[8.37] F.A. Gardner, "Charge-pump phase-locked loops," *IEEE Transctions on Communications*, vol. COM-28, pp. 1849–1858, Nov. 1980.

[8.38] F.A. Gardner, "Phase accuracy of charge-pump PLLs," *IEEE Transctions on Communications*, vol. COM-30, pp. 2362–2363, Oct. 1982.

[8.39] D.K. Jeong, G. Borriello, D.A. Hodges, and R.H. Katz, "Design of PLL-based clock generation circuits," *IEEE Journal of Solid-State Circuits*, vol. SC-22, no. 2, pp. 255–261, April 1987.

[8.40] M.G. Johnson and E.L. Hudson, "A variable delay line phase locked loop for CPU-coprocessor synchronization," *IEEE International Solid-State Circuits Conference (ISSCC'88)*, pp. 142–143, Feb. 17-19, 1988.

[8.41] K.D. Wagner and E.J. McCluskey, "Tuning, clock distribution and communication in VLSI high-speed chips," *Center for Reliable Computing Technical Report No. 84-5*, Stanford University, June 1984.

[8.42] S. Dhar, M.A. Franklin, and D.F. Wann, "Reduction of clock delays in VLSI structures," *IEEE International Conference on Computer Design: VLSI in Computers (ICCD'84)*, pp. 778–783, 1984.

[8.43] J. Beausang and A. Albicki, "A method to obtain an optimal clocking scheme for a digital system," *IEEE International Conference on Computer Design (ICCD'85)*, pp. 68–72, Oct. 1985.

[8.44] E. Friedman, W. Marking, E. Iodice, and S. Powell, "Parameterized buffer cells integrated into an automated layout system," *IEEE Custom Integrated Circuits Conference (CICC'85)*, pp. 389-392, May 1985.

[8.45] E.G. Friedman and S. Powell, "Design and analysis of a hierarchical clock distribution system for synchronous standard cell/macrocell VLSI," *IEEE Journal of Solid-State Circuits*, vol. SC-21, pp. 240–246, April 1986.

[8.46] E.G. Friedman, "Performance limitations in synchronous digital systems," PhD dissertation, University of California, Irvine, CA, 1989.

[8.47] A.L. Fisher and H.T. Kung, "Synchronizing large VLSI processor arrays," *IEEE Transactions on Computers*, vol. c-34, pp. 734–740, Aug. 1985.

[8.48] H.C. Lin and L.W. Linholm, "An optimized output stage for MOS Integrated Circuits," *IEEE Journal of Solid-State Circuits*, vol. SC-10, No. 2, pp. 106–109, April 1975.

[8.49] R.C. Jeager, "Comments on 'An optimized output stage for MOS integrated circuits'," *IEEE Journal of Solid-State Circuits*, vol. SC-10, pp. 185–186, June 1975.

[8.50] C. Mead and M. Rem, "Minimum propagation delays in VLSI," *IEEE Journal of Solid-State Circuits*, vol. SC-17, pp. 773–775, Aug. 1982.

[8.51] C.W. Ho, D.A. Chance, C.H. Bajorek, and R.E. Acosta, "The thin film module as a high-performance semiconductor package," *IBM Journal of Research and Development*, vol. 26, pp. 286–296, May 1982.

[8.52] B.D. Popovic, *Introductory Engineering Electromagnetics*, Addison-Wesley, Reading, MA, 1971.

[8.53] W.R. Wisseman, H.M. Macksey, G.E. Brehm, and P. Saunier, "GaAs microwave devices and circuits with submicron electron-beam defined features," *Proceedings of IEEE*, vol. 71, pp. 667–675, May 1983.

[8.54] H. Hasegawa, M. Furukawa, and H. Yanai, "Properties of microstrip line on $Si\text{-}SiO_2$ system," *IEEE Transactions on Microwave Theory and Techniques,* vol. MTT-19, pp. 869–881, Nov. 1971.

[8.55] T. Sakurai and T. Tamaru, "Simple formulas for two- and three-dimensional capacitances," *IEEE Transactions on Electron Devices,* vol. ED-30, pp. 183–185, Feb. 1983.

Chapter 9

[9.1] H.B. Bakoglu and J.D. Meindl, "A system level circuit model for multi- and single-chip CPUs," *IEEE International Solid State Circuits Conference (ISSCC'87),* pp. 308–309, New York, Feb. 1987.

[9.2] R.M. Russell, "The CRAY-1 computer system," *Communications of ACM,* vol. 21, no. 1, pp. 63–72 , Jan. 1978.

[9.3] "How Cray kept the 'biggest and fastest' title," *Electronics,* p. 92, Sept. 9, 1985.

[9.4] "Designers give packaging the credit for Cray-3 speed," *Electronic Engineering Times,* p. 10, Oct. 12, 1987.

[9.5] "Cray spins Cray-3 out to start-up," *Electronic Engineering Times,* issue 539, p. 1, May 22, 1989.

[9.6] A.J. Blodgett, "Microlectronic packaging", *Scientific American,* pp. 86–96, July 1983.

[9.7] A.J. Blodgett and D.R. Barbour, "Thermal conduction module: a high-performance multilayer ceramic package," *IBM Journal of Research Development,* pp. 30–36, Jan. 1982.

[9.8] R.F. Bonner, J.A. Asselta, and F.W. Haining, "Advanced printed-circuit board design for high-performance computer applications," *IBM Journal of Research and Development,* pp. 297–305, May 1982.

[9.9] D.P. Seraphim, "A new set of printed-circuit technologies for the IBM 3081 processor unit," *IBM Journal of Research and Development,* pp. 37–44, Jan. 1982.

[9.10] J.R. Lineback, "CMOS gates key to 'affordable' supercomputer," *Electronics Week,* pp. 17–19, Oct. 29, 1984.

[9.11] P.M. Russo, "VLSI impact on microprocessor evolution, usage and system design," *IEEE Journal of Solid-State Circuits,* pp. 397–406, vol. SC-15, no. 4, Aug. 1980.

[9.12] S.P. Morse, B.W. Ravenel, S. Mazor, and W.B. Pohlman, "Intel microprocessors—8008 to 8086," *IEEE Computer,* pp. 42–60, Oct. 1980.

[9.13] H.D. Toong and A. Gupta, "An architectural comparison of contemporary 16-bit microprocessors," *IEEE Micro,* pp. 26–38, May 1981.

[9.14] H.D. Toong and A. Gupta, "An architectural comparison of 32-bit microprocessors," *IEEE Micro,* pp. 9–22, Feb. 1983.

[9.15] D.A. Patterson and C.H. Sequin, "A VLSI RISC," *IEEE Computer,* pp. 8–22, Sept. 1982.

[9.16] B.T. Murphy, "Microcomputers: trends, technologies and design strategies," *IEEE Journal of Solid-State Circuits,* pp. 236–244, vol. SC-18, no. 3, June 1983.

[9.17] W.N. Johnson, "A VLSI superminicomputer CPU," *IEEE International Solid-State Circuits Conference (ISSCC'84),* pp. 174–175, San Francisco, Feb. 1984.

[9.18] C. Rowen, S.A. Przbylski, N.P. Jouppi, T.R. Gross, J.D. Shott,and J.L. Hennessy, "A pipelined 32b NMOS microprocessor," *IEEE International Solid-State Circuits Conference (ISSCC'84),* pp. 180–181, San Francisco, Feb. 1984.

[9.19] H. Sachs and W. Hollingsworth, "A high-performance 846,000 transistor UNIX engine—the Fairchild Clipper," *IEEE International Conference on Computer Design: VLSI in Computers (ICCD'85),* Oct. 1985.

[9.20] J.W. Prak, "High performance technology, circuits, and packaging for the 80386," *IEEE International Conference on Computer Design: VLSI in Computers (ICCD'86),* pp. 161–168, Oct. 1986.

[9.21] *80386 High Performance microprocessor with integrated memory management,* Intel Corp., Oct. 1985.

[9.22] K.A. El-Ayat and R.K. Agarwal, "The Intel 80386—architecture and implementation," *IEEE Micro,* pp. 4–22, Dec. 1985.

[9.23] "Intel i486 introduced; integrates MMU, FPU," *Electronics News,* vol. 35, no. 1754, p. 1, April 17, 1989.

[9.24] P.M. Kogge, *The Architecture of Pipelined Computers*, McGraw-Hill, New York, 1981.

[9.25] D.J. Lilja, "Reducing the branch penalty in pipelined processors," *IEEE Computer*, vol. 21, no. 7, pp. 47–55, July 1988.

[9.26] B.S. Landman and R.L. Russo, "On a pin versus block relationship for partitions of logic graphs," *IEEE Transactions on Computers*, vol. C-20, pp. 1469–1479, Dec. 1971.

[9.27] W.R. Heller, W.E. Donath, and W.F. Mikhail, "Prediction of wiring space requirements for LSI," *Proceedings of the 14th Design Automation Conference*, pp. 32–43, New Orleans, 1977.

[9.28] W.E. Donath, "Equivalance of memory to random logic," *IBM Journal of Research and Development*, vol. 18, pp. 401–407, Sept. 1974.

[9.29] W.E. Donath, "Placement and average interconnection lengths of computer logic," *IEEE Transactions on Circuits and Systems*, CAS-26, pp. 272–277, April 1979.

[9.30] W.E. Donath, "Wire length distribution for placements of computer logic," *IBM Journal of Research and Development*, vol. 25, pp. 152–155, May 1981.

[9.31] D.C. Schmidt, "Circuit pack parameter estimation using Rent's rule," *IEEE Transactions on Computer Aided Design of Integrated Circuits and Systems*, CAD-1, pp. 186–192, Oct. 1982.

[9.32] D.K. Ferry, "Interconnection Lengths and VLSI," *IEEE Circuits and Devices*, pp. 39–42, July 1985.

[9.33] R.H. Dennard, F.H. Gaensslen, H.N. Yu, V.L. Rideout, E. Bassous, and A.R. LeBlanc, "Design of ion implanted MOSFET's with very small physical dimensions," *IEEE Journal of Solid-State Circuits*, vol. SC-9, pp. 256–268, Oct. 1974.

[9.34] A. Moshen, "Device and circuit design for VLSI," *Caltech Conference on VLSI*, pp. 31–54, Jan. 1979.

[9.35] J.D. Meindl, K.N. Ratnakumar, L. Gerzberg, and K.C. Saraswat, "Circuit scaling limits for ultra-large scale integration," *IEEE International Solid-State Circuits Conference (ISSCC'81)*, pp. 36–37, Feb. 1981.

[9.36] VLSI Laboratory, Texas Instruments Inc., "Technology and design challenges of MOS VLSI," *IEEE Journal of Solid-State Circuits*, vol. SC-17, pp. 442–448, June 1982.

[9.37] P.M. Solomon, "A comparison of semiconductor devices for high speed logic," *Proceedings of IEEE*, vol. 70, pp. 489–509, May 1982.

[9.38] H.E. Oldham and S.L. Partridge, "A comparative study of CMOS processes for VLSI applications," *IEEE Transactions on Electron Devices*, vol. ED-29, pp. 1593–1598, Oct. 1982.

[9.39] S.S. Liu et al., "HMOS III technology," *IEEE Journal of Solid-State Circuits*, vol. SC-17, pp. 810–814, Oct. 1982.

[9.40] A. Reisman, "Device circuit and technology scaling to micron and submicron dimensions," *Proceedings of IEEE*, pp. 550–565, May 1983.

[9.41] R.J. Bayruns et al., "Delay analysis of Si NMOS Gbit/s logic circuits," *IEEE Journal of Solid-State Circuits*, vol. SC-19, pp. 755–764, Oct. 1984.

[9.42] G. Baccarani, M.R. Wordeman, and R.H. Dennard, "Generalized scaling theory and its application to a 1/4 micrometer MOSFET design," *IEEE Transactions on Electron Devices*, pp. 452–462, April 1984.

[9.43] K.C. Saraswat and F. Mohammadi, "Effect of scaling of interconnections on the time delay of VLSI circuits," *IEEE Journal of Solid-State Circuits*, vol. SC-17, pp. 275–280, April 1982.

[9.44] H.B. Bakoglu and J.D. Meindl, "Optimal interconnect circuits for VLSI," *IEEE International Solid-State Circuits Conference (ISSCC'84)*, pp. 164–165, San Francisco, Feb. 1984.

[9.45] H.B. Bakoglu and J.D. Meindl, "Optimal interconnection circuits for VLSI," *IEEE Transactions on Electron Devices*, vol. ED-32, no. 5, pp. 903–909, May 1985.

[9.46] C.A. Neugebauer, "Approaching wafer scale integration from the packaging point of view," *IEEE International Conference on Computer Design: VLSI in Computers (ICCD'84)*, pp. 115–120, Oct. 1984.

[9.47] D.L. Peltzer, "Wafer-scale integration: the limits of VLSI," *VLSI Design*, pp. 43–47, Sept. 1983.

[9.48] A.S. Bergendahl, B.J. Donlan, J.F. McDonald, R.H. Steinvorth, and G.F. Taylor, "A thick film lift-off technique for high frequency interconnection in wafer scale integration," *IEEE VLSI Multilevel Interconnection Conference*, pp. 154–162, June 1985.

[9.49] C.W. Ho, D.A. Chance, C.H. Bajorek, and R.E. Acosta, "The thin-film module as a high-performance semiconductor package," *IBM Journal of Research and Development*, vol. 26, no. 3, pp. 286–296, May 1982.

[9.50] G.F. Taylor, B.J. Donlan, J.F. McDonald, A.S. Bergendahl, and R.H. Steinvorth, "The wafer transmission module-wafer scale integration packaging," *IEEE 1985 Custom Integrated Circuits Conference (CICC'85)*, pp. 55–58, 1985.

[9.51] F.C. Chong, C.W. Ho, K. Liu, and S. Westbrook, "A high density multi-chip memory module," *Wescon/85 Professional Program Session Record 7: High Density/High Performance Semiconductor Packaging*, pp. 1–6, Nov. 1985.

[9.52] B.D. Popovic, *Introductory Engineering Electromagnetics*, Addison-Wesley Publishing Co., Reading, Mass., 1971, Chap. 14.

[9.53] R.W. Keyes, "The wire-limited logic chip," *IEEE Journal of Solid-State Circuits*, vol. SC-17, pp. 1232-1233, Dec. 1982.

[9.54] R.H. Krambeck, C.M. Lee, and H.S. Law, "High-speed compact circuits with CMOS," *IEEE Journal of Solid-State Circuits*, vol. SC-17, no. 3, pp. 614–619, June 1982.

[9.55] S. Waser and M.J. Flynn, *Introduction to Arithmetic for Digital Systems Designers*, CBS College Publishing, New York, 1982.

[9.56] D.B. Tuckerman and R.F.W. Pease, "High-performance heat sinking for VLSI," *IEEE Electron Device Letters*, vol. EDL-2, pp. 126–129, May 1981.

[9.57] H.C. Lin and L.W. Linholm, "An optimized output stage for MOS integrated circuits," *IEEE Journal of Solid-State Circuits*, vol. SC-10, no. 2, pp. 106–109, April 1975.

[9.58] R.C. Jeager, "Comments on 'An optimized output stage for MOS integrated circuits,' " *IEEE Journal of Solid-State Circuits*, vol SC-10, pp. 185–186, June 1975.

[9.59] C. Mead and M. Rem, "Minimum propagation delays in VLSI," *IEEE Journal of Solid-State Circuits*, vol. SC-17, pp. 773–775, Aug. 1982.

[9.60] E.T. Lewis, "Optimization of device area and overall delay for CMOS VLSI design," *Proceedings of IEEE*, vol. 72, pp. 670–689, June 1984.

[9.61] G. Moore, "VLSI: some fundamental challenges," *IEEE Spectrum*, vol. 16, p. 30, 1979.

[9.62] D. Bursky, "Advanced CISC processors catch up to RISC speeds," *Electronic Design*, vol. 37, no. 16, pp. 41–48, July 27, 1989.

[9.63] A. Barna, *VHSIC (Very High Speed Integrated Circuits) Technologies and Tradeoffs*, pp. 19–45, Wiley, New York, 1981.

[9.64] S. Konaka, Y. Yamamoto, and T. Sakai, "A 30 ps Si bipolar IC using super self-aligned process technology," *Extended Abstracts of the 16th (1984 International) Conference on Solid State Devices and Materials*, pp. 209–212, Kobe, Japan, 1984.

[9.65] P.M. Pitner, "Molecular beam epitaxy of ultra-high speed heterojunction bipolar transistor ICs," Ph.D. dissertation, Stanford University, 1987.

[9.66] B.T. Murphy, "Cost-size optima of monolithic integrated circuits," *Proceedings of IEEE*, vol. 52, pp. 1537–1545, Dec. 1964.

[9.67] J.E. Price, "A new look at yield of integrated circuits," *Proceedings of IEEE*, vol. 58, pp. 1290-1291, Aug. 1970.

INDEX